Springer Undergraduate Mathematics Series

T0225867

More information about this series at http://www.springer.com/series/3423

Karin Erdmann • Thorsten Holm

Algebras and Representation Theory

 Springer

Karin Erdmann
Mathematical Institute
University of Oxford
Oxford, United Kingdom

Thorsten Holm
Fakultät für Mathematik und Physik
Institut für Algebra, Zahlentheorie und
Diskrete Mathematik
Leibniz Universität Hannover
Hannover, Germany

ISSN 1615-2085   ISSN 2197-4144   (electronic)
Springer Undergraduate Mathematics Series
ISBN 978-3-319-91997-3   ISBN 978-3-319-91998-0   (eBook)
https://doi.org/10.1007/978-3-319-91998-0

Library of Congress Control Number: 2018950191

Mathematics Subject Classification (2010): 16-XX, 16G10, 16G20, 16D10, 16D60, 16G60, 20CXX

This Springer imprint is published by the registered company Springer Nature Switzerland AG
The registered company address is: Gewerbestrasse 11, 6330 Cham, Switzerland

Introduction

Representation theory is a beautiful subject which has numerous applications in mathematics and beyond. Roughly speaking, representation theory investigates how algebraic systems can act on vector spaces. When the vector spaces are finite-dimensional this allows one to explicitly express the elements of the algebraic system by matrices, and hence one can exploit basic linear algebra to study abstract algebraic systems. For example, one can study symmetry via group actions, but more generally one can also study processes which cannot be reversed. Algebras and their representations provide a natural framework for this. The idea of letting algebraic systems act on vector spaces is so general that there are many variations. A large part of these fall under the heading of representation theory of associative algebras, and this is the main focus of this text.

Examples of associative algebras which already appear in basic linear algebra are the spaces of $n \times n$-matrices with coefficients in some field K, with the usual matrix operations. Another example is provided by polynomials over some field, but there are many more. In general, roughly speaking, an associative algebra A is a ring which also is a vector space over some field K such that scalars commute with all elements of A. We start by introducing algebras and basic concepts, and describe many examples. In particular, we discuss group algebras, division algebras, and path algebras of quivers. Next, we introduce modules and representations of algebras and study standard concepts, such as submodules, factor modules and module homomorphisms.

A module is simple (or irreducible) if it does not have any submodules, except zero and the whole space. The first part of the text is motivated by simple modules. They can be seen as building blocks for arbitrary modules, this is made precise by the Jordan–Hölder theorem. It is therefore a fundamental problem to understand all simple modules of an algebra. The next question is then how the simple modules can be combined to form new modules. For some algebras, every module is a direct sum of simple modules, and in this case, the algebra is called semisimple. We study these in detail. In addition, we introduce the Jacobson radical of an algebra, which, roughly speaking, measures how far an algebra is away from being semisimple. The Artin–Wedderburn theorem completely classifies semisimple algebras. Given an arbitrary algebra, in general it is difficult to decide whether or not it is semisimple. However, when the algebra is the group algebra of a finite group G, Maschke's theorem answers this, namely the group algebra KG is semisimple if and only the characteristic of the field K does not divide the order of the group. We give a proof, and we discuss some applications.

If an algebra is not semisimple, one has to understand indecomposable modules instead of just simple modules. The second half of the text focusses on these. Any finite-dimensional module is a direct sum of indecomposable modules. Even more, such a direct sum decomposition is essentially unique, this is known as the Krull–Schmidt theorem. It shows that it is enough to understand indecomposable modules of an algebra. This suggests the definition of the representation type of an algebra. This is said to be finite if the algebra has only finitely many indecomposable modules, and is infinite otherwise. In general it is difficult to determine which algebras have finite representation type. However for group algebras of finite groups, there is a nice answer which we present.

The rest of the book studies quivers and path algebras of quivers. The main goal is to classify quivers whose path algebras have finite representation type. This is Gabriel's theorem, proved in the 1970s, which has led to a wide range of new directions, in algebra and also in geometry and other parts of mathematics. Gabriel proved that a path algebra KQ of a quiver Q has finite representation type if and only if the underlying graph of Q is the disjoint union of Dynkin diagrams of types A, D or E. In particular, this theorem shows that the representation type of KQ is independent of the field K and is determined entirely by the underlying graph of Q. Our aim is to give an elementary account of Gabriel's theorem, and this is done in several chapters. We introduce representations of quivers; they are the same as modules for the path algebra of the quiver, but working with representations has additional combinatorial information. We devote one chapter to the description of the graphs relevant for the proof of Gabriel's theorem, and to the development of further tools related to these graphs. Returning to representations, we introduce reflections of quivers and of representations, which are crucial to show that the representation type does not depend on the orientation of arrows. Combining the various tools allows us to prove Gabriel's Theorem, for arbitrary fields.

This text is an extended version of third year undergraduate courses which we gave at Oxford, and at Hannover. The aim is to give an elementary introduction, we assume knowledge of results from linear algebra, and on basic properties of rings and groups. Apart from this, we have tried to make the text self-contained. We have included a large number of examples, and also exercises. We are convinced that they are essential for the understanding of new mathematical concepts. In each section, we give sample solutions for some of the exercises, which can be found in the appendix. We hope that this will help to make the text also suitable for independent self-study.

Oxford, UK   Karin Erdmann
Hannover, Germany   Thorsten Holm
2018

Contents

Chapter 1
Algebras

We begin by defining associative algebras over fields, and we give a collection of typical examples. Roughly speaking, an algebra is a ring which is also a vector space such that scalars commute with everything. One example is given by the $n \times n$-matrices over some field, with the usual matrix addition and multiplication. We will introduce many more examples of algebras in this section.

1.1 Definition and Examples

We start by recalling the definition of a ring: A ring is a non-empty set R together with an addition $+ : R \times R \to R$, $(r, s) \mapsto r+s$ and a multiplication $\cdot : R \times R \to R$, $(r, s) \mapsto r \cdot s$ such that the following axioms are satisfied for all $r, s, t \in R$:

(R1) (Associativity of +) $r + (s + t) = (r + s) + t$.
(R2) (Zero element) There exists an element $0_R \in R$ such that $r + 0_R = r = 0_R + r$.
(R3) (Additive inverses) For every $r \in R$ there is an element $-r \in R$ such that $r + (-r) = 0_R$.
(R4) (Commutativity of +) $r + s = s + r$.
(R5) (Distributivity) $r \cdot (s + t) = r \cdot s + r \cdot t$ and $(r + s) \cdot t = r \cdot t + s \cdot t$.
(R6) (Associativity of $\cdot$) $r \cdot (s \cdot t) = (r \cdot s) \cdot t$.
(R7) (Identity element) There is an element $1_R \in R \setminus \{0\}$ such that $1_R \cdot r = r = r \cdot 1_R$.

Moreover, a ring R is called *commutative* if $r \cdot s = s \cdot r$ for all $r, s \in R$.

As usual, the multiplication in a ring is often just written as rs instead of $r \cdot s$; we will follow this convention from now on.

Note that axioms (R1)–(R4) say that $(R, +)$ is an abelian group. We assume by Axiom (R7) that all rings have an identity element; usually we will just write 1 for

© Springer International Publishing AG, part of Springer Nature 2018
K. Erdmann, T. Holm, *Algebras and Representation Theory*, Springer
Undergraduate Mathematics Series, https://doi.org/10.1007/978-3-319-91998-0_1

1_R. Axiom (R7) also implies that 1_R is not the zero element. In particular, a ring has at least two elements.

We list some common examples of rings.

(1) The integers $\mathbb{Z}$ form a ring. Every field is also a ring, such as the rational numbers $\mathbb{Q}$, the real numbers $\mathbb{R}$, the complex numbers $\mathbb{C}$, or the residue classes $\mathbb{Z}_p$ of integers modulo p where p is a prime number.
(2) The $n \times n$-matrices $M_n(K)$, with entries in a field K, form a ring with respect to matrix addition and matrix multiplication.
(3) The ring $K[X]$ of polynomials over a field K where X is a variable. Similarly, the ring of polynomials in two or more variables, such as $K[X, Y]$.

Examples (2) and (3) are not just rings but also vector spaces. There are many more rings which are vector spaces, and this has led to the definition of an algebra.

Definition 1.1.

(i) An *algebra* A over a field K (or a K-*algebra*) is a ring, with addition and multiplication

$$(a, b) \mapsto a + b \text{ and } (a, b) \mapsto ab \text{ for } a, b \in A$$

which is also a vector space over K, with the above addition and with scalar multiplication

$$(\lambda, a) \mapsto \lambda \cdot a \text{ for } \lambda \in K, a \in A,$$

and satisfying the following axiom

$$\lambda \cdot (ab) = (\lambda \cdot a)b = a(\lambda \cdot b) \text{ for all } \lambda \in K, a, b \in A. \tag{Alg}$$

(ii) The *dimension* of a K-algebra A is the dimension of A as a K-vector space. The K-algebra A is *finite-dimensional* if A is finite-dimensional as a K-vector space.
(iii) The algebra is *commutative* if it is commutative as a ring.

Remark 1.2.

(1) The condition (Alg) relating scalar multiplication and ring multiplication roughly says that *scalars commute with everything*.
(2) Spelling out the various axioms, we have already listed above the axioms for a ring. To say that A is a vector space over K means that for all $a, b \in A$ and $\lambda, \mu \in K$ we have

  (i) $\lambda \cdot (a + b) = \lambda \cdot a + \lambda \cdot b$;
  (ii) $(\lambda + \mu) \cdot a = \lambda \cdot a + \mu \cdot a$;
  (iii) $(\lambda \mu) \cdot a = \lambda \cdot (\mu \cdot a)$;
  (iv) $1_K \cdot a = a$.

(3) Strictly speaking, in Definition 1.1 we should say that A is an *associative algebra*, as the underlying multiplication in the ring is associative. There are other types of algebras, which we discuss at the end of Chap. 2.

(4) Since A is a vector space, and 1_A is a non-zero vector, it follows that the map $\lambda \mapsto \lambda \cdot 1_A$ from K to A is injective. We use this map to identify K as a subset of A. Similar to the convention for ring multiplication, for scalar multiplication we will usually also just write λa instead of $\lambda \cdot a$.

We describe now some important algebras which will appear again later.

Example 1.3.

(1) The field K is a commutative K-algebra, of dimension 1.

(2) The field $\mathbb{C}$ is also an algebra over $\mathbb{R}$, of dimension 2, with $\mathbb{R}$-vector space basis $\{1, i\}$, where $i^2 = -1$. More generally, if K is a subfield of a larger field L, then L is an algebra over K where addition and (scalar) multiplication are given by the addition and multiplication in the field L.

(3) The space of $n \times n$-matrices $M_n(K)$ with matrix addition and matrix multiplication form a K-algebra. It has dimension n^2; the *matrix units* E_{ij} for $1 \le i, j \le n$ form a K-basis. Here E_{ij} is the matrix which has entry 1 at position (i, j), and all other entries are 0. This algebra is not commutative for $n \ge 2$. For example we have $E_{11} E_{12} = E_{12}$ but $E_{12} E_{11} = 0$.

(4) Polynomial rings $K[X]$, or $K[X, Y]$, are commutative K-algebras. They are not finite-dimensional.

(5) Let V be a K-vector space, and consider the K-linear maps on V

$$\operatorname{End}_K(V) := \{\alpha : V \to V \mid \alpha \text{ is } K\text{-linear}\}.$$

This is a K-algebra, if one takes as multiplication the composition of maps, and where the addition and scalar multiplication are pointwise, that is

$$(\alpha + \beta)(v) = \alpha(v) + \beta(v) \text{ and } (\lambda \alpha)(v) = \lambda(\alpha(v))$$

for all $\alpha, \beta \in \operatorname{End}_K(V)$, $\lambda \in K$ and $v \in V$.

Exercise 1.1. For the $n \times n$-matrices $M_n(K)$ in Example 1.3 (3) check that the axioms for a K-algebra from Definition 1.1 are satisfied.

Remark 1.4. In order to perform computations in a K-algebra A, very often one fixes a K-vector space basis, say $\{b_1, b_2, \ldots\}$. It then suffices to know the products

$$b_r b_s \quad (r, s \ge 1). \tag{*}$$

Then one expresses arbitrary elements a and a' in A as finite linear combinations of elements in this basis, and uses the distributive laws to compute the product of a and a'. We refer to (*) as 'multiplication rules'.

In Example 1.3 where $A = \mathbb{C}$ and $K = \mathbb{R}$, taking the basis $\{1_{\mathbb{C}}, i\}$ one gets the usual multiplication rules for complex numbers.

For the $n \times n$-matrices $M_n(K)$ in Example 1.3, it is convenient to take the basis of matrix units $\{E_{ij} \mid 1 \leq i, j \leq n\}$ since products of two such matrices are either zero, or some other matrix of the same form, see Exercise 1.10.

Given some algebras, there are several general methods to construct new ones. We describe two such methods.

Definition 1.5. If $A_1, \ldots, A_n$ are K-algebras, their *direct product* is defined to be the algebra with underlying space

$$A_1 \times \ldots \times A_n := \{(a_1, \ldots, a_n) \mid a_i \in A_i \text{ for } i = 1, \ldots, n\}$$

and where addition and multiplication are componentwise. It is a straightforward exercise to check that the axioms of Definition 1.1 are satisfied.

Definition 1.6. If A is any K-algebra, then the *opposite algebra* A^{op} of A has the same underlying K-vector space as A, and the multiplication in A^{op}, which we denote by $*$, is given by reversing the order of the factors, that is

$$a * b := ba \quad \text{for } a, b \in A.$$

This is again a K-algebra, and $(A^{op})^{op} = A$.

There are three types of algebras which will play a special role in the following, namely division algebras, group algebras, and path algebras of quivers. We will study these now in more detail.

1.1.1   Division Algebras

A commutative ring is a field precisely when every non-zero element has an inverse with respect to multiplication. More generally, there are algebras in which every non-zero element has an inverse, and they need not be commutative.

Definition 1.7. An algebra A (over a field K) is called a *division algebra* if every non-zero element $a \in A$ is invertible, that is, there exists an element $b \in A$ such that $ab = 1_A = ba$. If so, we write $b = a^{-1}$. Note that if A is finite-dimensional and $ab = 1_A$ then it follows that $ba = 1_A$; see Exercise 1.8.

Division algebras occur naturally, we will see this later. Clearly, every field is a division algebra. There is a famous example of a division algebra which is not a field, this was discovered by Hamilton.

Example 1.8. The algebra $\mathbb{H}$ of *quaternions* is the 4-dimensional algebra over $\mathbb{R}$ with basis elements $1, i, j, k$ and with multiplication defined by

$$i^2 = j^2 = k^2 = -1$$

and

$$ij = k, \ ji = -k, \ jk = i, \ kj = -i, \ ki = j, \ ik = -j$$

and extending to linear combinations. That is, an arbitrary element of $\mathbb{H}$ has the form $a + bi + cj + dk$ with $a, b, c, d \in \mathbb{R}$, and the product of two elements in $\mathbb{H}$ is given by

$$(a_1 + b_1 i + c_1 j + d_1 k) \cdot (a_2 + b_2 i + c_2 j + d_2 k) =$$
$$(a_1 a_2 - b_1 b_2 - c_1 c_2 - d_1 d_2) + (a_1 b_2 + b_1 a_2 + c_1 d_2 - d_1 c_2)i$$
$$+ (a_1 c_2 - b_1 d_2 + c_1 a_2 + d_1 b_2)j + (a_1 d_2 + b_1 c_2 - c_1 b_2 + d_1 a_2)k.$$

It might be useful to check this formula, see Exercise 1.11.

One can check directly that the multiplication in $\mathbb{H}$ is associative, and that it satisfies the distributive law. But this will follow more easily later from a different construction of $\mathbb{H}$, see Example 1.27.

This 4-dimensional $\mathbb{R}$-algebra $\mathbb{H}$ is a division algebra. Indeed, take a general element of $\mathbb{H}$, of the form $u = a + bi + cj + dk$ with $a, b, c, d \in \mathbb{R}$. Let $\bar{u} := a - bi - cj - dk \in \mathbb{H}$, then we compute $\bar{u}u$, using the above formula, and get

$$\bar{u}u = (a^2 + b^2 + c^2 + d^2) \cdot 1 = u\bar{u}.$$

This is non-zero for any $u \neq 0$, and from this, one can write down the inverse of any non-zero element u.

Remark 1.9. We use the notation i, and this is justified: The subspace $\{a + bi \mid a, b \in \mathbb{R}\}$ of $\mathbb{H}$ really is $\mathbb{C}$, indeed from the multiplication rules in $\mathbb{H}$ we get

$$(a_1 + b_1 i) \cdot (a_2 + b_2 i) = a_1 a_2 - b_1 b_2 + (a_1 b_2 + b_1 a_2)i.$$

So the multiplication in $\mathbb{H}$ agrees with the one in $\mathbb{C}$.

However, $\mathbb{H}$ is not a $\mathbb{C}$-algebra, since axiom (Alg) from Definition 1.1 is not satisfied, for example we have

$$i(j1) = ij = k \neq -k = ji = j(i1).$$

The subset $\{\pm 1, \pm i, \pm j, \pm k\}$ of $\mathbb{H}$ forms a group under multiplication, this is known as the quaternion group.

1.1.2   Group Algebras

Let G be a group and K a field. We define a vector space over K which has basis the set $\{g \mid g \in G\}$, and we call this vector space KG. This space becomes a K-algebra if one defines the product on the basis by taking the group multiplication, and extends it to linear combinations. We call this algebra KG the *group algebra*.

Thus an arbitrary element of KG is a finite linear combination of the form $\sum_{g \in G} \alpha_g g$ with $\alpha_g \in K$. We can write down a formula for the product of two elements, following the recipe in Remark 1.4. Let $\alpha = \sum_{g \in G} \alpha_g g$ and $\beta = \sum_{h \in G} \beta_h h$ be two elements in KG; then their product has the form

$$\alpha\beta = \sum_{x \in G} \left(\sum_{gh=x} \alpha_g \beta_h \right) x.$$

Since the multiplication in the group is associative, it follows that the multiplication in KG is associative. Furthermore, one checks that the multiplication in KG is distributive. The identity element of the group algebra KG is given by the identity element of G.

Note that the group algebra KG is finite-dimensional if and only if the group G is finite, in which case the dimension of KG is equal to the order of the group G. The group algebra KG is commutative if and only if the group G is abelian.

Example 1.10. Let G be the cyclic group of order 3, generated by y, so that $G = \{1_G, y, y^2\}$ and $y^3 = 1_G$. Then we have

$$(a_0 1_G + a_1 y + a_2 y^2)(b_0 1_G + b_1 y + b_2 y^2) = c_0 1_G + c_1 y + c_2 y^2,$$

with

$$c_0 = a_0 b_0 + a_1 b_2 + a_2 b_1, \ c_1 = a_0 b_1 + a_1 b_0 + a_2 b_2, \ c_2 = a_0 b_2 + a_1 b_1 + a_2 b_0.$$

1.1.3   Path Algebras of Quivers

Path algebras of quivers are a class of algebras with an easy multiplication formula, and they are extremely useful for calculating examples. They also have connections to other parts of mathematics. The underlying basis of a path algebra is the set of paths in a finite directed graph. It is customary in representation theory to call such a graph a quiver. We assume throughout that a quiver has finitely many vertices and finitely many arrows.

Definition 1.11. A *quiver* Q is a finite directed graph. We sometimes write $Q = (Q_0, Q_1)$, where Q_0 is the set of vertices and Q_1 is the set of arrows.

We assume that Q_0 and Q_1 are finite sets. For any arrow $\alpha \in Q_1$ we denote by $s(\alpha) \in Q_0$ its starting point and by $t(\alpha) \in Q_0$ its end point.

A non-trivial *path* in Q is a sequence $p = \alpha_r \ldots \alpha_2 \alpha_1$ of arrows $\alpha_i \in Q_1$ such that $t(\alpha_i) = s(\alpha_{i+1})$ for all $i = 1, \ldots, r - 1$. Note that our convention is to read paths from right to left. The number r of arrows is called the *length* of p, and we denote by $s(p) = s(\alpha_1)$ the starting point, and by $t(p) = t(\alpha_r)$ the end point of p.

For each vertex $i \in Q_0$ we also need to have a trivial path of length 0, which we call e_i, and we set $s(e_i) = i = t(e_i)$.

We call a path p in Q an *oriented cycle* if p has positive length and $s(p) = t(p)$.

Definition 1.12. Let K be a field and Q a quiver. The *path algebra* KQ of the quiver Q over the field K has underlying vector space with basis given by all paths in Q.

The multiplication in KQ is defined on the basis by concatenation of paths (if possible), and extended linearly to linear combinations. More precisely, for two paths $p = \alpha_r \ldots \alpha_1$ and $q = \beta_s \ldots \beta_1$ in Q we set

$$p \cdot q = \begin{cases} \alpha_r \ldots \alpha_1 \beta_s \ldots \beta_1 & \text{if } t(\beta_s) = s(\alpha_1), \\ 0 & \text{otherwise.} \end{cases}$$

Note that for the trivial paths e_i, where i is a vertex in Q, we have that $p \cdot e_i = p$ for $i = s(p)$ and $p \cdot e_i = 0$ for $i \neq s(p)$; similarly $e_i \cdot p = p$ for $i = t(p)$ and 0 otherwise. In particular we have $e_i \cdot e_i = e_i$.

The multiplication in KQ is associative since the concatenation of paths is associative, and it is distributive, by definition of products for arbitrary linear combinations. We claim that the identity element of KQ is given by the sum of trivial paths, that is

$$1_{KQ} = \sum_{i \in Q_0} e_i.$$

In fact, for every path p in Q we have

$$p \cdot \left(\sum_{i \in Q_0} e_i \right) = \sum_{i \in Q_0} p \cdot e_i = p \cdot e_{s(p)} = p = e_{t(p)} \cdot p = \left(\sum_{i \in Q_0} e_i \right) \cdot p,$$

and by distributivity it follows that $\alpha \cdot \left(\sum_{i \in Q_0} e_i \right) = \alpha = \left(\sum_{i \in Q_0} e_i \right) \cdot \alpha$ for every $\alpha \in KQ$.

Example 1.13.

(1) We consider the quiver Q of the form $1 \xleftarrow{\alpha} 2$. The path algebra KQ has dimension 3, the basis consisting of paths is $\{e_1, e_2, \alpha\}$. The multiplication table

for KQ is given by

$\cdot$	e_1	α	e_2
e_1	e_1	α	0
α	0	0	α
e_2	0	0	e_2

(2) Let Q be the *one-loop quiver* with one vertex v and one arrow α with $s(\alpha) = v = t(\alpha)$, that is

$$v \;\circlearrowleft\; \alpha$$

Then the path algebra KQ has as basis the set $\{1, \alpha, \alpha^2, \alpha^3, \ldots\}$, and it is not finite-dimensional.

(3) A quiver can have multiple arrows between two vertices. This is the case for the *Kronecker quiver*

$$1 \underset{\beta}{\overset{\alpha}{\rightrightarrows}} 2$$

(4) Examples of quivers where more than two arrows start or end at a vertex are the *three-subspace quiver*

$$
\begin{array}{c}
2 \\
\downarrow \\
1 \longrightarrow 4 \longleftarrow 3
\end{array}
$$

or similarly with a different orientation of the arrows

$$
\begin{array}{c}
2 \\
\uparrow \\
1 \longleftarrow 4 \longrightarrow 3
\end{array}
$$

(5) Quivers can have oriented cycles, for example

$$
\begin{array}{ccc}
2 & \longrightarrow & 5 \\
\uparrow & & \downarrow \\
1 \longleftarrow & 4 & \longleftarrow 3
\end{array}
$$

Exercise 1.2. Let K be a field.

(a) Show that the path algebra KQ in part (5) of Example 1.13 is not finite-dimensional.
(b) Show that the path algebra KQ of a quiver Q is finite-dimensional if and only if Q does not contain any oriented cycles.

1.2 Subalgebras, Ideals and Factor Algebras

In analogy to the 'subspaces' of vector spaces, and 'subgroups' of groups, we should define the notion of a 'subalgebra'. Suppose A is a K-algebra, then, roughly speaking, a subalgebra B of A is a subset of A which is itself an algebra with respect to the operations in A. This is made precise in the following definition.

Definition 1.14. Let K be a field and A a K-algebra. A subset B of A is called a K-*subalgebra* (or just subalgebra) of A if the following holds:

(i) B is a K-subspace of A, that is, for every $\lambda, \mu \in K$ and $b_1, b_2 \in B$ we have $\lambda b_1 + \mu b_2 \in B$.
(ii) B is closed under multiplication, that is, for all $b_1, b_2 \in B$, the product $b_1 b_2$ belongs to B.
(iii) The identity element 1_A of A belongs to B.

Exercise 1.3. Let A be a K-algebra and $B \subseteq A$ a subset. Show that B is a K-subalgebra of A if and only if B itself is a K-algebra with the operations induced from A.

Remark 1.15. Suppose B is given as a subset of some algebra A, with the same addition, scalar multiplication, and multiplication. To decide whether or not B is an algebra, there is no need to check all the axioms of Definition 1.1. Instead, it is enough to verify conditions (i) to (iii) of Definition 1.14.

We consider several examples.

Example 1.16. Let K be a field.

(1) The K-algebra $M_n(K)$ of $n \times n$-matrices over K has many important subalgebras.

    (i) The upper triangular matrices

$$T_n(K) := \{a = (a_{ij}) \in M_n(K) \mid a_{ij} = 0 \text{ for } i > j\}$$

    form a subalgebra of $M_n(K)$. Similarly, one can define lower triangular matrices and they also form a subalgebra of $M_n(K)$.

(ii) The diagonal matrices

$$D_n(K) := \{a = (a_{ij}) \in M_n(K) \mid a_{ij} = 0 \text{ for } i \neq j\}$$

form a subalgebra of $M_n(K)$.

(iii) The *three-subspace algebra* is the subalgebra of $M_4(K)$ given by

$$\left\{ \begin{pmatrix} a_1 & b_1 & b_2 & b_3 \\ 0 & a_2 & 0 & 0 \\ 0 & 0 & a_3 & 0 \\ 0 & 0 & 0 & a_4 \end{pmatrix} \mid a_i, b_j \in K \right\}.$$

(iv) There are also subalgebras such as

$$\left\{ \begin{pmatrix} a & b & 0 & 0 \\ c & d & 0 & 0 \\ 0 & 0 & x & y \\ 0 & 0 & z & u \end{pmatrix} \mid a, b, c, d, x, y, z, u \in K \right\} \subseteq M_4(K).$$

(2) The $n \times n$-matrices $M_n(\mathbb{Z}) \subseteq M_n(\mathbb{R})$ over the integers are closed under addition and multiplication, but $M_n(\mathbb{Z})$ is not an $\mathbb{R}$-subalgebra of $M_n(\mathbb{R})$ since it is not an $\mathbb{R}$-subspace.

(3) The subset

$$\left\{ \begin{pmatrix} 0 & b \\ 0 & 0 \end{pmatrix} \mid b \in K \right\} \subseteq T_2(K)$$

is not a K-subalgebra of $T_2(K)$ since it does not contain the identity element.

(4) Let A be a K-algebra. For any element $a \in A$ define A_a to be the K-span of $\{1_A, a, a^2, \ldots\}$. That is, A_a is the space of polynomial expressions in a. This is a K-subalgebra of A, and it is always commutative. Note that if A is finite-dimensional then so is A_a.

(5) Let $A = A_1 \times A_2$, the direct product of two algebras. Then $A_1 \times \{0\}$ is not a subalgebra of A since it does not contain the identity element of A.

(6) Let H be a subgroup of a group G. Then the group algebra KH is a subalgebra of the group algebra KG.

(7) Let KQ be the path algebra of the quiver $1 \xleftarrow{\alpha} 2 \xleftarrow{\beta} 3$. We can consider the 'subquiver', Q' given by $1 \xleftarrow{\alpha} 2$. The path algebra KQ' is not a subalgebra of KQ since it does not contain the identity element $1_{KQ} = e_1 + e_2 + e_3$ of KQ.

Exercise 1.4. Verify that the three-subspace algebra from Example 1.16 is a subalgebra of $M_4(K)$, hence is a K-algebra.

In addition to subalgebras, there are also ideals, and they are needed when one wants to define factor algebras.

Definition 1.17. If A is a K-algebra then a subset I is a *left ideal* of A provided $(I, +)$ is a subgroup of $(A, +)$ such that $ax \in I$ for all $x \in I$ and $a \in A$. Similarly, I is a *right ideal* of A if $(I, +)$ is a subgroup such that $xa \in I$ for all $x \in I$ and $a \in A$. A subset I of A is a *two-sided ideal* (or just *ideal*) of A if it is both a left ideal and a right ideal.

Remark 1.18.

(1) The above definition works verbatim for rings instead of algebras.
(2) For commutative algebras A the notions of left ideal, right ideal and two-sided ideal clearly coincide. However, for non-commutative algebras, a left ideal might not be a right ideal, and vice versa; see the examples below.
(3) An ideal I (left or right or two-sided) of an algebra A is by definition closed under multiplication, and also a subspace, see Lemma 1.20 below. However, an ideal is in general not a subalgebra, since the identity element need not be in I. Actually, a (left or right or two-sided) ideal $I \subseteq A$ contains 1_A if and only if $I = A$. In addition, a subalgebra is in general not an ideal.

Exercise 1.5. Assume B is a subalgebra of A. Show that B is a left ideal (or right ideal) if and only if $B = A$.

Example 1.19.

(1) Every K-algebra A has the trivial two-sided ideals $\{0\}$ and A. In the sequel, we will usually just write 0 for the ideal $\{0\}$.
(2) Let A be a K-algebra. For every $z \in A$ the subset $Az = \{az \mid a \in A\}$ is a left ideal of A. Similarly, $zA = \{za \mid a \in A\}$ is a right ideal of A for every $z \in A$. These are called the *principal* (left or right) *ideals* generated by z. As a notation for principal ideals in commutative algebras we often use $Az = (z)$.
(3) For non-commutative algebras, Az need not be a two-sided ideal. For instance, let $A = M_n(K)$ for some $n \geq 2$ and consider the matrix unit $z = E_{ii}$ for some $i \in \{1, \ldots, n\}$. Then the left ideal Az consists of the matrices with non-zero entries only in column i, whereas the right ideal zA consists of those matrices with non-zero entries in row i. In particular, $Az \neq zA$, and both are not two-sided ideals of $M_n(K)$.
(4) The only two-sided ideals in $M_n(K)$ are the trivial ideals. In fact, suppose that $I \neq 0$ is a two-sided ideal in $M_n(K)$, and let $a = (a_{ij}) \in I \setminus \{0\}$, say $a_{k\ell} \neq 0$. Then for every $r, s \in \{1, \ldots, n\}$ we have that $E_{rk} a E_{\ell s} \in I$ since I is a two-sided ideal. On the other hand,

$$E_{rk} a E_{\ell s} = E_{rk} \left(\sum_{i, j=1}^{n} a_{ij} E_{ij} \right) E_{\ell s} = \left(\sum_{j=1}^{n} a_{kj} E_{rj} \right) E_{\ell s} = a_{k\ell} E_{rs}.$$

Since $a_{k\ell} \neq 0$ we conclude that $E_{rs} \in I$ for all r, s and hence $I = M_n(K)$.
(5) Consider the K-algebra $T_n(K)$ of upper triangular matrices. The K-subspace of strict upper triangular matrices $I_1 := \text{span}\{E_{ij} \mid 1 \leq i < j \leq n\}$ forms a two-sided ideal. More generally, for any $d \in \mathbb{N}$ the subspace

$I_d := \text{span}\{E_{ij} \mid d \leq j - i\}$ is a two-sided ideal of $T_n(K)$. Note that by definition, $I_d = 0$ for $d \geq n$.

(6) Let Q be a quiver and $A = KQ$ be its path algebra over a field K. Take the trivial path e_i for a vertex $i \in Q_0$. Then by Example (2) above we have the left ideal Ae_i. This is spanned by all paths in Q with starting vertex i. Similarly, the right ideal $e_i A$ is spanned by all paths in Q ending at vertex i.

A two-sided ideal of KQ is for instance given by the subspace $(KQ)^{\geq 1}$ spanned by all paths in Q of non-zero length (that is, by all paths in Q except the trivial paths). More generally, for every $d \in \mathbb{N}$ the linear combinations of all paths in Q of length at least d form a two-sided ideal $(KQ)^{\geq d}$.

Exercise 1.6. Let Q be a quiver, i a vertex in Q and $A = KQ$ the path algebra (over some field K). Find a condition on Q so that the left ideal Ae_i is a two-sided ideal.

Let A be a K-algebra and let $I \subset A$ be a proper two-sided ideal. Recall from basic algebra that the cosets $A/I := \{a + I \mid a \in A\}$ form a ring, the *factor ring*, with addition and multiplication defined by

$$(a + I) + (b + I) := a + b + I, \quad (a + I)(b + I) := ab + I$$

for $a, b \in A$. Note that these operations are well-defined, that is, they are independent of the choice of the representatives of the cosets, because I is a two-sided ideal. Moreover, the assumption $I \neq A$ is needed to ensure that the factor ring has an identity element; see Axiom (R7) in Sect. 1.1.

For K-algebras we have some extra structure on the factor rings.

Lemma 1.20. *Let A be a K-algebra. Then the following holds.*

(a) Every left (or right) ideal I of A is a K-subspace of A.

(b) If I is a proper two-sided ideal of A then the factor ring A/I is a K-algebra, the factor algebra *of A with respect to I.*

Proof. (a) Let I be a left ideal. By definition, $(I, +)$ is an abelian group. We need to show that if $\lambda \in K$ and $x \in I$ then $\lambda x \in I$. But $\lambda 1_A \in A$, and we obtain

$$\lambda x = \lambda(1_A x) = (\lambda 1_A)x \in I,$$

since I is a left ideal. The same argument works if I is a right ideal, by axiom (Alg) in Definition 1.1.

(b) We have already recalled above that the cosets A/I form a ring. Moreover, by part (a), I is a K-subspace and hence A/I is also a K-vector space with (well-defined) scalar multiplication $\lambda(a+I) = \lambda a + I$ for all $\lambda \in K$ and $a \in A$. According to Definition 1.1 it only remains to show that axiom (Alg) holds. But this property is inherited from A; explicitly, let $\lambda \in K$ and $a, b \in A$, then

$$\lambda((a + I)(b + I)) = \lambda(ab + I) = \lambda(ab) + I = (\lambda a)b + I$$

$$= (\lambda a + I)(b + I) = (\lambda(a + I))(b + I).$$

Similarly, using that $\lambda(ab) = a(\lambda b)$ by axiom (Alg), one shows that $\lambda((a+I)(b+I)) = (a+I)(\lambda(b+I))$. □

Example 1.21. Consider the algebra $K[X]$ of polynomials in one variable over a field K. Recall from a course on basic algebra that every non-zero ideal I of $K[X]$ is of the form $K[X]f = (f)$ for some non-zero polynomial $f \in K[X]$ (that is, $K[X]$ is a principal ideal domain). The factor algebra $A/I = K[X]/(f)$ is finite-dimensional. More precisely, we claim that it has dimension equal to the degree d of the polynomial f and that a K-basis of $K[X]/(f)$ is given by the cosets $1 + (f), X + (f), \ldots, X^{d-1} + (f)$. In fact, if $g \in K[X]$ then division with remainder in $K[X]$ (polynomial long division) yields $g = qf + r$ with polynomials $q, r \in K[X]$ and r has degree less than d, the degree of f. Hence

$$g + (f) = r + (f) \in \text{span}\{1 + (f), X + (f), \ldots, X^{d-1} + (f)\}.$$

On the other hand, considering degrees one checks that this spanning set of $K[X]/(f)$ is also linearly independent.

1.3 Algebra Homomorphisms

As with any algebraic structure (like vector spaces, groups, rings) one needs to define and study maps between algebras which 'preserve the structure'.

Definition 1.22. Let A and B be K-algebras. A map $\phi : A \to B$ is a *K-algebra homomorphism* (or homomorphism of K-algebras) if

 (i) ϕ is a K-linear map of vector spaces,
 (ii) $\phi(ab) = \phi(a)\phi(b)$ for all $a, b \in A$,
(iii) $\phi(1_A) = 1_B$.

The map $\phi : A \to B$ is a K-algebra *isomorphism* if it is a K-algebra homomorphism and is in addition bijective. If so, then the K-algebras A and B are said to be isomorphic, and one writes $A \cong B$. Note that the inverse of an algebra isomorphism is also an algebra isomorphism, see Exercise 1.14.

Remark 1.23.

(1) To check condition (ii) of Definition 1.22, it suffices to take for a, b any two elements in some fixed basis. Then it follows for arbitrary elements of A as long as ϕ is K-linear.
(2) Note that the definition of an algebra homomorphism requires more than just being a homomorphism of the underlying rings. Indeed, a ring homomorphism between K-algebras is in general not a K-algebra homomorphism.

 For instance, consider the complex numbers $\mathbb{C}$ as a $\mathbb{C}$-algebra. Let $\phi : \mathbb{C} \to \mathbb{C}, \phi(z) = \bar{z}$ be the complex conjugation map. By the usual rules for complex conjugation ϕ satisfies axioms (ii) and (iii) from Definition 1.22, that is, ϕ is a ring homomorphism. But ϕ does not satisfy axiom (i), since for example

$$\phi(ii) = \phi(i^2) = \phi(-1) = -1 \neq 1 = i(-i) = i\phi(i).$$

So ϕ is not a $\mathbb{C}$-algebra homomorphism. However, if one considers $\mathbb{C}$ as an algebra over $\mathbb{R}$, then complex conjugation is an $\mathbb{R}$-algebra homomorphism. In fact, for $r \in \mathbb{R}$ and $z \in \mathbb{C}$ we have

$$\phi(rz) = \overline{rz} = \bar{r}\bar{z} = r\bar{z} = r\phi(z).$$

We list a few examples, some of which will occur frequently. For each of these, we recommend checking that the axioms of Definition 1.22 are indeed satisfied.

Example 1.24. Let K be a field.

(1) Let Q be the one-loop quiver with one vertex v and one arrow α such that $s(\alpha) = v = t(\alpha)$. As pointed out in Example 1.13, the path algebra KQ has a basis consisting of $1, \alpha, \alpha^2, \ldots$. The multiplication is given by $\alpha^i \cdot \alpha^j = \alpha^{i+j}$. From this we can see that the polynomial algebra $K[X]$ and KQ are isomorphic, via the homomorphism defined by sending $\sum_i \lambda_i X^i$ to $\sum_i \lambda_i \alpha^i$. That is, we substitute α into the polynomial $\sum_i \lambda_i X^i$.

(2) Let A be a K-algebra. For every element $a \in A$ we consider the 'evaluation map'

$$\varphi_a : K[X] \to A , \quad \sum_i \lambda_i X^i \mapsto \sum_i \lambda_i a^i.$$

This is a homomorphism of K-algebras.

(3) Let A be a K-algebra and $I \subset A$ a proper two-sided ideal, with factor algebra A/I. Then the canonical map $\pi : A \to A/I$, $a \mapsto a + I$ is a surjective K-algebra homomorphism.

(4) Let $A = A_1 \times \ldots \times A_n$ be a direct product of K-algebras. Then for each $i \in \{1, \ldots, n\}$ the projection map

$$\pi_i : A \to A_i , \quad (a_1, \ldots, a_n) \mapsto a_i$$

is a surjective K-algebra homomorphism. However, note that the embeddings

$$\iota_i : A_i \to A , \quad a_i \mapsto (0, \ldots, 0, a_i, 0, \ldots, 0)$$

are not K-algebra homomorphisms when $n \geq 2$ since the identity element 1_{A_i} is not mapped to the identity element $1_A = (1_{A_1}, \ldots, 1_{A_n})$.

(5) Let $A = T_n(K)$ be the algebra of upper triangular matrices. Denote by $B = K \times \ldots \times K$ the direct product of n copies of K. Define $\phi : A \to B$ by setting

$$\phi \begin{pmatrix} a_{11} & & * \\ & \ddots & \\ 0 & & \ddots \\ & & a_{nn} \end{pmatrix} = (a_{11}, \ldots, a_{nn}).$$

Then ϕ is a homomorphism of K-algebras.

(6) Consider the matrix algebra $M_n(K)$ where $n \geq 1$. Then the opposite algebra $M_n(K)^{op}$ (as defined in Definition 1.6) is isomorphic to the algebra $M_n(K)$. In fact, consider the map given by transposition of matrices

$$\tau : M_n(K) \to M_n(K)^{op} , \quad m \mapsto m^t.$$

Clearly, this is a K-linear map, and it is bijective (since τ^2 is the identity). Moreover, for all matrices $m_1, m_2 \in M_n(K)$ we have $(m_1 m_2)^t = m_2^t m_1^t$ by a standard result from linear algebra, that is,

$$\tau(m_1 m_2) = (m_1 m_2)^t = m_2^t m_1^t = \tau(m_1) * \tau(m_2).$$

Finally, τ maps the identity matrix to itself, hence τ is an algebra homomorphism.

(7) When writing linear transformations of a finite-dimensional vector space as matrices with respect to a fixed basis, one basically proves that the algebra of linear transformations is isomorphic to the algebra of square matrices. We recall the proof, partly as a reminder, but also since we will later need a generalization.

Suppose V is an n-dimensional vector space over the field K. Then the K-algebras $\operatorname{End}_K(V)$ and $M_n(K)$ are isomorphic.

In fact, we fix a K-basis of V. Suppose α is a linear transformation of V, let $M(\alpha)$ be the matrix of α with respect to the fixed basis. Then define a map

$$\psi : \operatorname{End}_K(V) \to M_n(K), \quad \psi(\alpha) := M(\alpha).$$

From linear algebra it is known that ψ is K-linear and that it preserves the multiplication, that is, $M(\beta)M(\alpha) = M(\beta \circ \alpha)$. The map ψ is also injective. Suppose $M(\alpha) = 0$, then by definition α maps the fixed basis to zero, but then $\alpha = 0$. The map ψ is surjective, because every $n \times n$-matrix defines a linear transformation of V.

(8) We consider the algebra $T_2(K)$ of upper triangular 2×2-matrices. This algebra is of dimension 3 and has a basis of matrix units E_{11}, E_{12} and E_{22}. Their products can easily be computed (for instance using the formula in Exercise 1.10), and they are collected in the multiplication table below. Let us now compare the algebra $T_2(K)$ with the path algebra KQ for the quiver

$1 \xleftarrow{\alpha} 2$ which has appeared already in Example 1.13. The multiplication tables for these two algebras are given as follows

$\cdot$	E_{11}	E_{12}	E_{22}
E_{11}	E_{11}	E_{12}	0
E_{12}	0	0	E_{12}
E_{22}	0	0	E_{22}

$\cdot$	e_1	α	e_2
e_1	e_1	α	0
α	0	0	α
e_2	0	0	e_2

From this it easily follows that the assignment $E_{11} \mapsto e_1$, $E_{12} \mapsto \alpha$, and $E_{22} \mapsto e_2$ defines a K-algebra isomorphism $T_2(K) \to KQ$.

Remark 1.25. The last example can be generalized. For every $n \in \mathbb{N}$ the K-algebra $T_n(K)$ of upper triangular $n \times n$-matrices is isomorphic to the path algebra of the quiver

$$1 \longleftarrow 2 \longleftarrow \ldots \longleftarrow n-1 \longleftarrow n$$

See Exercise 1.18.

In general, homomorphisms and isomorphisms are very important when comparing different algebras. Exactly as for rings we have an isomorphism theorem for algebras. Note that the kernel and the image of an algebra homomorphism are just the kernel and the image of the underlying K-linear map.

Theorem 1.26 (Isomorphism Theorem for Algebras). *Let K be a field and let A and B be K-algebras. Suppose $\phi : A \to B$ is a K-algebra homomorphism. Then the kernel $\ker(\phi)$ is a two-sided ideal of A, and the image $\mathrm{im}(\phi)$ is a subalgebra of B. Moreover:*

(a) If I is an ideal of A and $I \subseteq \ker(\phi)$ then we have a surjective algebra homomorphism

$$\bar{\phi} : A/I \to \mathrm{im}(\phi), \quad \bar{\phi}(a + I) = \phi(a).$$

(b) The map $\bar{\phi}$ is injective if and only if $I = \ker(\phi)$. Hence ϕ induces an isomorphism

$$\Psi : A/\ker(\phi) \to \mathrm{im}(\phi), \quad a + \ker(\phi) \mapsto \phi(a).$$

Proof. From linear algebra we know that the kernel $\ker(\phi) = \{a \in A \mid \phi(a) = 0\}$ is a subspace of A, and $\ker(\phi) \neq A$ since $\phi(1_A) = 1_B$ (see Definition 1.22). If $a \in A$ and $x \in \ker(\phi)$ then we have

$$\phi(ax) = \phi(a)\phi(x) = \phi(a) \cdot 0 = 0 = 0 \cdot \phi(a) = \phi(x)\phi(a) = \phi(xa),$$

that is, $ax \in \ker(\phi)$ and $xa \in \ker(\phi)$ and $\ker(\phi)$ is a two-sided ideal of A.

We check that $\text{im}(\phi)$ is a subalgebra of B (see Definition 1.14). Since ϕ is a K-linear map, the image $\text{im}(\phi)$ is a subspace. It is also closed under multiplication and contains the identity element; in fact, we have

$$\phi(a)\phi(b) = \phi(ab) \in \text{im}(\phi) \quad \text{and} \quad 1_B = \phi(1_A) \in \text{im}(\phi)$$

since ϕ is an algebra homomorphism.

(a) If $a + I = a' + I$ then $a - a' \in I$ and since $I \subseteq \text{ker}(\phi)$ we have

$$0 = \phi(a - a') = \phi(a) - \phi(a').$$

Hence $\bar{\phi}$ is well defined, and its image is obviously equal to $\text{im}(\phi)$. It remains to check that $\bar{\phi}$ is an algebra homomorphism. It is known from linear algebra (and easy to check) that the map is K-linear. It takes the identity $1_A + I$ to the identity element of B since ϕ is an algebra homomorphism. To see that it preserves products, let $a, b \in A$; then

$$\bar{\phi}((a + I)(b + I)) = \bar{\phi}(ab + I) = \phi(ab) = \phi(a)\phi(b) = \bar{\phi}(a + I)\bar{\phi}(b + I).$$

(b) We see directly that $\text{ker}(\bar{\phi}) = \text{ker}(\phi)/I$. The homomorphism $\bar{\phi}$ is injective if and only if its kernel is zero, that is, $\text{ker}(\phi) = I$. The last part follows. $\qquad \square$

Example 1.27.

(1) Consider the following evaluation homomorphism of $\mathbb{R}$-algebras (see Example 1.24),

$$\Phi : \mathbb{R}[X] \to \mathbb{C}, \quad \Phi(f) = f(i)$$

where $i^2 = -1$. In order to apply the isomorphism theorem we have to determine the kernel of Φ. Clearly, the ideal $(X^2 + 1) = \mathbb{R}[X](X^2 + 1)$ generated by $X^2 + 1$ is contained in the kernel. On the other hand, if $g \in \text{ker}(\Phi)$, then division with remainder in $\mathbb{R}[X]$ yields polynomials h and r such that $g = h(X^2 + 1) + r$, where r is of degree ≤ 1. Evaluating at i gives $r(i) = 0$, but r has degree ≤ 1, so r is the zero polynomial, that is, $g \in (X^2 + 1)$. Since Φ is surjective by definition, the isomorphism theorem gives

$$\mathbb{R}[X]/(X^2 + 1) \cong \mathbb{C}$$

as $\mathbb{R}$-algebras.

(2) Let G be a cyclic group of order n, generated by the element $a \in G$. For a field K consider the group algebra KG; this is a K-algebra of dimension n. Similar to the previous example we consider the surjective evaluation homomorphism

$$\Phi : K[X] \to KG, \quad \Phi(f) = f(a).$$

From the isomorphism theorem we know that

$$K[X]/\ker(\Phi) \cong \mathrm{im}(\Phi) = KG.$$

The ideal generated by the polynomial $X^n - 1$ is contained in the kernel of Φ, since a^n is the identity element in G (and then also in KG). Thus, the algebra on the left-hand side has dimension at most n, see Example 1.21. On the other hand, KG has dimension n. From this we can conclude that $\ker(\Phi) = (X^n - 1)$ and that as K-algebras we have

$$K[X]/(X^n - 1) \cong KG.$$

(3) Let A be a finite-dimensional K-algebra and $a \in A$, and let A_a be the subalgebra of A as in Example 1.16. Let $t \in \mathbb{N}_0$ such that $\{1, a, a^2, \ldots a^t\}$ is linearly independent and $a^{t+1} = \sum_{i=0}^{t} \lambda_i a^i$ for $\lambda_i \in K$. The polynomial

$$m_a := X^{t+1} - \sum_{i=0}^{t} \lambda_i X^i \in K[X]$$

is called the minimal polynomial of a. It is the (unique) monic polynomial of smallest degree such that the evaluation map (see Example 1.24) at a is zero. By the same arguments as in the first two examples we have

$$K[X]/(m_a) \cong A_a.$$

(4) Suppose $A = A_1 \times \ldots \times A_r$, the direct product of K-algebras $A_1 \ldots, A_r$. Then for every $i \in \{1, \ldots, r\}$ the projection $\pi_i : A \to A_i$, $(a_1, \ldots, a_r) \mapsto a_i$, is an algebra homomorphism, and it is surjective. By definition the kernel has the form

$$\ker(\pi_i) = A_1 \times \ldots \times A_{i-1} \times 0 \times A_{i+1} \times \ldots \times A_r.$$

By the isomorphism theorem we have $A/\ker(\pi_i) \cong A_i$. We also see from this that $A_1 \times \ldots \times A_{i-1} \times 0 \times A_{i+1} \times \ldots \times A_r$ is a two-sided ideal of A.

(5) We consider the following map on the upper triangular matrices

$$\Phi : T_n(K) \to T_n(K), \quad \begin{pmatrix} a_{11} & & * \\ & \ddots & \\ 0 & & a_{nn} \end{pmatrix} \mapsto \begin{pmatrix} a_{11} & & 0 \\ & \ddots & \\ 0 & & a_{nn} \end{pmatrix},$$

which sets all non-diagonal entries to 0. It is easily checked that Φ is a homomorphism of K-algebras. By definition, the kernel is the two-sided ideal

$$U_n(K) := \{a = (a_{ij}) \in M_n(K) \mid a_{ij} = 0 \text{ for } i \geq j\}$$

of strict upper triangular matrices and the image is the K-algebra $D_n(K)$ of diagonal matrices. Hence the isomorphism theorem yields that

$$T_n(K)/U_n(K) \cong D_n(K)$$

as K-algebras. Note that, moreover, we have that $D_n(K) \cong K \times \ldots \times K$, the n-fold direct product of copies of K.

(6) We want to give an alternative description of the $\mathbb{R}$-algebra $\mathbb{H}$ of quaternions from Sect. 1.1.1. To this end, consider the map

$$\Phi : \mathbb{H} \to M_4(\mathbb{R}) \, , \, a + bi + cj + dk \mapsto \begin{pmatrix} a & b & c & d \\ -b & a & -d & c \\ -c & d & a & -b \\ -d & -c & b & a \end{pmatrix}.$$

Using the formula from Example 1.8 for the product of two elements from $\mathbb{H}$ one can check that Φ is an $\mathbb{R}$-algebra homomorphism, see Exercise 1.11.

Looking at the first row of the matrices in the image, it is immediate that Φ is injective. Therefore, the algebra $\mathbb{H}$ is isomorphic to the subalgebra $\mathrm{im}(\Phi)$ of $M_4(\mathbb{R})$. Since we know (from linear algebra) that matrix multiplication is associative and that the distributivity law holds in $M_4(\mathbb{R})$, we can now conclude with no effort that the multiplication in $\mathbb{H}$ is associative and distributive.

Exercise 1.7. Explain briefly how examples (1) and (2) in Example 1.27 are special cases of (3).

1.4   Some Algebras of Small Dimensions

One might like to know how many K-algebras there are of a given dimension, up to isomorphism. In general there might be far too many different algebras, but for small dimensions one can hope to get a complete overview. We fix a field K, and we consider K-algebras of dimension at most 2. For these, there are some restrictions.

Proposition 1.28. *Let K be a field.*

(a) Every 1-dimensional K-algebra is isomorphic to K.
(b) Every 2-dimensional K-algebra is commutative.

Proof. (a) Let A be a 1-dimensional K-algebra. Then A must contain the scalar multiples of the identity element, giving a subalgebra $U := \{\lambda 1_A \, | \, \lambda \in K\} \subseteq A$. Then $U = A$, since A is 1-dimensional. Moreover, according to axiom (Alg) from Definition 1.1 the product in U is given by $(\lambda 1_A)(\mu 1_A) = (\lambda\mu)1_A$ and hence the map $A \to K, \lambda 1_A \mapsto \lambda$, is an isomorphism of K-algebras.

(b) Let A be a 2-dimensional K-algebra. We can choose a basis which contains the identity element of A, say $\{1_A, b\}$ (use from linear algebra that every linearly independent subset can be extended to a basis). The basis elements clearly commute; but then also any linear combinations of basis elements commute, and therefore A is commutative.   □

We consider now algebras of dimension 2 over the real numbers $\mathbb{R}$. The aim is to classify these, up to isomorphism. The method will be to find suitable bases, leading to 'canonical' representatives of the isomorphism classes. It will turn out that there are precisely three $\mathbb{R}$-algebras of dimension 2, see Proposition 1.29 below.

So we take a 2-dimensional $\mathbb{R}$-algebra A, and we choose a basis of A containing the identity, say $\{1_A, b\}$, as in the above proof of Proposition 1.28. Then b^2 must be a linear combination of the basis elements, so there are scalars $\gamma, \delta \in \mathbb{R}$ such that $b^2 = \gamma 1_A + \delta b$. We consider the polynomial $X^2 - \delta X - \gamma \in \mathbb{R}[X]$ and we complete squares,

$$X^2 - \delta X - \gamma = (X - \delta/2)^2 - (\gamma + (\delta/2)^2).$$

Let $b' := b - (\delta/2)1_A$, this is an element in the algebra A, and we also set $\rho = \gamma + (\delta/2)^2$, which is a scalar from $\mathbb{R}$. Then we have

$$b'^2 = b^2 - \delta b + (\delta/2)^2 1_A = \gamma 1_A + (\delta/2)^2 1_A = \rho 1_A.$$

Note that $\{1_A, b'\}$ is still an $\mathbb{R}$-vector space basis of A. Then we rescale this basis by setting

$$\tilde{b} := \begin{cases} \sqrt{|\rho|^{-1}}\, b' & \text{if } \rho \neq 0, \\ b' & \text{if } \rho = 0. \end{cases}$$

Then the set $\{1_A, \tilde{b}\}$ also is an $\mathbb{R}$-vector space basis of A, and now we have $\tilde{b}^2 \in \{0, \pm 1_A\}$.

This leaves only three possible forms for the algebra A. We write A_j for the algebra in which $\tilde{b}^2 = j1_{A_j}$ for $j = 0, 1, -1$. We want to show that no two of these three algebras are isomorphic. For this we use Exercise 1.15.

(1) The algebra A_0 has a non-zero element with square zero, namely $\tilde{b}$. By Exercise 1.15, any algebra isomorphic to A_0 must also have such an element.

(2) The algebra A_1 does not have a non-zero element whose square is zero: Suppose $a^2 = 0$ for $a \in A_1$ and write $a = \lambda 1_{A_1} + \mu \tilde{b}$ with $\lambda, \mu \in \mathbb{R}$. Then, using that $\tilde{b}^2 = 1_{A_1}$, we have

$$0 = a^2 = (\lambda^2 + \mu^2)1_{A_1} + 2\lambda\mu\tilde{b}.$$

Since 1_{A_1} and $\tilde{b}$ are linearly independent, it follows that $2\lambda\mu = 0$ and $\lambda^2 = -\mu^2$. So $\lambda = 0$ or $\mu = 0$, which immediately forces $\lambda = \mu = 0$, and therefore $a = 0$, as claimed.

This shows that the algebra A_1 is not isomorphic to A_0, by Exercise 1.15.
(3) Consider the algebra A_{-1}. This occurs in nature, namely $\mathbb{C}$ is such an $\mathbb{R}$-algebra, taking $\tilde{b} = i$. In fact, we can see directly that A_{-1} is a field; take an arbitrary non-zero element $\alpha 1_{A_{-1}} + \beta \tilde{b} \in A_{-1}$ with $\alpha, \beta \in \mathbb{R}$. Then we have

$$(\alpha 1_{A_{-1}} + \beta \tilde{b})(\alpha 1_{A_{-1}} - \beta \tilde{b}) = (\alpha^2 + \beta^2) 1_{A_{-1}}$$

and since α and β are not both zero, we can write down the inverse of the above non-zero element with respect to multiplication.

Clearly A_0 and A_1 are not fields, since they have zero divisors, namely $\tilde{b}$ for A_0, and $(\tilde{b} - 1)(\tilde{b} + 1) = 0$ in A_1. So A_{-1} is not isomorphic to A_0 or A_1, again by Exercise 1.15.

We can list a 'canonical representative' for each of the three isomorphism classes of algebras. For $j \in \{0, \pm 1\}$ consider the $\mathbb{R}$-algebra

$$\mathbb{R}[X]/(X^2 - j) = \text{span}\{1 + (X^2 - j), \tilde{X} := X + (X^2 - j)\}$$

and observe that $\mathbb{R}[X]/(X^2 - j) \cong A_j$ for $j \in \{0, \pm 1\}$, where an isomorphism maps $\tilde{X}$ to $\tilde{b}$.

To summarize, we have proved the following classification of 2-dimensional $\mathbb{R}$-algebras.

Proposition 1.29. *Up to isomorphism, there are precisely three 2-dimensional algebras over $\mathbb{R}$. Any 2-dimensional algebra over $\mathbb{R}$ is isomorphic to precisely one of*

$$\mathbb{R}[X]/(X^2), \quad \mathbb{R}[X]/(X^2 - 1), \quad \mathbb{R}[X]/(X^2 + 1).$$

Example 1.30. Any 2-dimensional $\mathbb{R}$-algebra will be isomorphic to one of these. For example, which one is isomorphic to the 2-dimensional $\mathbb{R}$-algebra $\mathbb{R} \times \mathbb{R}$? The element $(-1, -1) \in \mathbb{R} \times \mathbb{R}$ squares to the identity element, so $\mathbb{R} \times \mathbb{R}$ is isomorphic to $A_1 \cong \mathbb{R}[X]/(X^2 - 1)$.

There is an alternative explanation for those familiar with the Chinese Remainder Theorem for polynomial rings; in fact, this yields

$$\mathbb{R}[X]/(X^2 - 1) \cong \mathbb{R}[X]/(X - 1) \times \mathbb{R}[X]/(X + 1)$$

and for any scalar, $\mathbb{R}[X]/(X - \lambda) \cong \mathbb{R}$: apply the isomorphism theorem with the evaluation map $\varphi_\lambda : \mathbb{R}[X] \to \mathbb{R}$.

Remark 1.31. One might ask what happens for different fields. For the complex numbers we can adapt the above proof and see that there are precisely two 2-dimensional $\mathbb{C}$-algebras, up to isomorphism, see Exercise 1.26. However, the situation for the rational numbers is very different: there are infinitely many non-isomorphic 2-dimensional algebras over $\mathbb{Q}$, see Exercise 1.27.

Remark 1.32. In this book we focus on algebras over fields. One can also define algebras where for K one takes a commutative ring, instead of a field. With this, large parts of the constructions in this chapter work as well, but generally the situation is more complicated. Therefore we will not discuss these here.

In addition, as we have mentioned, our algebras are *associative* algebras, that is, for any a, b, c in the algebra we have

$$(ab)c = a(bc).$$

There are other kinds of algebras which occur in various contexts.

Sometimes one defines an 'algebra' to be a vector space V over a field, together with a product $(v, w) \mapsto vw : V \times V \to V$ which is bilinear. Then one can impose various identities, and obtain various types of algebras. Perhaps the most common algebras, apart from associative algebras, are the Lie algebras, which are algebras as above where the product is written as a bracket $[v, w]$ and where one requests that for any $v \in V$, the product $[v, v] = 0$, and in addition that the 'Jacobi identity' must hold, that is for any $v, w, z \in V$

$$[[v, w], z] + [[w, z], v] + [[z, v], w] = 0.$$

The properties of such Lie algebras are rather different; see the book by Erdmann and Wildon in this series for a thorough treatment of Lie algebras at an undergraduate level.[1]

EXERCISES

1.8. Assume A is a finite-dimensional algebra over K. If $a, b \in A$ and $ab = 1_A$ then also $ba = 1_A$. That is, left inverses are automatically two-sided inverses. (Hint: Observe that $S(x) = ax$ and $T(x) = bx$ for $x \in A$ define linear maps of A and that $S \circ T = \text{id}_A$. Apply a result from linear algebra.)

1.9. Let $A = \text{End}_K(V)$ be the algebra of K-linear maps of the infinite-dimensional K-vector space V with basis $\{v_i \mid i \in \mathbb{N}\}$. Define elements a, b in A on the basis, and extend to finite linear combinations, as follows: Take $b(v_i) = v_{i+1}$ for $i \geq 1$, and

$$a(v_i) := \begin{cases} v_{i-1} & i > 1, \\ 0 & i = 1. \end{cases}$$

Check that $a \circ b = 1_A$ and show also that $b \circ a$ is not the identity of A.

---

[1] K. Erdmann, M.J. Wildon, *Introduction to Lie Algebras*. Springer Undergraduate Mathematics Series. Springer-Verlag London, Ltd., London, 2006. x+251 pp.

1.10. Let K be a field and let $E_{ij} \in M_n(K)$ be the matrix units as defined in Example 1.3, that is, E_{ij} has an entry 1 at position (i, j) and all other entries are 0. Show that for all $i, j, k, \ell \in \{1, \ldots, n\}$ we have

$$E_{ij} E_{k\ell} = \delta_{jk} E_{i\ell} = \begin{cases} E_{i\ell} & \text{if } j = k, \\ 0 & \text{if } j \neq k, \end{cases}$$

where δ_{jk} is the Kronecker symbol, that is, $\delta_{jk} = 1$ if $j = k$ and 0 otherwise.

1.11. Let $\mathbb{H}$ be the $\mathbb{R}$-algebra of quaternions from Sect. 1.1.1.

(a) Compute the product of two arbitrary elements $a_1 + b_1 i + c_1 j + d_1 k$ and $a_2 + b_2 i + c_2 j + d_2 k$ in $\mathbb{H}$, that is, verify the formula given in Example 1.8.

(b) Verify that the following set of matrices

$$A := \left\{ \begin{pmatrix} a & b & c & d \\ -b & a & -d & c \\ -c & d & a & -b \\ -d & -c & b & a \end{pmatrix} \mid a, b, c, d \in \mathbb{R} \right\} \subseteq M_4(\mathbb{R})$$

forms a subalgebra of the $\mathbb{R}$-algebra $M_4(\mathbb{R})$.

(c) Prove that the algebra A from part (b) is isomorphic to the algebra $\mathbb{H}$ of quaternions, that is, fill in the details in Example 1.27.

1.12. Let $\mathbb{H}$ be the $\mathbb{R}$-algebra of quaternions. By using the formula in Example 1.8, determine all roots in $\mathbb{H}$ of the polynomial $X^2 + 1$. In particular, verify that there are infinitely many such roots.

1.13. Let K be a field. Which of the following subsets of $M_3(K)$ are K-subalgebras? Each asterisk indicates an arbitrary entry from K, not necessarily the same.

$$\begin{pmatrix} * & * & 0 \\ 0 & * & * \\ 0 & 0 & * \end{pmatrix}, \quad \begin{pmatrix} * & * & * \\ 0 & * & 0 \\ 0 & 0 & * \end{pmatrix}, \quad \begin{pmatrix} * & 0 & * \\ 0 & * & 0 \\ 0 & 0 & * \end{pmatrix},$$

$$\begin{pmatrix} * & * & * \\ 0 & * & 0 \\ 0 & * & * \end{pmatrix}, \quad \begin{pmatrix} * & * & * \\ 0 & * & * \\ 0 & 0 & 0 \end{pmatrix}, \quad \begin{pmatrix} * & 0 & * \\ 0 & * & 0 \\ * & 0 & * \end{pmatrix}.$$

1.14. Assume A and B are K-algebras, and $\phi : A \to B$ is a K-algebra isomorphism. Show that then the inverse ϕ^{-1} of ϕ is a K-algebra isomorphism from B to A.

1.15. Suppose $\phi : A \to B$ is an isomorphism of K-algebras. Show that the following holds.

(i) If $a \in A$ then $a^2 = 0$ if and only if $\phi(a)^2 = 0$.

(ii) $a \in A$ is a zero divisor if and only if $\phi(a) \in B$ is a zero divisor.

(iii) A is commutative if and only if B is commutative.

(iv) A is a field if and only if B is a field.

1.16. For a field K we consider the following 3-dimensional subspaces of $M_3(K)$:

$$A_1 = \left\{ \begin{pmatrix} x & 0 & 0 \\ 0 & y & 0 \\ 0 & 0 & z \end{pmatrix} \mid x, y, z \in K \right\} ; A_2 = \left\{ \begin{pmatrix} x & y & z \\ 0 & x & 0 \\ 0 & 0 & x \end{pmatrix} \mid x, y, z \in K \right\} ;$$

$$A_3 = \left\{ \begin{pmatrix} x & y & z \\ 0 & x & y \\ 0 & 0 & x \end{pmatrix} \mid x, y, z \in K \right\} ; A_4 = \left\{ \begin{pmatrix} x & y & 0 \\ 0 & z & 0 \\ 0 & 0 & x \end{pmatrix} \mid x, y, z \in K \right\} ;$$

$$A_5 = \left\{ \begin{pmatrix} x & 0 & y \\ 0 & x & z \\ 0 & 0 & x \end{pmatrix} \mid x, y, z \in K \right\} .$$

(i) For each i, verify that A_i is a K-subalgebra of $M_3(K)$. Is A_i commutative?

(ii) For each i, determine the set $\bar{A}_i := \{\alpha \in A_i \mid \alpha^2 = 0\}$.

(iii) For each pair of algebras above, determine whether they are isomorphic, or not. Hint: Apply Exercise 1.15.

1.17. Find all quivers whose path algebras have dimension at most 5 (up to an obvious notion of isomorphism). For this, note that quivers don't have to be connected and that they are allowed to have multiple arrows.

1.18. Show that for every $n \in \mathbb{N}$ the K-algebra $T_n(K)$ of upper triangular $n \times n$-matrices is isomorphic to the path algebra of the quiver

$$1 \longleftarrow 2 \longleftarrow \ldots \longleftarrow n - 1 \longleftarrow n$$

1.19. Let K be a field.

(a) Consider the following sets of matrices in $M_3(K)$ (where each asterisk indicates an arbitrary entry from K, not necessarily the same)

$$A := \begin{pmatrix} * & * & 0 \\ 0 & * & 0 \\ 0 & * & * \end{pmatrix} , \qquad B := \begin{pmatrix} * & 0 & 0 \\ * & * & 0 \\ * & 0 & * \end{pmatrix} .$$

Show that A and B are subalgebras of $M_3(K)$.

If possible, for each of the algebras A and B find an isomorphism to the path algebra of a quiver.

(b) Find subalgebras of $M_3(K)$ which are isomorphic to the path algebra of
the quiver $1 \longrightarrow 2 \longleftarrow 3$.

1.20. Let K be a field. Consider the following set of upper triangular matrices

$$A := \left\{ \begin{pmatrix} a & 0 & c \\ 0 & a & d \\ 0 & 0 & b \end{pmatrix} \mid a, b, c, d \in K \right\} \subseteq T_3(K).$$

(a) Show that A is a subalgebra of $T_3(K)$.
(b) Show that A is isomorphic to the path algebra of the Kronecker quiver,
as defined in Example 1.13.

1.21. For a field K let A be the three-subspace algebra as in Example 1.16, that is,

$$A = \left\{ \begin{pmatrix} a_1 & b_1 & b_2 & b_3 \\ 0 & a_2 & 0 & 0 \\ 0 & 0 & a_3 & 0 \\ 0 & 0 & 0 & a_4 \end{pmatrix} \mid a_i, b_j \in K \right\}.$$

(a) The three-subspace quiver is the quiver in Example 1.13 where all arrows
point towards the branch vertex. Show that the path algebra of the three-
subspace quiver is isomorphic to the three-subspace algebra. (Hint: It
might be convenient to label the branch vertex as vertex 1.)
(b) Determine the opposite algebra A^{op}. Is A isomorphic to A^{op}? Is it
isomorphic to the path algebra of some quiver?

1.22. This exercise gives a criterion by which one can sometimes deduce that a
certain algebra cannot be isomorphic to the path algebra of a quiver.

(a) An element e in a K-algebra A is called an *idempotent* if $e^2 = e$. Note
that 0 and 1_A are always idempotent elements. Show that if $\phi : A \to B$
is an isomorphism of K-algebras and $e \in A$ is an idempotent, then
$\phi(e) \in B$ is also an idempotent.
(b) Suppose that A is a K-algebra of dimension > 1 which has only 0 and 1_A
as idempotents. Then A is not isomorphic to the path algebra of a quiver.
(Hint: consider the trivial paths in the quiver.)
(c) Show that every division algebra A has no idempotents other than 0 and
1_A; deduce that if A has dimension > 1 then A cannot be isomorphic to
the path algebra of a quiver. In particular, this applies to the $\mathbb{R}$-algebra $\mathbb{H}$
of quaternions.
(d) Show that the factor algebra $K[X]/(X^2)$ is not isomorphic to the path
algebra of a quiver.

1.23. Let K be a field and A a K-algebra. Recall from Definition 1.6 that the opposite algebra A^{op} has the same underlying vector space as A, but a new multiplication

$$* : A^{op} \times A^{op} \to A^{op} , \quad a * b = ba$$

where on the right-hand the side the product is given by the multiplication in A.

(a) Show that A^{op} is again a K-algebra.

(b) Let $\mathbb{H}$ be the $\mathbb{R}$-algebra of quaternions (see Example 1.8). Show that the map

$$\varphi : \mathbb{H} \to \mathbb{H}^{op} , \quad a + bi + cj + dk \mapsto a - bi - cj - dk$$

is an $\mathbb{R}$-algebra isomorphism.

(c) Let G be a group and KG the group algebra. Show that the K-algebras KG and $(KG)^{op}$ are isomorphic.

(d) Let Q be a quiver and KQ its path algebra. Show that the opposite algebra $(KQ)^{op}$ is isomorphic to the path algebra $K\overline{Q}$, where $\overline{Q}$ is the quiver obtained from Q by reversing all arrows.

1.24. Consider the following 2-dimensional $\mathbb{R}$-subalgebras of $M_2(\mathbb{R})$ and determine to which algebra of Proposition 1.29 they are isomorphic:

(i) $D_2(\mathbb{R})$, the diagonal matrices;

(ii) $A := \left\{ \begin{pmatrix} a & b \\ 0 & a \end{pmatrix} \mid a, b \in \mathbb{R} \right\}$;

(iii) $B := \left\{ \begin{pmatrix} a & b \\ -b & a \end{pmatrix} \mid a, b \in \mathbb{R} \right\}$.

1.25. Let K be any field, and let A be a 2-dimensional K-algebra with basis $\{1_A, a\}$. Hence $A = A_a$ as in Example 1.16. Let $a^2 = \gamma 1_A + \delta a$, where $\gamma, \delta \in K$. As in Example 1.27, the element a has minimal polynomial $m_a := X^2 - \delta X - \gamma$, and

$$A = A_a \cong K[X]/(m_a).$$

(a) By applying the Chinese Remainder Theorem, show that if m_a has two distinct roots in K, then A is isomorphic to the direct product $K \times K$ of K-algebras.

(b) Show also that if $m_a = (X - \lambda)^2$ for $\lambda \in K$, then A is isomorphic to the algebra $K[T]/(T^2)$.

(c) Show that if m_a is irreducible in $K[X]$, then A is a field, containing K as a subfield.

(d) Explain briefly why the algebras in (a) and (b) are not isomorphic, and also not isomorphic to any of the algebras in (c).

1.26. Show that there are precisely two 2-dimensional algebras over $\mathbb{C}$, up to isomorphism.

1.27. Consider 2-dimensional algebras over $\mathbb{Q}$. Show that the algebras $\mathbb{Q}[X]/(X^2 - p)$ and $\mathbb{Q}[X]/(X^2 - q)$ are not isomorphic if p and q are distinct prime numbers.

1.28. Let $K = \mathbb{Z}_2$, the field with two elements.

(a) Let B be the set of matrices

$$B = \left\{ \begin{pmatrix} a & b \\ b & a+b \end{pmatrix} \mid a, b \in \mathbb{Z}_2 \right\}.$$

Show that B is a subalgebra of $M_2(\mathbb{Z}_2)$. Moreover, show that B is a field with four elements.

(b) Find all 2-dimensional $\mathbb{Z}_2$-algebras, up to isomorphism. (Hint: there are three different algebras.)

1.29. This exercise shows that every finite-dimensional algebra is isomorphic to a subalgebra of some matrix algebra. Assume A is an n-dimensional algebra over a field K.

(a) For $a \in A$, define the map $l_a : A \to A, l_a(x) = ax$.

  (i) Show that l_a is K-linear, hence is an element of $\mathrm{End}_K(A)$.
  (ii) Show that $l_{\lambda a + \mu b} = \lambda l_a + \mu l_b$ for $\lambda, \mu \in K$ and $a, b \in A$.
  (iii) Show that $l_{ab} = l_a \circ l_b$ for $a, b \in A$.

(b) By part (a), the map $\psi : A \to \mathrm{End}_K(A)$, where $\psi(a) = l_a$, is an algebra homomorphism. Show that it is injective.

(c) Fix a K-basis of A and write each l_a as a matrix with respect to such a fixed basis. Deduce that A is isomorphic to a subalgebra of $M_n(K)$.

(d) Let Q be the quiver $1 \overset{\alpha}{\longleftarrow} 2$, so that the path algebra KQ is 3-dimensional. Identify KQ with a subalgebra of $M_3(K)$.

Chapter 2
Modules and Representations

Representation theory studies how algebras can act on vector spaces. The fundamental notion is that of a module, or equivalently (as we shall see), that of a representation. Perhaps the most elementary way to think of modules is to view them as generalizations of vector spaces, where the role of scalars is played by elements in an algebra, or more generally, in a ring.

2.1 Definition and Examples

A vector space over a field K is an abelian group V together with a scalar multiplication $K \times V \to V$, satisfying the usual axioms. If one replaces the field K by a ring R, then one gets the notion of an R-module. Although we mainly deal with algebras over fields in this book, we slightly broaden the perspective in this chapter by defining modules over rings. We always assume that rings contain an identity element.

Definition 2.1. Let R be a ring with identity element 1_R. A *left R-module* (or just R-module) is an abelian group $(M, +)$ together with a map

$$R \times M \to M, \quad (r, m) \mapsto r \cdot m$$

such that for all $r, s \in R$ and all $m, n \in M$ we have

(i) $(r + s) \cdot m = r \cdot m + s \cdot m$;
(ii) $r \cdot (m + n) = r \cdot m + r \cdot n$;
(iii) $r \cdot (s \cdot m) = (rs) \cdot m$;
(iv) $1_R \cdot m = m$.

K. Erdmann, T. Holm, *Algebras and Representation Theory*, Springer Undergraduate Mathematics Series, https://doi.org/10.1007/978-3-319-91998-0_2

Exercise 2.1. Let R be a ring (with zero element 0_R and identity element 1_R) and M an R-module with zero element 0_M. Show that the following holds for all $r \in R$ and $m \in M$:

(i) $0_R \cdot m = 0_M$;
(ii) $r \cdot 0_M = 0_M$;
(ii) $-(r \cdot m) = (-r) \cdot m = r \cdot (-m)$, in particular $-m = (-1_R) \cdot m$.

Remark 2.2. Completely analogous to Definition 2.1 one can define *right* R-modules, using a map $M \times R \to M$, $(m, r) \mapsto m \cdot r$. When the ring R is not commutative the behaviour of left modules and of right modules can be different; for an illustration see Exercise 2.22. We will consider only left modules, since we are mostly interested in the case when the ring is a K-algebra, and scalars are usually written to the left.

Before dealing with elementary properties of modules we consider a few examples.

Example 2.3.

(1) When $R = K$ is a field, then R-modules are exactly the same as K-vector spaces. Thus, modules are a true generalization of the concept of a vector space.
(2) Let $R = \mathbb{Z}$, the ring of integers. Then every abelian group can be viewed as a $\mathbb{Z}$-module: If $n \geq 1$ then $n \cdot a$ is set to be the sum of n copies of a, and $(-n) \cdot a := -(n \cdot a)$, and $0_{\mathbb{Z}} \cdot a = 0$. With this, conditions (i) to (iv) in Definition 2.1 hold in any abelian group.
(3) Let R be a ring (with 1). Then every left ideal I of R is an R-module, with R-action given by ring multiplication. First, as a left ideal, $(I, +)$ is an abelian group. The properties (i)–(iv) hold even for arbitrary elements in R.
(4) A very important special case of (3) is that every ring R is an R-module, with action given by ring multiplication.
(5) Suppose $M_1, \ldots, M_n$ are R-modules. Then the cartesian product

$$M_1 \times \ldots \times M_n := \{(m_1, \ldots, m_n) \mid m_i \in M_i\}$$

is an R-module if one defines the addition and the action of R componentwise, so that

$$r \cdot (m_1, \ldots, m_n) := (rm_1, \ldots, rm_n) \text{ for } r \in R \text{ and } m_i \in M_i.$$

The module axioms follow immediately from the fact that they hold in $M_1, \ldots, M_n$.

We will almost always study modules when the ring is a K-algebra A. In this case, there is a range of important types of A-modules, which we will now introduce.

Example 2.4. Let K be a field.

(1) If A is a subalgebra of the algebra of $n \times n$-matrices $M_n(K)$, or a subalgebra of the algebra $\text{End}_K(V)$ of K-linear maps on a vector space V (see Example 1.3), then A has a *natural module*, which we will now describe.

  (i) Let A be a subalgebra of $M_n(K)$, and let $V = K^n$, the space of column vectors, that is, of $n \times 1$-matrices. By properties of matrix multiplication, multiplying an $n \times n$-matrix by an $n \times 1$-matrix gives an $n \times 1$-matrix, and this satisfies axioms (i) to (iv). Hence V is an A-module, the natural A-module. Here A could be all of $M_n(K)$, or the algebra of upper triangular $n \times n$-matrices, or any other subalgebra of $M_n(K)$.

  (ii) Let V be a vector space over the field K. Assume that A is a subalgebra of the algebra $\text{End}_K(V)$ of all K-linear maps on V (see Example 1.3). Then V becomes an A-module, where the action of A is just applying the linear maps to the vectors, that is, we set

$$A \times V \to V, \ (\varphi, v) \mapsto \varphi \cdot v := \varphi(v).$$

To check the axioms, let $\varphi, \psi \in A$ and $v, w \in V$, then we have

$$(\varphi + \psi) \cdot v = (\varphi + \psi)(v) = \varphi(v) + \psi(v) = \varphi \cdot v + \psi \cdot v$$

by the definition of the sum of two maps, and similarly

$$\varphi \cdot (v + w) = \varphi(v + w) = \varphi(v) + \varphi(w) = \varphi \cdot v + \varphi \cdot w$$

since φ is K-linear. Moreover,

$$\varphi \cdot (\psi \cdot v) = \varphi(\psi(v)) = (\varphi\psi) \cdot v$$

since the multiplication in $\text{End}_K(V)$ is given by composition of maps, and clearly we have $1_A \cdot v = \text{id}_V(v) = v$.

(2) Let $A = KQ$ be the path algebra of a quiver Q. For a fixed vertex i, let M be the span of all paths in Q starting at i. Then $M = Ae_i$, which is a left ideal of A and hence is an A-module (see Example 2.3).

(3) Let $A = KG$ be the group algebra of a group G. The *trivial KG-module* has underlying vector space K, and the action of A is defined by

$$g \cdot x = x \ \text{ for all } g \in G \text{ and } x \in K$$

and extended linearly to the entire group algebra KG. The module axioms are trivially satisfied.

(4) Let B be an algebra and A a subalgebra of B. Then every B-module M can be viewed as an A-module with respect to the given action. The axioms are then satisfied since they even hold for elements in the larger algebra B. We have already used this, when describing the natural module for subalgebras of $M_n(K)$, or of $\mathrm{End}_K(V)$.

(5) Let A, B be K-algebras and suppose $\varphi : A \to B$ is a K-algebra homomorphism. If M is a B-module, then M also becomes an A-module by setting

$$A \times M \to M \ , \ (a, m) \mapsto a \cdot m := \varphi(a)m$$

where on the right we use the given B-module structure on M. It is straightforward to check the module axioms.

Exercise 2.2. Explain briefly why example (4) is a special case of example (5).

We will almost always focus on the case when the ring is an algebra over a field K. However, for some of the general properties it is convenient to see these for rings. In this chapter we will write R and M if we are working with an R-module for a general ring, and we will write A and V if we are working with an A-module where A is a K-algebra.

Assume A is a K-algebra, then we have the following important observation, namely all A-modules are automatically vector spaces.

Lemma 2.5. *Let K be a field and A a K-algebra. Then every A-module V is a K-vector space.*

Proof. Recall from Remark 1.2 that we view K as a subset of A, by identifying $\lambda \in K$ with $\lambda 1_A \in A$. Restricting the action of A on V gives us a map $K \times V \to V$. The module axioms (i)–(iv) from Definition 2.1 are then just the K-vector space axioms for V.   □

Remark 2.6. Let A be a K-algebra. To simplify the construction of A-modules, or to check the axioms, it is usually enough to deal with elements of a fixed K-basis of A, recall Remark 1.4. Sometimes one can simplify further. For example if $A = K[X]$, it has basis X^r for $r \geq 1$. Because of axiom (iii) in Definition 2.1 it suffices to define and to check the action of X as this already determines the action of arbitrary basis elements.

Similarly, since an A-module V is always a K-vector space, it is often convenient (and enough) to define actions on a K-basis of V and also check axioms using a basis.

2.2 Modules for Polynomial Algebras

In this section we will completely describe the modules for algebras of the form $K[X]/(f)$ where $f \in K[X]$ is a polynomial. We first recall the situation for the case $f = 0$, that is, modules for the polynomial algebra $K[X]$.

Definition 2.7. Let K be a field and V a K-vector space. For any K-linear map $\alpha : V \to V$ we can use this and turn V into a $K[X]$-module by setting

$$g \cdot v := g(\alpha)(v) = \sum_i \lambda_i \alpha^i(v) \quad \text{(for } g = \sum_i \lambda_i X^i \in K[X] \text{ and } v \in V).$$

Here $\alpha^i = \alpha \circ \ldots \circ \alpha$ is the i-fold composition of maps. We denote this $K[X]$-module by V_α.

Checking the module axioms (i)–(iv) from Definition 2.1 is straightforward. For example, consider condition (iii),

$$g \cdot (h \cdot v) = g(\alpha)(h \cdot v) = g(\alpha)(h(\alpha)(v)) = ((gh)(\alpha))(v) = (gh) \cdot v.$$

Verifying the other axioms is similar and is left as an exercise. Note that, to define a $K[X]$-module structure on a vector space, one only has to specify the action of X, see Remark 2.6.

Example 2.8. Let $K = \mathbb{R}$, and take $V = \mathbb{R}^2$, the space of column vectors. Let $\alpha : V \to V$ be the linear map with matrix

$$\begin{pmatrix} 0 & 1 \\ 0 & 0 \end{pmatrix}$$

with respect to the standard basis of unit vectors of $\mathbb{R}^2$. According to Definition 2.7, V becomes a module for $\mathbb{R}[X]$ by setting $X \cdot v := \alpha(v)$. Here $\alpha^2 = 0$, so if $g = \sum_i \lambda_i X^i \in \mathbb{R}[X]$ is a general polynomial then

$$g \cdot v = g(\alpha)(v) = \sum_i \lambda_i \alpha^i(v) = \lambda_0 v + \lambda_1 \alpha(v).$$

The definition of V_α is more than just an example. We will now show that every $K[X]$-module is equal to V_α for some K-linear map α.

Proposition 2.9. *Let K be a field and let V be a $K[X]$-module. Then $V = V_\alpha$, where $\alpha : V \to V$ is the K-linear map given by $\alpha(v) := X \cdot v$ for $v \in V$.*

Proof. We first want to show that the map α defined in the statement is K-linear. In fact, for every $\lambda, \mu \in K$ and $v, w \in V$ we have

$$\alpha(\lambda v + \mu w) = X \cdot (\lambda v + \mu w) = X \cdot (\lambda v) + X \cdot (\mu w)$$
$$= X \cdot (\lambda 1_{K[X]} \cdot v) + X \cdot (\mu 1_{K[X]} \cdot w)$$
$$= (X\lambda 1_{K[X]}) \cdot v + (X\mu 1_{K[X]}) \cdot w$$
$$= (\lambda 1_{K[X]} X) \cdot v + (\mu 1_{K[X]} X) \cdot w$$
$$= \lambda(X \cdot v) + \mu(X \cdot w) = \lambda\alpha(v) + \mu\alpha(w).$$

To prove that $V = V_\alpha$ it remains to check that for any polynomial $g \in K[X]$ we have $g \cdot v = g(\alpha)(v)$. In fact, by induction on r one sees that $X^r \cdot v = \alpha^r(v)$ for each r, and then one uses linearity to prove the claim.   □

The following result relates $K[X]$-modules with modules for factor algebras $K[X]/I$. This is important since for $I \neq 0$, these factor algebras are finite-dimensional, and many finite-dimensional algebras occuring 'in nature' are of this form. Therefore, this result will be applied many times throughout the book. Recall from basic algebra that $K[X]$ is a principal ideal domain, so that every ideal of $K[X]$ can be generated by one element. Consider now an ideal of $K[X]$ of the form $(f) = K[X]f$ for some polynomial $f \in K[X]$. We assume that f has positive degree, to ensure that $K[X]/(f)$ is an algebra (that is, contains a non-zero identity element).

Theorem 2.10. *Let K be a field and $f \in K[X]$ a non-constant polynomial, and let A be the factor algebra $A = K[X]/(f)$. Then there is a bijection between the set of A-modules and the set of those $K[X]$-modules V_α which satisfy $f(\alpha) = 0$.*

Proof. We set $I = (f)$. Suppose first that V is an A-module. Then we can view V as a module for $K[X]$ by setting

$$g \cdot v = (g + I) \cdot v \ \text{ for all } v \in V \text{ and } g \in K[X].$$

(Note that this is a special case of Example 2.4, for the algebra homomorphism $K[X] \to A$, $g \mapsto g + I$.) Then as a $K[X]$-module, $V = V_\alpha$, where α is the linear map $v \mapsto X \cdot v$, by Proposition 2.9. It follows that for every $v \in V$ we have

$$f(\alpha)(v) = f \cdot v = (f + I) \cdot v = 0 \cdot v = 0$$

since $f \in I$, that is, $f(\alpha) = 0$, as desired.

Conversely, consider a $K[X]$-module V_α where $f(\alpha) = 0$. Then we view V_α as a module for A, by setting

$$(g + I) \cdot v := g(\alpha)(v) \ \text{ for all } g \in K[X] \text{ and } v \in V.$$

This is well-defined, that is, independent of the representatives of the cosets: if $g + I = h + I$ then $g - h \in I = (f)$ and thus $g - h = pf$ for some polynomial $p \in K[X]$. Therefore $(g - h)(\alpha) = (pf)(\alpha) = p(\alpha)f(\alpha) = 0$ by assumption, and hence $g(\alpha) = h(\alpha)$. Finally, one checks that the above action satisfies the module axioms from Definition 2.1.

Since we only changed the point of view, and did not do anything to the vector space and the action as such, the two constructions above are inverse to each other and hence give bijections as claimed in the theorem. □

Example 2.11. Let K be a field.

(1) Consider the K-algebras $K[X]/(X^n)$, where $n \in \mathbb{N}$. By Theorem 2.10, the $K[X]/(X^n)$-modules are in bijection with those $K[X]$-modules V_α such that $\alpha^n = 0$. In particular, the 1-dimensional $K[X]/(X^n)$ modules correspond to scalars $\alpha \in K$ such that $\alpha^n = 0$. But for every field K this has only one solution $\alpha = 0$, that is, there is precisely one 1-dimensional $K[X]/(X^n)$-module $V_0 = \text{span}\{v\}$ with zero action $X \cdot v = 0$. Note that this is independent of the field K.

(2) Let $A = K[X]/(X^n - 1)$, where $n \geq 1$. By Theorem 2.10, we have that A-modules are in bijection with the $K[X]$-modules V_α such that $\alpha^n = \text{id}_V$. In particular, a 1-dimensional A-module is given by the $K[X]$-modules V_α, where $V = \text{span}\{v\} \cong K$ is a 1-dimensional vector space and $\alpha : K \to K$ satisfies $\alpha^n = \text{id}_K$. That is, $\alpha \in K$ is an n-th root of unity. Hence the number of 1-dimensional A-modules depends on how many n-th roots of unity K contains.

When $K = \mathbb{C}$, we have n possibilities, namely $\alpha = (e^{2\pi i/n})^j$ for $j = 0, 1, \ldots, n - 1$.

Assume $K = \mathbb{R}$, then we only get one or two 1-dimensonal A-modules. If n is even then $\alpha = 1$ or -1, and if n is odd we only have $\alpha = 1$.

(3) In general, the K-algebra $K[X]/(f)$ has a 1-dimensional module if and only if the polynomial f has a root in K. So, for example, $\mathbb{R}[X]/(X^2 + 1)$ does not have a 1-dimensional module.

(4) Let C_n be a cyclic group of order n. We have seen in Example 1.27 that the group algebra KC_n is isomorphic to the factor algebra $K[X]/(X^n - 1)$. So we get information on KC_n from example (2).

2.3 Submodules and Factor Modules

In analogy to the vector space constructions, we expect there to be 'submodules' and 'factor modules' of R-modules. A submodule of an R-module M will be a subset U of M which itself is a module with the operations inherited from M. If so, then one can turn the factor group M/U into an R-module.

Definition 2.12. Let R be a ring and let M be some R-module. An R-*submodule* U of M is a subgroup $(U, +)$ which is closed under the action of R, that is, $r \cdot u \in U$ for all $r \in R$ and $u \in U$.

Example 2.13.

(1) Let $R = K$ be a field. Then K-modules are just K-vector spaces, and K-submodules are just K-subspaces.
(2) Let $R = \mathbb{Z}$ be the ring of integers. Then $\mathbb{Z}$-modules are nothing but abelian groups, and $\mathbb{Z}$-submodules are just (abelian) subgroups.
(3) Let R be a ring, considered as an R-module, with action given by ring multiplication. Then the R-submodules of R are precisely the left ideals of R.
(4) Every R-module M has the trivial submodules $\{0\}$ and M. We often just write 0 for the trivial submodule consisting only of the zero element.
(5) Let R be a ring and M an R-module. For every $m \in M$ the subset

$$Rm := \{r \cdot m \mid r \in R\}$$

is an R-submodule of M, the *submodule generated by m*.
(6) Assume M is an R-module, and U, V are R-submodules of M. Then the intersection $U \cap V$ is a submodule of M. Furthermore, let $U + V := \{u + v \mid u \in U, v \in V\}$. Then $U + V$ is a submodule of M.

Exercise 2.3. Prove the statements in Example 2.13 (6).

For modules of K-algebras, we have some specific examples of submodules.

Example 2.14. Let K be a field.

(1) Assume $A = K[X]$. We consider submodules of an A-module V_α. Recall V is a vector space, and α is a linear transformation of V, and the action of A is given by

$$g \cdot v = g(\alpha)(v) \quad (v \in V, g \in K[X]).$$

Then an A-submodule of V_α is the same as a subspace U of V such that $\alpha(U) \subseteq U$. For example, the kernel $U = \ker(\alpha)$, and the image $U = \mathrm{im}(\alpha)$, are A-submodules of V_α.
(2) Consider the matrix algebra $M_n(K)$, and let K^n be the natural A-module, see Example 2.4. We claim that the only $M_n(K)$-submodules of K^n are the trivial submodules 0 and K^n. In fact, let $U \subseteq K^n$ be a non-zero $M_n(K)$-submodule, then there is some $u = (u_1, \ldots, u_n)^t \in U \setminus \{0\}$, say $u_j \neq 0$ for some j. For every $i \in \{1, \ldots, n\}$ consider the matrix unit $E_{ij} \in M_n(K)$ with entry 1 at position (i, j) and 0 otherwise. Then $(u_j)^{-1} E_{ij} \cdot u$ is the ith unit vector and it is in U since U is an $M_n(K)$-submodule. So U contains all unit vectors and hence $U = K^n$, as claimed.

(3) Let $A = T_n(K)$, the algebra of upper triangular matrices. We can also consider K^n as an A-module. Then K^n has non-trivial submodules, for example there is a 1-dimensional submodule, spanned by $(1, 0, \ldots, 0)^t$. Exercise 2.14 determines all $T_n(K)$-submodules of K^n.

(4) Let Q be a quiver and $A = KQ$, the path algebra of Q. For any $r \geq 1$, let $A^{\geq r}$ be the subspace of A spanned by paths of length $\geq r$. Then $A^{\geq r}$ is a submodule of the A-module A. We have seen Ae_i if i is a vertex of Q, this is also a submodule of A. Then we also have the submodule $(Ae_i)^{\geq r} := Ae_i \cap A^{\geq r}$, by Example 2.13.

(5) Consider the 2-dimensional $\mathbb{R}$-algebra $A_0 = \text{span}\{1_{A_0}, \tilde{b}\}$ with $\tilde{b}^2 = 0$ as in Sect. 1.4, as an A-module. The 1-dimensional subspace spanned by $\tilde{b}$ is an A_0-submodule of A_0. Alternatively, $A_0 \cong \mathbb{R}[X]/(X^2)$ and the subspace $\text{span}\{X + (X^2)\}$ is an $\mathbb{R}[X]/(X^2)$-submodule of $\mathbb{R}[X]/(X^2)$.

On the other hand, consider the algebra $A_1 = \text{span}\{1_{A_1}, \tilde{b}\}$ with $\tilde{b}^2 = 1_{A_1}$ in the same section, then the subspace spanned by $\tilde{b}$ is not a submodule. But the space spanned by $\tilde{b} - 1_{A_1}$ is a submodule. Alternatively, $A_1 \cong \mathbb{R}[X]/(X^2 - 1)$; here the subspace $U_1 := \text{span}\{X + (X^2 - 1)\}$ is not a submodule since

$$X \cdot (X + (X^2 - 1)) = 1 + (X^2 - 1) \notin U_1,$$

but the subspace $U_2 := \text{span}\{X - 1 + (X^2 - 1)\}$ is a submodule since

$$X \cdot (X - 1 + (X^2 - 1)) = 1 - X + (X^2 - 1) \in U_2.$$

Exercise 2.4. Let $A = \mathbb{R}[X]/(X^2 + 1)$ (which is the algebra A_{-1} in Sect. 1.4). Why does A as an A-module not have any submodules except 0 and A?

Sometimes a module can be broken up into 'smaller pieces'; the fundamental notion for such phenomena is that of the direct sum of submodules. Very often we will only need finite direct sums, but when dealing with semisimple modules we will also need arbitrary direct sums. For clarity, we will give the definition in both situations separately.

Definition 2.15. Let R be a ring and let M be an R-module.

(a) Let $U_1, U_2, \ldots, U_t$ be R-submodules of M. We say that M is the *direct sum* of $U_1, \ldots, U_t$, denoted $M = U_1 \oplus U_2 \oplus \ldots \oplus U_t$, if the following two conditions are satisfied:

   (i) $M = U_1 + U_2 + \ldots + U_t$, that is, every element of M can be expressed as a sum of elements from the submodules U_i.
   (ii) For every j with $1 \leq j \leq t$ we have $U_j \cap \sum_{i \neq j} U_i = 0$.

(b) We say that M is the *direct sum* of a family $(U_i)_{i \in I}$ of R-submodules (where I is an arbitrary index set), denoted $M = \bigoplus_{i \in I} U_i$, if the following two conditions are satisfied:

(i) $M = \sum_{i \in I} U_i$, that is, every element of M can be expressed as a finite sum of elements from the submodules U_i.

(ii) For every $j \in I$ we have $U_j \cap \sum_{i \neq j} U_i = 0$.

Remark 2.16.

(1) When the ring is a K-algebra A, conditions (i) and (ii) in the above definition state that M is, as a vector space, the direct sum of the subspaces $U_1, \ldots, U_t$ (or U_i where $i \in I$). One should keep in mind that each U_i must be an A-submodule of M.

(2) Note that for $R = K$ a field, every K-module (that is, K-vector space) can be decomposed as a direct sum of 1-dimensional K-submodules (that is, K-subspaces); this is just another formulation of the fact that every K-vector space has a basis.

Given an algebra A, expressing A as an A-module as a direct sum of submodules has numerous applications, we will see this later. The following exercises are examples.

Exercise 2.5. Let $A = M_n(K)$, the algebra of $n \times n$-matrices over a field K, considered as an A-module. For any $i \in \{1, \ldots, n\}$ we define $C_i \subseteq A$ to be the set of matrices which are zero outside the i-th column. Show that C_i is an A-submodule of A, and that

$$A = C_1 \oplus C_2 \oplus \ldots \oplus C_n$$

is a direct sum of these submodules.

Exercise 2.6. Let $A = KQ$, the path algebra of some quiver. Suppose the vertices of Q are $1, 2, \ldots, n$. Recall that the trivial paths $e_i \in KQ$ are orthogonal idempotents (that is, $e_i^2 = e_i$ and $e_i e_j = 0$ for $i \neq j$), and that $e_1 + e_2 + \ldots + e_n = 1_A$. Show that as an A-module, $A = Ae_1 \oplus Ae_2 \oplus \ldots \oplus Ae_n$, the direct sum of A-submodules.

Recall that we have seen direct products of finitely many modules for a ring (in Example 2.3). These products are necessary to construct a new module from given modules, which may not be related in any way. We will later need arbitrary direct products, and also some important submodules therein.

Definition 2.17. Let R be a ring and let $(M_i)_{i \in I}$ be a family of R-modules, where I is some index set.

(a) The cartesian product

$$\prod_{i \in I} M_i = \{(m_i)_{i \in I} \mid m_i \in M_i \text{ for all } i \in I\}$$

becomes an R-module with componentwise addition and R-action. This R-module is called the *direct product* of the family $(M_i)_{i \in I}$ of R-modules.

(b) The subset

$$\bigoplus_{i \in I} M_i = \{(m_i)_{i \in I} \mid m_i \in M_i \text{ for all } i \in I, \text{ only finitely many } m_i \text{ non-zero}\}$$

is an R-submodule of the direct product $\prod_{i \in I} M_i$, called the *direct sum* of the family $(M_i)_{i \in I}$ of R-modules.

Note that we now have two notions of a direct sum. In Definition 2.15 it is assumed that in $M = \bigoplus_{i \in I} U_i$ the U_i are submodules of the given R-module M, whereas in Definition 2.17 the modules M_i in $\bigoplus_{i \in I} M_i$ are not related and a priori are not contained in some given module. (Some books distinguish between these two constructions, calling them an 'internal' and 'external' direct sum.)

We now come to the important construction of factor modules. Suppose U is a submodule of an R-module M, then one knows from basic algebra that the cosets

$$M/U := \{m + U \mid m \in M\}$$

form an abelian group, with addition $(m + U) + (n + U) = m + n + U$. In the case, when M is an R-module and U is an R-submodule, this is actually an R-module in a natural way.

Proposition 2.18. *Let R be a ring, let M be an R-module and U an R-submodule of M. Then the cosets M/U form an R-module if one defines*

$$r \cdot (m + U) := rm + U \quad \text{for all } r \in R \text{ and } m \in M.$$

This module is called the factor module of M modulo U.

Proof. One has to check that the R-action is well-defined, that is, independent of the choice of representatives. If $m + U = m' + U$ then $m - m' \in U$ and then also $rm - rm' = r(m - m') \in U$ since U is an R-submodule. Therefore $rm + U = rm' + U$. Finally, the module axioms are inherited from those for M. $\square$

Example 2.19.

(1) Let R be a ring and I some left ideal of R, then R/I is a factor module of the R-module R. For example, if $R = \mathbb{Z}$ then every ideal is of the form $I = \mathbb{Z}d$ for $d \in \mathbb{Z}$, and this gives the well-known $\mathbb{Z}$-modules $\mathbb{Z}/\mathbb{Z}d$ of integers modulo d.

(2) Let $A = KQ$, the path algebra of a quiver Q, and let $M = Ae_i$ where i is a vertex of Q. This has the submodule $(Ae_i)^{\geq 1} = Ae_i \cap A^{\geq 1}$ (see Example 2.14). The factor module $Ae_i/(Ae_i)^{\geq 1}$ is 1-dimensional, spanned by the coset of e_i. Note that Ae_i may be infinite-dimensional.

2.4   Module Homomorphisms

We have introduced modules as a generalization of vector spaces. In this section we introduce the suitable maps between modules; these 'module homomorphisms' provide a direct generalization of the concept of a K-linear map of vector spaces.

Definition 2.20. Suppose R is a ring, and M and N are R-modules. A map $\phi : M \to N$ is an *R-module homomorphism* if for all $m, m_1, m_2 \in M$ and $r \in R$ we have

(i) $\phi(m_1 + m_2) = \phi(m_1) + \phi(m_2)$;
(ii) $\phi(rm) = r\phi(m)$.

An *isomorphism of R-modules* is an R-module homomorphism which is also bijective. The R-modules M and N are then called *isomorphic*; notation: $M \cong N$. Note that if ϕ is an isomorphism of modules then so is the inverse map; the proof is analogous to Exercise 1.14.

Remark 2.21. Assume that the ring in Definition 2.20 is a K-algebra A (for some field K).

(1) The above definition also says that every A-module homomorphism ϕ must be K-linear. Indeed, recall that we identified scalars $\lambda \in K$ with elements $\lambda 1_A$ in A. Then we have for $\lambda, \mu \in K$ and $m_1, m_2 \in M$ that

$$\phi(\lambda m_1 + \mu m_2) = \phi((\lambda 1_A)m_1 + (\mu 1_A)m_2)$$
$$= (\lambda 1_A)\phi(m_1) + (\mu 1_A)\phi(m_2)$$
$$= \lambda\phi(m_1) + \mu\phi(m_2).$$

(2) It suffices to check condition (ii) in Definition 2.20 for elements $r \in A$ in a fixed basis, or just for elements which generate A, such as $r = X$ in the case when $A = K[X]$.

Exercise 2.7. Suppose V is an A-module where A is a K-algebra. Let $\text{End}_A(V)$ be the set of all A-module homomorphisms $V \to V$, which is by Remark 2.21 a subset of $\text{End}_K(V)$. Check that it is a K-subalgebra.

Example 2.22. Let R be a ring.

(1) Suppose U is a submodule of an R-module M, then the 'canonical map' $\pi : M \to M/U$ defined by $\pi(m) = m + U$ is an R-module homomorphism.
(2) Suppose M is an R-module, and $m \in M$. Then there is an R-module homomorphism

$$\phi : R \to M, \quad \phi(r) = rm.$$

This is very common, and we call it a 'multiplication homomorphism'. There is a general version of this, also very common. Namely, suppose $m_1, m_2, \ldots, m_n$ are given elements in M. Now take the R-module $R^n := R \times R \times \ldots \times R$, and define

$$\psi : R^n \to M, \quad \psi(r_1, r_2, \ldots, r_n) = r_1 m_1 + r_2 m_2 + \ldots + r_n m_n.$$

You are encouraged to check that this is indeed an R-module homomorphism.

(3) Suppose $M = M_1 \times \ldots \times M_r$, the direct product of R-modules $M_1, \ldots, M_r$. Then for every $i \in \{1, \ldots, r\}$ the projection map

$$\pi_i : M \to M_i , \quad (m_1, \ldots, m_r) \mapsto m_i$$

is an R-module homomorphism. Similarly, the inclusion maps into the i-th coordinate

$$\iota_i : M_i \to M , \quad m_i \mapsto (0, \ldots, 0, m_i, 0, \ldots, 0)$$

are R-module homomorphisms. We recommend to check this as well.

Similarly, if $M = U \oplus V$, the direct sum of two submodules U and V, then the projection maps, and the inclusion maps, are R-module homomorphisms.

Now consider the case when the ring is an algebra, then we have several other types of module homomorphisms.

Example 2.23. Let K be a field.

(1) Consider the polynomial algebra $A = K[X]$. For a K-linear map between A-modules to be an A-module homomorphism it suffices that it commutes with the action of X. We have described the A-modules (see Sect. 2.2), so let V_α and W_β be A-modules. An A-module homomorphism is a K-linear map $\theta : V \to W$ such that $\theta(Xv) = X\theta(v)$ for all $v \in V$. On V_α, the element X acts by α, and on W_β, the action of X is given by β. So this means

$$\theta(\alpha(v)) = \beta(\theta(v)) \text{ for all } v \in V.$$

This holds for all v, so we have $\theta \circ \alpha = \beta \circ \theta$.

Conversely, if $\theta : V \to W$ is a linear map such that $\theta \circ \alpha = \beta \circ \theta$, then θ defines a module homomorphism. In particular, $V_\alpha \cong W_\beta$ are isomorphic A-modules if and only if there is an invertible linear map θ such that $\theta^{-1} \circ \beta \circ \theta = \alpha$. This condition means that α and β are similar as linear maps.

(2) Let $A = K[X]$ and let V_α, W_β be 1-dimensional A-modules. In this case, α and β are scalar multiples of the identity map. Then by the previous result, V_α and W_β are isomorphic if and only if $\alpha = \beta$. This follows from the fact that, since β is a scalar multiple of the identity map, it commutes with θ.

(3) Let $G = C_n$ be a cyclic group of order n. Then we have seen in Example 2.11
that the group algebra $\mathbb{C}G \cong \mathbb{C}[X]/(X^n - 1)$ has n one-dimensional modules
given by multiplication with the scalars $(e^{2\pi i/n})^j$ for $j = 0, 1, \ldots, n-1$. These
are pairwise non-isomorphic, by part (2) above. Thus, $\mathbb{C}G$ has precisely n one-
dimensional modules, up to isomorphism.

Exercise 2.8. Let $A = KQ$, where Q is a quiver. We have seen that A has for each
vertex i of Q a 1-dimensional module $S_i := Ae_i/(Ae_i)^{\geq 1}$ (see Example 2.19). Show
that if $i \neq j$ then S_i and S_j are not isomorphic.

In analogy to the isomorphism theorem for linear maps, or group homomorphisms, there are also isomorphism theorems for module homomorphisms.

Theorem 2.24 (Isomorphism Theorems).   *Let R be a ring. Then the following
hold.*

(a) *Suppose $\phi : M \to N$ is an R-module homomorphism. Then the kernel $\ker(\phi)$ is
an R-submodule of M and the image $\operatorname{im}(\phi)$ is an R-submodule of N. Moreover,
we have an isomorphism of R-modules*

$$M/\ker(\phi) \cong \operatorname{im}(\phi).$$

(b) *Suppose U, V are submodules of an R-module M, then the sum $U + V$ and the
intersection $U \cap V$ are also R-submodules of M. Moreover,*

$$U/(U \cap V) \cong (U + V)/V$$

are isomorphic as R-modules.

(c) *Let M be an R-module. Suppose $U \subseteq V \subseteq M$ are R-submodules, then V/U is
an R-submodule of M/U, and*

$$(M/U)/(V/U) \cong M/V$$

are isomorphic as R-modules.

Proof. Readers who have seen the corresponding theorem for abelian groups may
apply this, and just check that all maps involved are R-module homomorphisms.
But for completeness, we give a slightly more detailed proof here.
(a) Since ϕ is in particular a homomorphism of (abelian) groups, we know (or can
easily check) that $\ker(\phi)$ is a subgroup of M. Moreover, for every $m \in \ker(\phi)$ and
$r \in R$ we have

$$\phi(rm) = r\phi(m) = r \cdot 0 = 0$$

and $rm \in \ker(\phi)$. Similarly one checks that the image $\mathrm{im}(\phi)$ is a submodule of N. For the second statement we consider the map

$$\psi : M/\ker(\phi) \to \mathrm{im}(\phi), \quad m + \ker(\phi) \mapsto \phi(m).$$

This map is well-defined and injective: in fact, for any $m_1, m_2 \in M$ we have

$$m_1 + \ker(\phi) = m_2 + \ker(\phi) \iff m_1 - m_2 \in \ker(\phi) \iff \phi(m_1) = \phi(m_2).$$

By definition, ψ is surjective. It remains to check that this map is an R-module homomorphism. For every $m_1, m_2, m \in M$ and $r \in R$ we have

$$\psi((m_1 + \ker(\phi)) + (m_2 + \ker(\phi))) = \phi(m_1 + m_2) = \phi(m_1) + \phi(m_2)$$
$$= \psi(m_1 + \ker(\phi)) + \psi(m_2 + \ker(\phi)),$$

and

$$\psi(r(m + \ker(\phi))) = \phi(rm) = r\phi(m) = r\psi(m + \ker(\phi)).$$

(b) We have already seen in Example 2.13 that $U + V$ and $U \cap V$ are submodules. Then we consider the map

$$\psi : U \to (U + V)/V , \quad u \mapsto u + V.$$

From the addition and R-action on factor modules being defined on representatives, it is easy to check that ψ is an R-module homomorphism. Since every coset in $(U + V)/V$ is of the form $u + v + V = u + V$, the map ψ is surjective. Moreover, it follows directly from the definition that $\ker(\psi) = U \cap V$. So part (a) implies that

$$U/(U \cap V) \cong (U + V)/V.$$

(c) That V/U is an R-submodule of M/U follows directly from the fact that V is an R-submodule of M. We then consider the map

$$\psi : M/U \to M/V , \quad m + U \mapsto m + V.$$

Note that this map is well-defined since $U \subseteq V$ by assumption. One checks that ψ is an R-module homomorphism. By definition, ψ is surjective, and the kernel consists precisely of the cosets of the form $m + U$ with $m \in V$, that is, $\ker(\psi) = V/U$. So part (a) implies that

$$(M/U)/(V/U) \cong M/V,$$

as claimed. $\qquad\qquad\qquad\qquad\qquad\qquad\qquad\qquad\qquad\qquad\qquad\qquad\qquad\qquad$ $\square$

Example 2.25. Let R be a ring and M an R-module. For any $m \in M$ consider the R-module homomorphism

$$\phi : R \to M , \ \phi(r) = rm$$

from Example 2.22. The kernel

$$\mathrm{Ann}_R(m) := \{r \in R \mid rm = 0\}$$

is called the *annihilator* of m in R. By the isomorphism theorem we have that

$$R/\mathrm{Ann}_R(m) \cong \mathrm{im}(\phi) = Rm,$$

that is, the factor module is actually isomorphic to the submodule of M generated by m; this has appeared already in Example 2.13.

In the isomorphism theorem we have seen that factor modules occur very naturally in the context of module homomorphisms. We now describe the submodules of a factor module. This so-called submodule correspondence is very useful, as it allows one to translate between factor modules and modules. This is based on the following observation:

Proposition 2.26. *Let R be a ring and $\phi : M \to N$ an R-module homomorphism. Then for every R-submodule $W \subseteq N$ the preimage $\phi^{-1}(W) := \{m \in M \mid \phi(m) \in W\}$ is an R-submodule of M, which contains the kernel of ϕ.*

Proof. We first show that $\phi^{-1}(W)$ is a subgroup. It contains the zero element since $\phi(0) = 0 \in W$. Moreover, if $m_1, m_2 \in \phi^{-1}(W)$, then

$$\phi(m_1 \pm m_2) = \phi(m_1) \pm \phi(m_2) \in W$$

since W is a subgroup of N, that is, $m_1 \pm m_2 \in \phi^{-1}(W)$. Finally, let $r \in R$ and $m \in \phi^{-1}(W)$. Then $\phi(rm) = r\phi(m) \in W$ since W is an R-submodule, that is, $rm \in \phi^{-1}(W)$. The kernel of ϕ is mapped to zero, and $0 \in W$, hence the kernel of ϕ is contained in $\phi^{-1}(W)$.   □

Example 2.27. Let R be a ring and M an R-module, and let U be a submodule of M with factor module M/U. Then we have the natural homomorphism $\pi : M \to M/U, \ \pi(m) = m + U$. Proposition 2.26 with $\phi = \pi$ shows that for every submodule W of M/U, the preimage under π is a submodule of M containing U. Explicitly, this is the module

$$\widehat{W} := \pi^{-1}(W) = \{m \in M \mid m + U \in W\}.$$

This construction leads to the following submodule correspondence:

Theorem 2.28. *Let R be a ring. Suppose M is an R-module and U is an R-submodule of M. Then the map $W \mapsto \widehat{W}$ induces a bijection, inclusion preserving, between the set $\mathcal{F}$ of R-submodules of M/U and the set $\mathcal{S}$ of R-submodules of M that contain U. Its inverse takes V in $\mathcal{S}$ to $V/U \in \mathcal{F}$.*

Proof. If W is a submodule of M/U then $\widehat{W}$ belongs to the set $\mathcal{S}$, by Example 2.27. Take a module V which belongs to $\mathcal{S}$, then V/U is in $\mathcal{F}$, by part (c) of Theorem 2.24. To show that this gives a bijection, we must check that

(i) if W is in $\mathcal{F}$ then $\widehat{W}/U = W$,

(ii) if V is in $\mathcal{S}$ then $\widehat{(V/U)} = V$.

To prove (i), let $m \in \widehat{W}$, then by definition $m + U \in W$ so that $\widehat{W}/U \subseteq W$. Conversely if $w \in W$ then $w = m + U$ for some $m \in M$. Then $m \in \widehat{W}$ and therefore $w \in \widehat{W}/U$.

To prove (ii), note that $\widehat{(V/U)} = \{m \in M \mid m + U \in V/U\}$. First, let $m \in \widehat{(V/U)}$, that is, $m + U \in V/U$. Then $m + U = v + U$ for some $v \in V$, and therefore $m - v \in U$. But $U \subseteq V$ and therefore $m - v \in V$, and it follows that $m = (m - v) + v \in V$. This shows that $\widehat{(V/U)} \subseteq V$. Secondly, if $v \in V$ then $v + U \in V/U$, so that $v \in \widehat{(V/U)}$. This completes the proof of (ii). It is clear from the constructions that they preserve inclusions. $\square$

Example 2.29. For a field K, we consider the K-algebra $A = K[X]/(X^n)$ as an A-module. We apply Theorem 2.28 in the case $R = M = K[X]$ and $U = (X^n)$; then the $K[X]$-submodules of A are in bijection with the $K[X]$-submodules of $K[X]$ containing the ideal (X^n). The $K[X]$-submodules of $K[X]$ are precisely the (left) ideals. Since $K[X]$ is a principal ideal domain every ideal is of the form $(g) = K[X]g$ for some polynomial $g \in K[X]$; moreover, (g) contains the ideal (X^n) if and only if g divides X^n. Hence there are precisely $n + 1$ different $K[X]$-submodules of $A = K[X]/(X^n)$, namely all

$$(X^j)/(X^n) = \{X^j h + (X^n) \mid h \in K[X]\} \ (0 \le j \le n).$$

Note that the $K[X]$-action on A is the same as the A-action on A (both given by multiplication in $K[X]$); thus the $K[X]$-submodules of A are the same as the A-submodules of A and the above list gives precisely the submodules of A as an A-module.

Alternatively, we know that as a $K[X]$-module, $A = V_\alpha$, where $V = A$ as a vector space, and α is the linear map which is given as multiplication by X (see Sect. 2.2). The matrix of α with respect to the basis $\{1 + (X^n), X + (X^n), \ldots, X^{n-2} + (X^n), X^{n-1} + (X^n)\}$ is the Jordan block $J_n(0)$

of size n with zero diagonal entries, that is,

$$J_n(0) = \begin{pmatrix} 0 & 0 & \ldots\ldots & 0 \\ 1 & 0 & & \vdots \\ 0 & \ddots & \ddots & \vdots \\ \vdots & & \ddots & 0 & 0 \\ 0 & \ldots & 0 & 1 & 0 \end{pmatrix}.$$

A submodule is a subspace which is invariant under this matrix. With this, we also get the above description of submodules.

Exercise 2.9. Check the details in the above example.

Often, properties of module homomorphisms already give rise to direct sum decompositions. We give an illustration of this, which will actually be applied twice later.

Lemma 2.30. *Let A be a K-algebra, and let M, N, N' be non-zero A-modules. Suppose there are A-module homomorphisms $j : N \to M$ and $\pi : M \to N'$ such that the composition $\pi \circ j : N \to N'$ is an isomorphism. Then j is injective and π is surjective, and M is the direct sum of two submodules,*

$$M = \mathrm{im}(j) \oplus \ker(\pi).$$

Proof. The first part is clear. We must show that $\mathrm{im}(j) \cap \ker(\pi) = 0$ and that $M = \mathrm{im}(j) + \ker(\pi)$, see Definition 2.15.

Suppose $w \in \mathrm{im}(j) \cap \ker(\pi)$, so that $w = j(n)$ for some $n \in N$ and $\pi(w) = 0$. Then $0 = \pi(w) = (\pi \circ j)(n)$ from which it follows that $n = 0$ since $\pi \circ j$ is injective. Clearly, then also $w = 0$, as desired. This proves that the intersection $\mathrm{im}(j) \cap \ker(\pi)$ is zero.

Let $\phi : N' \to N$ be the inverse of $\pi \circ j$, so that we have $\pi \circ j \circ \phi = \mathrm{id}_{N'}$. Take $w \in M$ then

$$w = (j \circ \phi \circ \pi)(w) + (w - (j \circ \phi \circ \pi)(w)).$$

The first summand belongs to $\mathrm{im}(j)$, and the second summand is in $\ker(\pi)$ since

$$\pi(w - (j \circ \phi \circ \pi)(w)) = \pi(w) - (\pi \circ j \circ \phi \circ \pi)(w) = \pi(w) - \pi(w) = 0.$$

This proves that $w \in \mathrm{im}(j) + \ker(\pi)$, and hence $M = \mathrm{im}(j) + \ker(\pi)$. $\square$

2.5 Representations of Algebras

In basic group theory one learns that a group action of a group G on a set Ω is 'the same' as a group homomorphism from G into the group of all permutations of the set Ω.

In analogy, let A be a K-algebra; as we will see, an A-module V 'is the same' as an algebra homomorphism $\phi : A \to \text{End}_K(V)$, that is, a representation of A.

Definition 2.31. Let K be a field and A a K-algebra.

(a) A *representation of A over K* is a K-vector space V together with a K-algebra homomorphism $\theta : A \to \text{End}_K(V)$.
(b) A *matrix representation* of A is a K-algebra homomorphism $\theta : A \longrightarrow M_n(K)$, for some $n \geq 1$.
(c) Suppose in (a) that V is finite-dimensional. We may fix a basis in V and write linear maps of V as matrices with respect to such a basis. Then the representation of A becomes a K-algebra homomorphism $\theta : A \to M_n(K)$, that is, a matrix representation of A.

Example 2.32. Let K be a field.

(1) Let A be a K-subalgebra of $M_n(K)$, the algebra of $n \times n$-matrices, then the inclusion map is an algebra homomorphism and hence is a matrix representation of A. Similarly, let A be a subalgebra of the K-algebra $\text{End}_K(V)$ of K-linear maps on V, where V is a vector space over K. Then the inclusion map from A to $\text{End}_K(V)$ is an algebra homomorphism, hence it is a representation of A.
(2) Consider the polynomial algebra $K[X]$, and take a K-vector space V together with a fixed linear transformation α. We have seen in Example 1.24 that evaluation at α is an algebra homomorphism. Hence we have a representation of $K[X]$ given by

$$\theta : K[X] \to \text{End}_K(V), \quad \theta(f) = f(\alpha).$$

(3) Let $A = KG$, the group algebra of a finite group. Define $\theta : A \to M_1(K) = K$ by mapping each basis vector $g \in G$ to 1, and extend to linear combinations. This is a representation of A: by Remark 2.6 it is enough to check the conditions on a basis. This is easy, since $\theta(g) = 1$ for $g \in G$.

In Sect. 2.1 we introduced modules for an algebra as vector spaces on which the algebra acts by linear transformations. The following crucial result observes that modules and representations of an algebra are the same. Going from one notion to the other is a formal matter, nothing is 'done' to the modules or representations and it only describes two different views of the same thing.

Theorem 2.33. *Let K be a field and let A be a K-algebra.*

(a) *Suppose V is an A-module, with A-action $A \times V \to V$, $(a, v) \mapsto a \cdot v$. Then we have a representation of A given by*

$$\theta : A \to \operatorname{End}_K(V), \ \theta(a)(v) = a \cdot v \text{ for all } a \in A, v \in V.$$

(b) *Suppose $\theta : A \to \operatorname{End}_K(V)$ is a representation of A. Then V becomes an A-module by setting*

$$A \times V \to V, \ a \cdot v := \theta(a)(v) \text{ for all } a \in A, v \in V.$$

Roughly speaking, the representation θ corresponding to an A-module V describes how each element $a \in A$ acts linearly on the vector space V, and vice versa.

Proof. The proof basically consists of translating the module axioms from Definition 2.1 to the new language of representations from Definition 2.31, and vice versa. (a) We first have to show that $\theta(a) \in \operatorname{End}_K(V)$ for all $a \in A$. Recall from Lemma 2.5 that every A-module is also a K-vector space; then for all $\lambda, \mu \in K$ and $v, w \in V$ we have

$$\theta(a)(\lambda v + \mu w) = \theta(a)(\lambda 1_A \cdot v + \mu 1_A \cdot w) = (a\lambda 1_A) \cdot v + (a\mu 1_A) \cdot w$$
$$= \lambda(a \cdot v) + \mu(a \cdot w) = \lambda\theta(a)(v) + \mu\theta(a)(w),$$

where we again used axiom (Alg) from Definition 1.1.

It remains to check that θ is an algebra homomorphism. First, it has to be a K-linear map; for any $\lambda, \mu \in K, a, b \in A$ and $v \in V$ we have

$$\theta(\lambda a + \mu b)(v) = (\lambda a + \mu b) \cdot v = \lambda(a \cdot v) + \mu(b \cdot v) = (\lambda\theta(a) + \mu\theta(b))(v),$$

that is, $\theta(\lambda a + \mu b) = \lambda\theta(a) + \mu\theta(b)$. Next, for any $a, b \in A$ and $v \in V$ we get

$$\theta(ab)(v) = (ab) \cdot v = a \cdot (b \cdot v) = (\theta(a) \circ \theta(b))(v),$$

which holds for all $v \in V$, hence $\theta(ab) = \theta(a) \circ \theta(b)$. Finally, it is immediate from the definition that $\theta(1_A) = \operatorname{id}_V$.
(b) This is analogous to the proof of part (a), where each argument can be reversed.   □

Example 2.34. Let K be a field.

(1) Let V be a K-vector space and $A \subseteq \operatorname{End}_K(V)$ a subalgebra. As observed in Example 2.32 the inclusion map $\theta : A \to \operatorname{End}_K(V)$ is a representation. When V is interpreted as an A-module we obtain precisely the 'natural module' from Example 2.4 with A-action given by $A \times V \to V, (\varphi, v) \mapsto \varphi(v)$.

(2) The representation $\theta : K[X] \to \mathrm{End}_K(V)$, $\theta(f) = f(\alpha)$ for any K-linear map α on a K-vector space V when interpreted as a $K[X]$-module is precisely the $K[X]$-module V_α studied in Sect. 2.2.

(3) Let A be a K-algebra, then A is an A-module with A-action given by the multiplication in A. The corresponding representation of A is then given by

$$\theta : A \to \mathrm{End}_K(A) , \quad \theta(a)(x) = ax \text{ for all } a, x \in A.$$

This representation θ is called the *regular representation* of A. (See also Exercise 1.29.)

We now want to transfer the notion of module isomorphism (see Definition 2.20) to the language of representations. It will be convenient to have the following notion:

Definition 2.35. Let K be a field. Suppose we are given two representations θ_1, θ_2 of a K-algebra A where $\theta_1 : A \to \mathrm{End}_K(V_1)$ and $\theta_2 : A \to \mathrm{End}_K(V_2)$. Then θ_1 and θ_2 are said to be *equivalent* if there is a vector space isomorphism $\psi : V_1 \to V_2$ such that

$$\theta_1(a) = \psi^{-1} \circ \theta_2(a) \circ \psi \text{ for all } a \in A. \tag{$*$}$$

If $\theta_1 : A \to M_{n_1}(K)$ and $\theta_2 : A \to M_{n_2}(K)$ are matrix representations of A, this means that $\theta_1(a)$ and $\theta_2(a)$ should be *simultaneously similar*, that is, there exists an invertible $n_2 \times n_2$-matrix C (independent of a) such that $\theta_1(a) = C^{-1}\theta_2(a)C$ for all $a \in A$. (Note that for equivalent representations the dimensions of the vector spaces must agree, that is, we have $n_1 = n_2$.)

For example, let $A = K[X]$, and assume V_1, V_2 are finite-dimensional. Then $(*)$ holds for all $a \in K[X]$ if and only if $(*)$ holds for $a = X$, that is, if and only if the matrices $\theta_1(X)$ and $\theta_2(X)$ are similar.

Proposition 2.36. *Let K be a field and A a K-algebra. Then two representations $\theta_1 : A \to \mathrm{End}_K(V_1)$ and $\theta_2 : A \to \mathrm{End}_K(V_2)$ of A are equivalent if and only if the corresponding A-modules V_1 and V_2 are isomorphic.*

Proof. Suppose first that θ_1 and θ_2 are equivalent via the vector space isomorphism $\psi : V_1 \to V_2$. We claim that ψ is also an A-module homomorphism (and then it is an isomorphism). By definition $\psi : V_1 \to V_2$ is K-linear. Moreover, for $v \in V_1$ and $a \in A$ we have

$$\psi(a \cdot v) = \psi(\theta_1(a)(v)) = \theta_2(a)(\psi(v)) = a \cdot \psi(v).$$

Conversely, suppose $\psi : V_1 \to V_2$ is an A-module isomorphism. Then for all $a \in A$ and $v \in V_1$ we have

$$\psi(\theta_1(a)(v)) = \psi(a \cdot v) = a \cdot \psi(v) = \theta_2(a)(\psi(v)).$$

This is true for all $v \in V_1$, that is, for all $a \in A$ we have

$$\psi \circ \theta_1(a) = \theta_2(a) \circ \psi$$

and hence the representations θ_1 and θ_2 are equivalent.   □

Focussing on representations (rather than modules) gives some new insight. Indeed, an algebra homomorphism has a kernel which quite often is non-trivial. Factoring out the kernel, or an ideal contained in the kernel, gives a representation of a smaller algebra.

Lemma 2.37. *Let A be a K-algebra, and I a proper two-sided ideal of A, with factor algebra A/I. Then the representations of A/I are in bijection with the representations of A whose kernel contains I. The bijection takes a representation ψ of A/I to the representation $\psi \circ \pi$ where $\pi : A \to A/I$ is the natural homomorphism.*

Translating this to modules, there is a bijection between the set of A/I-modules and the set of those A-modules V for which $xv = 0$ for all $x \in I$ and $v \in V$.

Proof. We prove the statement about the representations. The second statement then follows directly by Theorem 2.33.

(i) Let $\psi : A/I \to \mathrm{End}_K(V)$ be a representation of the factor algebra A/I, then the composition $\psi \circ \pi : A \to \mathrm{End}_K(V)$ is an algebra homomorphism and hence a representation of A. By definition, $I = \ker(\pi)$ and hence $I \subseteq \ker(\psi \circ \pi)$.

(ii) Conversely, assume $\theta : A \to \mathrm{End}_K(V)$ is a representation of A such that $I \subseteq \ker(\theta)$. By Theorem 1.26, we have a well-defined algebra homomorphism $\bar{\theta} : A/I \to \mathrm{End}_K(V)$ defined by

$$\bar{\theta}(a + I) = \theta(a) \quad (a \in A)$$

and moreover $\theta = \bar{\theta} \circ \pi$. Then $\bar{\theta}$ is a representation of A/I.

In (i) we define a map between the two sets of representations, and in (ii) we show that it is surjective. Finally, let ψ, ψ' be representations of A/I such that $\psi \circ \pi = \psi' \circ \pi$. The map π is surjective, and therefore $\psi = \psi'$. Thus, the map in (i) is also injective.   □

Remark 2.38. In the above lemma, the A/I-module V is called *inflation* when it is viewed as an A-module. The two actions in this case are related by

$$(a + I)v = av \quad \text{for all } a \in A, v \in V.$$

We want to illustrate the above inflation procedure with an example from linear algebra.

Example 2.39. Let $A = K[X]$ be the polynomial algebra. We take a representation $\theta : K[X] \to \mathrm{End}_K(V)$ and let $\theta(X) =: \alpha$. The kernel is an ideal of $K[X]$, so it is of the form $\ker(\theta) = K[X]m = (m)$ for some polynomial $m \in K[X]$. Assume

$m \neq 0$, then we can take it to be monic. Then m is the minimal polynomial of α, as it is studied in linear algebra. That is, we have $m(\alpha) = 0$ and $f(\alpha) = 0$ if and only if f is a multiple of m in $K[X]$.

By the above Lemma 2.37, the representation θ can be viewed as a representation of the algebra $K[X]/I$ whenever I is an ideal of $K[X]$ contained in $\ker(\theta) = (m)$, that is, $I = (f)$ and f is a multiple of m.

2.5.1 Representations of Groups vs. Modules for Group Algebras

In this book we focus on representations of algebras. However, we explain in this section how the important theory of representations of groups can be interpreted in this context. Representations of groups have historically been the starting point for representation theory at the end of the 19th century. The idea of a representation of a group is analogous to that for algebras: one lets a group act by linear transformations on a vector space in a way compatible with the group structure. Group elements are invertible, therefore the linear transformations by which they act must also be invertible. The invertible linear transformations of a vector space form a group $GL(V)$, with respect to composition of maps, and it consists of the invertible elements in $\mathrm{End}_K(V)$. For an n-dimensional vector space, if one fixes a basis, one gets the group $GL_n(K)$, which are the invertible matrices in $M_n(K)$, a group with respect to matrix multiplication.

Definition 2.40. Let G be a group. A *representation* of G over a field K is a homomorphism of groups $\rho : G \to GL(V)$, where V is a vector space over K.

If V is finite-dimensional one can choose a basis and obtain a *matrix representation*, that is, a group homomorphism $\rho : G \to GL_n(K)$.

The next result explains that group representations are basically the same as representations for the corresponding group algebras (see Sect. 1.1.2). By Theorem 2.33 this can be rephrased by saying that a representation of a group is nothing but a module for the corresponding group algebra. Sometimes group homomorphisms are more useful, and sometimes it is useful that one can also use linear algebra.

Proposition 2.41. *Let G be a group and K a field.*

(a) *Every representation $\rho : G \to GL(V)$ of the group G over K extends to a representation $\theta_\rho : KG \to \mathrm{End}_K(V)$ of the group algebra KG given by $\sum_{g \in G} \alpha_g g \mapsto \sum_{g \in G} \alpha_g \rho(g)$.*
(b) *Conversely, given a representation $\theta : KG \to \mathrm{End}_K(V)$ of the group algebra KG, then the restriction $\rho_\theta : G \to GL(V)$ of θ to G is a representation of the group G over K.*

Proof. (a) Given a representation, that is, a group homomorphism $\rho : G \to GL(V)$, we must show that θ_ρ is an algebra homomorphism from KG to $\text{End}_K(V)$. First, θ_ρ is linear, by definition, and maps into $\text{End}_K(V)$. Next, we must check that $\theta_\rho(ab) = \theta_\rho(a)\theta_\rho(b)$. As mentioned in Remark 1.23, it suffices to take a, b in a basis of KG, so we can take $a, b \in G$ and then also $ab \in G$, and we have

$$\theta_\rho(ab) = \rho(ab) = \rho(a)\rho(b) = \theta_\rho(a)\theta_\rho(b).$$

In addition $\theta_\rho(1_{KG}) = \rho(1_G) = 1_{GL(V)} = \text{id}_V$.

(b) The elements of G form a subset (even a vector space basis) of the group algebra KG. Thus we can restrict θ to this subset and get a map

$$\rho_\theta : G \to GL(V) , \quad \rho_\theta(g) = \theta(g) \text{ for all } g \in G.$$

In fact, every $g \in G$ has an inverse g^{-1} in the group G; since θ is an algebra homomorphism it follows that

$$\theta(g)\theta(g^{-1}) = \theta(gg^{-1}) = \theta(1_{KG}) = \text{id}_V,$$

that is, $\rho_\theta(g) = \theta(g) \in GL(V)$ is indeed an invertible linear map. Moreover, for every $g, h \in G$ we have

$$\rho_\theta(gh) = \theta(gh) = \theta(g)\theta(h) = \rho_\theta(g)\rho_\theta(h)$$

and ρ_θ is a group homomorphism, as required. $\square$

Example 2.42. We consider a square in the plane, with corners $(\pm 1, \pm 1)$. Recall that the group of symmetries of the square is by definition the group of orthogonal transformations of $\mathbb{R}^2$ which leave the square invariant. This group is called *dihedral group of order* 8 and we denote it by D_4 (some texts call it D_8). It consists of four rotations (including the identity), and four reflections. Let r be the anti-clockwise rotation by $\pi/2$, and let s be the reflection in the x-axis. One can check that $D_4 = \{s^i r^j \mid 0 \le i \le 1, 0 \le j \le 3\}$. We define this group as a subgroup of $GL(\mathbb{R}^2)$. Therefore the inclusion map $D_4 \longrightarrow GL(\mathbb{R}^2)$ is a group homomorphism, hence is a representation. We can write the elements in this group as matrices with respect to the standard basis. Then

$$\rho(r) = \begin{pmatrix} 0 & -1 \\ 1 & 0 \end{pmatrix} \quad \text{and } \rho(s) = \begin{pmatrix} 1 & 0 \\ 0 & -1 \end{pmatrix}.$$

Then the above group homomorphism translates into a matrix representation

$$\rho : D_4 \to GL_2(\mathbb{R}) , \quad \rho(s^i r^j) = \rho(s)^i \rho(r)^j.$$

By Proposition 2.41 this yields a representation θ_ρ of the group algebra $\mathbb{R}D_4$. Interpreting representations of algebras as modules (see Theorem 2.33) this means that $V = \mathbb{R}^2$ becomes an $\mathbb{R}D_4$-module where each $g \in D_4$ acts on V by applying the matrix $\rho(g)$ to a vector.

Assume N is a normal subgroup of G with factor group G/N, then it is reasonable to expect that representations of G/N may be related to some representations of G. Indeed, we have the canonical map $\pi : G \to G/N$ where $\pi(g) = gN$, which is a group homomorphism. In fact, using this we have an analogue of inflation for algebras, described in Lemma 2.37.

Lemma 2.43. *Let G be a group and N a normal subgroup of G with factor group G/N, and consider representations over the field K. Then the representations of G/N are in bijection with the representations of G whose kernel contains N. The bijection takes a representation θ of G/N to the representation $\theta \circ \pi$ where $\pi : G \to G/N$ is the canonical map.*

Translating this to modules, this gives a bijection between the set of $K(G/N)$-modules and the set of those KG-modules V for which $n \cdot v = v$ for all $n \in N$ and $v \in V$.

Proof. This is completely analogous to the proof of Lemma 2.37, so we leave it as an exercise. In fact, Exercise 2.24 shows that it is the same. □

2.5.2 Representations of Quivers vs. Modules for Path Algebras

The group algebra KG has basis the elements of the group G, which allows us to relate KG-modules and representations of G. The path algebra KQ of a quiver has basis the paths of Q. In analogy, we can define a representation of a quiver Q, which allows us to relate KQ-modules and representations of Q. We will introduce this now, and later we will study it in more detail.

Roughly speaking, a representation is as follows. A quiver consists of vertices and arrows, and if we want to realize it in the setting of vector spaces, we represent vertices by vector spaces, and arrows by linear maps, so that when arrows can be composed, the corresponding maps can also be composed.

Definition 2.44. Let $Q = (Q_0, Q_1)$ be a quiver. A *representation* $\mathcal{V}$ of Q over a *field* K is a set of K-vector spaces $\{V(i) \mid i \in Q_0\}$ together with K-linear maps $V(\alpha) : V(i) \to V(j)$ for each arrow α from i to j. We sometimes also write $\mathcal{V}$ as a tuple $\mathcal{V} = ((V(i))_{i \in Q_0}, (V(\alpha))_{\alpha \in Q_1})$.

Example 2.45.

(1) Let Q be the one-loop quiver as in Example 1.13 with one vertex 1, and one arrow α with starting and end point 1. Then a representation $\mathcal{V}$ of Q consists of a K-vector space $V(1)$ and a K-linear map $V(\alpha) : V(1) \to V(1)$.

(2) Let Q be the quiver $1 \xrightarrow{\alpha} 2$. Then a representation V consists of two K-vector
    spaces $V(1)$ and $V(2)$ and a K-linear map $V(\alpha) : V(1) \to V(2)$.

Consider the second example. We can construct from this a module for the path
algebra $KQ = \text{span}\{e_1, e_2, \alpha\}$. Take as the underlying space $V := V(1) \times V(2)$,
a direct product of K-vector spaces (a special case of a direct product of modules
as in Example 2.3 and Definition 2.17). Let e_i act as the projection onto $V(i)$ with
kernel $V(j)$ for $j \neq i$. Then define the action of α on V by

$$\alpha((v_1, v_2)) := V(\alpha)(v_1)$$

(where $v_i \in V(i)$). Conversely, if we have a KQ-module V then we can turn it into
a representation of Q by setting

$$V(1) := e_1 V, \ \ V(2) = e_2 V$$

and $V(\alpha) : e_1 V \to e_2 V$ is given by (left) multiplication by α.

The following result shows how this can be done in general, it says that
representations of a quiver Q over a field K are 'the same' as modules for the path
algebra KQ.

Proposition 2.46. *Let K be a field and Q a quiver, with vertex set Q_0 and arrow
set Q_1.*

(a) *Let $V = ((V(i))_{i \in Q_0}, (V(\alpha))_{\alpha \in Q_1})$ be a representation of Q over K. Then
    the direct product $V := \prod_{i \in Q_0} V(i)$ becomes a KQ-module as follows: let
    $v = (v_i)_{i \in Q_0} \in V$ and let $p = \alpha_r \ldots \alpha_1$ be a path in Q starting at vertex
    $s(p) = s(\alpha_1)$ and ending at vertex $t(p) = t(\alpha_r)$. Then we set*

$$p \cdot v = (0, \ldots, 0, V(\alpha_r) \circ \ldots \circ V(\alpha_1)(v_{s(p)}), 0, \ldots, 0)$$

*where the (possibly) non-zero entry is in position $t(p)$. In particular, if $r = 0$,
then $e_i \cdot v = (0, \ldots, 0, v_i, 0, \ldots, 0)$. This action is extended to all of KQ by
linearity.*
(b) *Let V be a KQ-module. For any vertex $i \in Q_0$ we set*

$$V(i) = e_i V = \{e_i \cdot v \mid v \in V\};$$

for any arrow $i \xrightarrow{\alpha} j$ in Q_1 we set

$$V(\alpha) : V(i) \to V(j), \ \ e_i \cdot v \mapsto \alpha \cdot (e_i \cdot v) = \alpha \cdot v.$$

*Then $V = ((V(i))_{i \in Q_0}, (V(\alpha))_{\alpha \in Q_1})$ is a representation of the quiver Q
over K.*
(c) *The constructions in (a) and (b) are inverse to each other.*

Proof. (a) We check that the module axioms from Definition 2.1 are satisfied. Let $p = \alpha_r \ldots \alpha_1$ and q be paths in Q. Since the KQ-action is defined on the basis of KQ and then extended by linearity, the distributivity $(p+q) \cdot v = p \cdot v + q \cdot v$ holds by definition. Moreover, let $v, w \in V$; then

$$p \cdot (v + w) = (0, \ldots, 0, V(\alpha_r) \circ \ldots \circ V(\alpha_1)(v_{s(p)} + w_{s(p)}), 0, \ldots, 0)$$

$$= p \cdot v + p \cdot w$$

since all the maps $V(\alpha_i)$ are K-linear. Since the multiplication in KQ is defined by concatenation of paths (see Sect. 1.1.3), it is immediate that $p \cdot (q \cdot v) = (pq) \cdot v$ for all $v \in V$ and all paths p, q in Q, and then by linearity also for arbitrary elements in KQ. Finally, the identity element is $1_{KQ} = \sum_{i \in Q_0} e_i$, the sum of all trivial paths (see Sect. 1.1.3); by definition, e_i acts by picking out the ith component, then for all $v \in V$ we have

$$1_{KQ} \cdot v = \sum_{i \in Q_0} e_i \cdot v = v.$$

(b) According to Definition 2.44 we have to confirm that the $V(i) = e_i V$ are K-vector spaces and that the maps $V(\alpha)$ are K-linear. The module axioms for V imply that for every $v, w \in V$ and $\lambda \in K$ we have

$$e_i \cdot v + e_i \cdot w = e_i \cdot (v + w) \in e_i V = V(i)$$

and also

$$\lambda(e_i \cdot v) = (\lambda 1_{KQ} e_i) \cdot v = (e_i \lambda 1_{KQ}) \cdot v = e_i \cdot (\lambda v) \in e_i V.$$

So $V(i)$ is a K-vector space for every $i \in Q_0$.

Finally, let us check that $V(\alpha)$ is a K-linear map for every arrow $i \xrightarrow{\alpha} j$ in Q. Note first that $\alpha e_i = \alpha$, so that the map $V(\alpha)$ is indeed given by $e_i \cdot v \mapsto \alpha \cdot v$. Then for all $\lambda, \mu \in K$ and all $v, w \in V(i)$ we have

$$V(\alpha)(\lambda v + \mu w) = \alpha \cdot (\lambda v + \mu w) = \alpha \cdot (\lambda 1_{KQ} \cdot v) + \alpha \cdot (\mu 1_{KQ} \cdot w)$$

$$= (\alpha \lambda 1_{KQ}) \cdot v + (\alpha \mu 1_{KQ}) \cdot w = (\lambda 1_{KQ} \alpha) \cdot v + (\mu 1_{KQ} \alpha) \cdot w$$

$$= \lambda 1_{KQ} \cdot (\alpha \cdot v) + \mu 1_{KQ} \cdot (\alpha \cdot w) = \lambda V(\alpha)(v) + \mu V(\alpha)(w).$$

So $V(\alpha)$ is a K-linear map.
(c) It is straightforward to check from the definitions that the two constructions are inverse to each other; we leave the details to the reader. $\qquad\square$

Example 2.47. We consider the quiver $1 \xrightarrow{\alpha} 2$. The 1-dimensional vector space span$\{e_2\}$ is a KQ-module (more precisely, a KQ-submodule of the path algebra KQ). We interpret it as a representation of the quiver Q by $0 \xrightarrow{0}$ span$\{e_2\}$. Also the two-dimensional vector space span$\{e_1, \alpha\}$ is a KQ-module. As a representation of Q this takes the form span$\{e_1\} \xrightarrow{V(\alpha)}$ span$\{\alpha\}$, where $V(\alpha)(e_1) = \alpha$. Often, the vector spaces are only considered up to isomorphism, then the latter representation takes the more concise form $K \xrightarrow{\mathrm{id}_K} K$.

EXERCISES

2.10. Let R be a ring. Suppose M is an R-module with submodules U, V and W. We have seen in Exercise 2.3 that the sum $U + V := \{u + v \mid u \in U, v \in V\}$ and the intersection $U \cap V$ are submodules of M.

   (a) Show by means of an example that it is not in general the case that

$$U \cap (V + W) = (U \cap V) + (U \cap W).$$

   (b) Show that $U \setminus V$ is never a submodule. Show also that the union $U \cup V$ is a submodule if and only if $U \subseteq V$ or $V \subseteq U$.

2.11. Let $A = M_2(K)$, the K-algebra of 2×2-matrices over a field K. Take the A-module $M = A$, and for $i = 1, 2$ define U_i to be the subset of matrices where all entries not in the i-th column are zero. Moreover, let
$$U_3 := \left\{ \begin{pmatrix} a & a \\ b & b \end{pmatrix} \mid a, b \in K \right\}.$$

   (a) Check that each U_i is an A-submodule of M.
   (b) Verify that for $i \neq j$, the intersection $U_i \cap U_j$ is zero.
   (c) Show that M is not the direct sum of U_1, U_2 and U_3.

2.12. For a field K we consider the factor algebra $A = K[X]/(X^4 - 2)$. In each of the following cases find the number of 1-dimensional A-modules (up to isomorphism); moreover, describe explicitly the action of the coset of X on the module.

   (i) $K = \mathbb{Q}$; (ii) $K = \mathbb{R}$; (iii) $K = \mathbb{C}$; (iv) $K = \mathbb{Z}_3$; (v) $K = \mathbb{Z}_7$.

2.13. Let $A = \mathbb{Z}_p[X]/(X^n - 1)$, where p is a prime number. We investigate 1-dimensional A-modules, that is, the roots of $X^n - 1$ in $\mathbb{Z}_p$ (see Theorem 2.10).

   (a) Show that $\bar{1} \in \mathbb{Z}_p$ is always a root.

(b) Show that the number of 1-dimensional A-modules is d where d is the greatest common divisor of n and $p - 1$. (Hint: You might use that the non-zero elements in $\mathbb{Z}_p$ are precisely the roots of the polynomial $X^{p-1} - 1$.)

2.14. For a field K we consider the natural $T_n(K)$-module K^n, where $T_n(K)$ is the algebra of upper triangular $n \times n$-matrices.

(a) In K^n consider the K-subspaces

$$V_i := \{(\lambda_1, \ldots, \lambda_n)^t \mid \lambda_j = 0 \text{ for all } j > i\},$$

where $0 \le i \le n$. Show that these are precisely the $T_n(K)$-submodules of K^n.

(b) For any $0 \le j < i \le n$ consider the factor module $V_{i,j} := V_i/V_j$. Show that these give $\frac{n(n+1)}{2}$ pairwise non-isomorphic $T_n(K)$-modules.

(c) For each standard basis vector $e_i \in K^n$, where $1 \le i \le n$, determine the annihilator $\mathrm{Ann}_{T_n(K)}(e_i)$ (see Example 2.25) and identify the factor module $T_n(K)/\mathrm{Ann}_{T_n(K)}(e_i)$ with a $T_n(K)$-submodule of K^n, up to isomorphism.

2.15. Let R be a ring and suppose $M = U \times V$, the direct product of R-modules U and V. Check that $\tilde{U} := \{(u, 0) \mid u \in U\}$ is a submodule of M, isomorphic to U. Write down a similar submodule $\tilde{V}$ of M isomorphic to V, and show that $M = \tilde{U} \oplus \tilde{V}$, the direct sum of submodules.

2.16. Let K be a field. Assume M, M_1, M_2 are K-vector spaces and $\alpha_i : M_i \to M$ are K-linear maps. The *pull-back* of (α_1, α_2) is defined to be

$$E := \{(m_1, m_2) \in M_1 \times M_2 \mid \alpha_1(m_1) + \alpha_2(m_2) = 0\}.$$

(a) Check that E is a K-subspace of $M_1 \times M_2$.

(b) Assume that M, M_1, M_2 are finite-dimensional K-vector spaces and that $M = \mathrm{im}(\alpha_1) + \mathrm{im}(\alpha_2)$. Show that then for the vector space dimensions we have $\dim_K E = \dim_K M_1 + \dim_K M_2 - \dim_K M$. (Hint: Show that the map $(m_1, m_2) \mapsto \alpha_1(m_1) + \alpha_2(m_2)$ from $M_1 \times M_2$ to M is surjective, and has kernel E.)

(c) Now assume that M, M_1, M_2 are A-modules where A is some K-algebra, and that α_1 and α_2 are A-module homomorphisms. Show that then E is an A-submodule of $M_1 \times M_2$.

2.17. Let K be a field. Assume W, M_1, M_2 are K-vector spaces and $\beta_i : W \to M_i$ are K-linear maps. The *push-out* of (β_1, β_2) is defined to be

$$F := (M_1 \times M_2)/C,$$

where $C := \{(\beta_1(w), \beta_2(w)) \mid w \in W\}$.

(a) Check that C is a K-subspace of $M_1 \times M_2$, hence F is a K-vector space.

(b) Assume W, M_1, M_2 are finite-dimensional and $\ker(\beta_1) \cap \ker(\beta_2) = 0$. Show that then

$$\dim_K F = \dim_K M_1 + \dim_K M_2 - \dim_K W.$$

(Hint: Show that the linear map $w \mapsto (\beta_1(w), \beta_2(w))$ from W to C is an isomorphism.)

(c) Assume that W, M_1, M_2 are A-modules where A is some K-algebra and that β_1, β_2 are A-module homomorphisms. Show that then C and hence F are A-modules.

2.18. Let E be the pull-back as in Exercise 2.16 and assume $M = \text{im}(\alpha_1) + \text{im}(\alpha_2)$. Now take the push-out F as in Exercise 2.17 where $W = E$ with the same M_1, M_2, and where $\beta_i : W \to M_i$ are the maps

$$\beta_1(m_1, m_2) := m_1, \quad \beta_2(m_1, m_2) := m_2 \quad (\text{where } (m_1, m_2) \in W = E).$$

(a) Check that $\ker(\beta_1) \cap \ker(\beta_2) = 0$.

(b) Show that the vector space C in the construction of the pushout is equal to E.

(c) Show that the pushout F is isomorphic to M. (Hint: Consider the map from $M_1 \times M_2$ to M defined by $(m_1, m_2) \mapsto \alpha_1(m_1) + \alpha_2(m_2)$.)

2.19. Let K be a field and KG be the group algebra where G is a finite group. Recall from Example 2.4 that the trivial KG-module is the 1-dimensional module with action $g \cdot x = x$ for all $g \in G$ and $x \in K$, linearly extended to all of KG. Show that the corresponding representation $\theta : KG \to \text{End}_K(K)$ satisfies $\theta(a) = \text{id}_K$ for all $a \in KG$. Check that this is indeed a representation.

2.20. Let G be the group of symmetries of a regular pentagon, that is, the group of orthogonal transformations of $\mathbb{R}^2$ which leave the pentagon invariant. That is, G is the dihedral group of order 10, a subgroup of $GL(\mathbb{R}^2)$. As a group, G is generated by the (counterclockwise) rotation by $\frac{2\pi i}{5}$, which we call r, and a reflection s; the defining relations are $r^5 = \text{id}_{\mathbb{R}^2}$, $s^2 = \text{id}_{\mathbb{R}^2}$ and $s^{-1}rs = r^{-1}$. Consider the group algebra $\mathbb{C}G$ of G over the complex numbers, and suppose that $\omega \in \mathbb{C}$ is some 5-th root of unity. Show that the matrices

$$\rho(r) = \begin{pmatrix} \omega & 0 \\ 0 & \omega^{-1} \end{pmatrix}, \quad \rho(s) = \begin{pmatrix} 0 & 1 \\ 1 & 0 \end{pmatrix}$$

satisfy the defining relations for G, hence give rise to a group representation $\rho : G \to GL_2(\mathbb{C})$, and a 2-dimensional $\mathbb{C}G$-module.

2.21. Let K be a field. We consider the quiver Q given by $1 \xleftarrow{\alpha} 2 \xleftarrow{\beta} 3$ and the path algebra KQ as a KQ-module.

    (a) Let $V := \text{span}\{e_2, \alpha\} \subseteq KQ$. Explain why $V = KQe_2$ and hence V is a KQ-submodule of KQ.

    (b) Find a K-basis of the KQ-submodule $W := KQ\beta$ generated by β.

    (c) Express the KQ-modules V and W as a representation of the quiver Q. Are V and W isomorphic as KQ-modules?

2.22. Let $A = KQ$ where Q is the following quiver: $1 \xrightarrow{\alpha} 2 \xleftarrow{\beta} 3$. This exercise illustrates that A as a left module and A as a right module have different properties.

    (a) As a left module $A = Ae_1 \oplus Ae_2 \oplus Ae_3$ (see Exercise 2.6). For each Ae_i, find a K-basis, and verify that each of Ae_1 and Ae_3 is 2-dimensional, and Ae_2 is 1-dimensional.

    (b) Show that the only 1-dimensional A-submodule of Ae_1 is $\text{span}\{\alpha\}$. Deduce that Ae_1 cannot be expressed as $Ae_1 = U \oplus V$ where U and V are non-zero A-submodules.

    (c) Explain briefly why the same holds for Ae_3.

    (d) As a right A-module, $A = e_1A \oplus e_2A \oplus e_3A$ (by the same reasoning as in Exercise 2.6). Verify that e_1A and e_3A are 1-dimensional.

2.23. Assume $A = K[X]$ and let $f = gh$ where g and h are polynomials in A. Then $Af = (f) \subseteq (g) = Ag$ and the factor module is

$$(g)/(f) = \{rg + (f) \mid r \in K[X]\}.$$

Show that $(g)/(f)$ is isomorphic to $K[X]/(h)$ as a $K[X]$-module.

2.24. Let G be a group and N a normal subgroup of G, and let $A = KG$, the group algebra over the field K.

    (a) Show that the space $I = \text{span}\{g_1 - g_2 \mid g_1g_2^{-1} \in N\}$ is a 2-sided ideal of A.

    (b) Show that the A-modules on which I acts as zero are precisely the A-modules V such that $n \cdot v = v$ for all $n \in N$ and $v \in V$.

    (c) Explain briefly how this connects Lemmas 2.37 and 2.43.

Chapter 3
Simple Modules and the Jordan–Hölder Theorem

In this section we introduce simple modules for algebras over a field. Simple modules can be seen as the building blocks of arbitrary modules. We will make this precise by introducing and studying composition series, in particular we will prove the fundamental Jordan–Hölder theorem. This shows that it is an important problem to classify, if possible, all simple modules of a (finite-dimensional) algebra. We discuss tools to find and to compare simple modules of a fixed algebra. Furthermore, we determine the simple modules for algebras of the form $K[X]/(f)$ where f is a non-constant polynomial, and also for finite-dimensional path algebras of quivers.

3.1 Simple Modules

Throughout this chapter, K is a field. Let A be a K-algebra.

Definition 3.1. An A-module V is called *simple* if V is non-zero, and if it does not have any A-submodules other than 0 and V.

We start by considering some examples; in fact, we have already seen simple modules in the previous chapter.

Example 3.2.

(1) Every A-module V such that $\dim_K V = 1$ is a simple module. In fact, it does not have any K-subspaces except for 0 and V and therefore it cannot have any A-submodules except for 0 or V.

(2) Simple modules can have arbitrarily large dimensions. As an example, let $A = M_n(K)$, and take $V = K^n$ to be the natural module. Then we have seen in Example 2.14 that the only $M_n(K)$-submodules of K^n are the trivial submodules 0 and K^n. That is, K^n is a simple $M_n(K)$-module, and it is n-dimensional.

© Springer International Publishing AG, part of Springer Nature 2018
K. Erdmann, T. Holm, *Algebras and Representation Theory*, Springer
Undergraduate Mathematics Series, https://doi.org/10.1007/978-3-319-91998-0_3

However, if we consider $V = K^n$ as a module for the upper triangular matrix algebra $T_n(K)$ then V is not simple when $n \geq 2$, see Exercise 2.14.

(3) Consider the symmetry group D_4 of a square, that is, the dihedral group of order 8. Let $A = \mathbb{R}D_4$ be the group algebra over the real numbers. We have seen in Example 2.42 that there is a matrix representation $\rho : D_4 \to GL_2(\mathbb{R})$ such that

$$\rho(r) = \begin{pmatrix} 0 & -1 \\ 1 & 0 \end{pmatrix} \quad \text{and} \quad \rho(s) = \begin{pmatrix} 1 & 0 \\ 0 & -1 \end{pmatrix},$$

where r is the rotation by $\pi/2$ and s the reflection about the x-axis. The corresponding A-module is $V = \mathbb{R}^2$, the action of every $g \in D_4$ is given by applying the matrix $\rho(g)$ to (column) vectors.

We claim that $V = \mathbb{R}^2$ is a simple $\mathbb{R}D_4$-module. Suppose, for a contradiction, that V has a non-zero submodule U and $U \neq V$. Then U is 1-dimensional, say U is spanned by a vector u. Then $\rho(r)u = \lambda u$ for some $\lambda \in \mathbb{R}$, which means that $u \in \mathbb{R}^2$ is an eigenvector of $\rho(r)$. But the matrix $\rho(r)$ does not have a real eigenvalue, a contradiction.

Since $GL_2(\mathbb{R})$ is a subgroup of $GL_2(\mathbb{C})$, we may view $\rho(g)$ for $g \in D_4$ also as elements in $GL_2(\mathbb{C})$. This gives the 2-dimensional module $V = \mathbb{C}^2$ for the group algebra $\mathbb{C}D_4$. In Exercise 3.13 it will be shown that this module is simple. Note that this does not follow from the argument we used in the case of the $\mathbb{R}D_4$-module $\mathbb{R}^2$.

(4) Let K be a field and D a division algebra over K, see Definition 1.7. We view D as a D-module (with action given by multiplication in D). Then D is a simple D-module: In fact, let $0 \neq U \subseteq D$ be a D-submodule, and take an element $0 \neq u \in U$. Then $1_D = u^{-1}u \in U$ and hence if $d \in D$ is arbitrary, we have that $d = d1_D \in U$. Therefore $U = D$, and D is a simple D-module.

We describe now a method which allows us to show that a given A-module V is simple. For any element $v \in V$ set $Av := \{av \mid a \in A\}$. This is an A-submodule of V, the submodule of V generated by v, see Example 2.13.

Lemma 3.3. *Let A be a K-algebra and let V be a non-zero A-module. Then V is simple if and only if for each $v \in V \setminus \{0\}$ we have $Av = V$.*

Proof. First suppose that V is simple, and take an arbitrary element $0 \neq v \in V$. We know that Av is a submodule of V, and it contains $v = 1_A v$, and so Av is non-zero and therefore $Av = V$ since V is simple.

Conversely, suppose U is a non-zero submodule of V. Then there is some non-zero $u \in U$. Since U is a submodule, we have $Au \subseteq U$, but by the hypothesis, $V = Au \subseteq U \subseteq V$ and hence $U = V$.   □

A module isomorphism takes a simple module to a simple module, see Exercise 3.3; this is perhaps not surprising.

We would like to understand when a factor module of a given module is simple. This can be answered by using the submodule correspondence (see Theorem 2.28).

Lemma 3.4. *Let A be a K-algebra. Suppose V is an A-module and U is an A-submodule of V with $U \neq V$. Then the following conditions are equivalent:*

(i) The factor module V/U is simple.
(ii) U is a maximal submodule of V, that is, if $U \subseteq W \subseteq V$ are A-submodules then $W = U$ or $W = V$.

Proof. This follows directly from the submodule correspondence in Theorem 2.28.   □

Suppose A is an algebra and I is a two-sided ideal of A with $I \neq A$. Then we have the factor algebra $B = A/I$. According to Lemma 2.37, any B-module M can be viewed as an A-module, with action $am = (a + I)m$ for $m \in M$; this A-module is called the inflation of M to A, see Remark 2.38. This can be used as a method to find simple A-modules:

Lemma 3.5. *Let A be a K-algebra, and let $B = A/I$ where I is an ideal of A with $I \neq A$. If S is a simple B-module, then the inflation of S to A is a simple A-module.*

Proof. The submodules of S as an A-module are inflations of submodules of S as a B-module, since the inflation only changes the point of view. Since S has no submodules other than S and 0 as a B-module, it also has no submodules other than S and 0 as an A-module, hence is a simple A-module.   □

3.2 Composition Series

Roughly speaking, a composition series of a module breaks the module up into 'simple pieces'. This will make precise in what sense the simple modules are the building blocks for arbitrary modules.

Definition 3.6. Let A be a K-algebra and V an A-module. A *composition series* of V is a finite chain of A-submodules

$$0 = V_0 \subset V_1 \subset V_2 \subset \ldots \subset V_n = V$$

such that the factor modules V_i / V_{i-1} are simple, for all $1 \leq i \leq n$. The *length* of the composition series is n, the number of factor modules appearing. We refer to the V_i as the *terms* of the composition series.

Example 3.7.

(1) The zero module has a composition series $0 = V_0 = 0$ of length 0. If V is a simple module then $0 = V_0 \subset V_1 = V$ is a composition series, of length 1.
(2) Assume we have a composition series as in Definition 3.6. If V_k is one of the terms, then V_k 'inherits' the composition series

$$0 = V_0 \subset V_1 \subset \ldots \subset V_k.$$

(3) Let $K = \mathbb{R}$ and take A to be the 2-dimensional algebra over $\mathbb{R}$, with basis $\{1_A, \beta\}$ such that $\beta^2 = 0$ (see Proposition 1.29); an explicit realisation would be $A = \mathbb{R}[X]/(X^2)$. Take the A-module $V = A$, and let V_1 be the space spanned by β, then V_1 is a submodule. Since V_1 and V/V_1 are 1-dimensional, they are simple (see Example 3.2). Hence V has a composition series

$$0 = V_0 \subset V_1 \subset V_2 = V.$$

(4) Let $A = M_n(K)$ and take the A-module $V = A$. In Exercise 2.5 we have considered the A-submodules C_i consisting of the matrices with zero entries outside the i-th column, where $1 \leq i \leq n$. In Exercise 3.1 it is shown that every A-module C_i is isomorphic to the natural A-module K^n. In particular, each A-module C_i is simple (see Example 3.2). On the other hand we have a direct sum decomposition $A = C_1 \oplus C_2 \oplus \ldots \oplus C_n$ and therefore we have a finite chain of submodules

$$0 \subset C_1 \subset C_1 \oplus C_2 \subset \ldots \subset C_1 \oplus \ldots \oplus C_{n-1} \subset A.$$

Each factor module is simple: By the isomorphism theorem (see Theorem 2.24)

$$C_1 \oplus \ldots \oplus C_k / C_1 \oplus \ldots \oplus C_{k-1} \cong C_k / C_k \cap (C_1 \oplus \ldots \oplus C_{k-1}) = C_k / \{0\} \cong C_k.$$

This shows that the above chain is a composition series.

(5) Let $A = T_2(K)$, the 2×2 upper triangular matrices over K, and consider the A-module $V = A$. One checks that the sets

$$V_1 := \left\{ \begin{pmatrix} a & 0 \\ 0 & 0 \end{pmatrix} \mid a \in K \right\} \quad \text{and} \quad V_2 := \left\{ \begin{pmatrix} a & b \\ 0 & 0 \end{pmatrix} \mid a, b \in K \right\}$$

are A-submodules of V. The chain $0 = V_0 \subset V_1 \subset V_2 \subset V$ is a composition series: each factor module V_i / V_{i-1} is 1-dimensional and hence simple.

(6) Let $A = KQ$ be the path algebra of the quiver $1 \overset{\alpha}{\longrightarrow} 2$ and let V be the A-module $V = A$. Then the following chains are composition series for V:

$$0 \subset \operatorname{span}\{e_2\} \subset \operatorname{span}\{e_2, \alpha\} \subset V.$$

$$0 \subset \operatorname{span}\{\alpha\} \subset \operatorname{span}\{\alpha, e_1\} \subset V.$$

In each case, the factor modules are 1-dimensional and hence are simple A-modules.

Exercise 3.1. Let $A = M_n(K)$, and let $C_i \subseteq A$ be the space of matrices which are zero outside the i-th column. Show that C_i is isomorphic to the natural module $V = K^n$ of column vectors. *Hint:* Show that placing $v \in V$ into the i-th column of a matrix and extending by zeros is a module homomorphism $V \to C_i$.

Remark 3.8. Not every module has a composition series. For instance, take $A = K$ to be the 1-dimensional algebra over a field K. Then A-modules are K-vector spaces, and A-submodules are K-subspaces. Therefore it follows from Definition 3.1 that the simple K-modules are precisely the 1-dimensional K-vector spaces. This means that an infinite-dimensional K-vector space does not have a composition series since a composition series is by definition a *finite* chain of submodules, see Definition 3.6. On the other hand, we will now see that for any algebra, finite-dimensional modules have a composition series.

Lemma 3.9. *Let A be a K-algebra. Every finite-dimensional A-module V has a composition series.*

Proof. This is proved by induction on the dimension of V. If $\dim_K V = 0$ or $\dim_K V = 1$ then we are done by Example 3.7.

So assume now that $\dim_K V > 1$. If V is simple then, again by Example 3.7, V has a composition series. Otherwise, V has proper non-zero submodules. So we can choose a proper submodule $0 \neq U \subset V$ of largest possible dimension. Then U must be a maximal submodule of V and hence the factor module V/U is a simple A-module, by Lemma 3.4. Since $\dim_K U < \dim_K V$, by the induction hypothesis U has a composition series, say

$$0 = U_0 \subset U_1 \subset U_2 \subset \ldots \subset U_k = U.$$

Since V/U is simple, it follows that

$$0 = U_0 \subset U_1 \subset U_2 \subset \ldots \subset U_k = U \subset V$$

is a composition series of V. $\qquad\square$

In Example 3.7 we have seen that a term of a composition series inherits a composition series. This is a special case of the following result, which holds for arbitrary submodules.

Proposition 3.10. *Let A be a K-algebra, and let V be an A-module. If V has a composition series, then every submodule $U \subseteq V$ also has a composition series.*

Proof. Take any composition series for V, say

$$0 = V_0 \subset V_1 \subset V_2 \subset \ldots \subset V_{n-1} \subset V_n = V.$$

Taking intersections with the submodule U yields a chain of A-submodules of U,

$$0 = V_0 \cap U \subseteq V_1 \cap U \subseteq V_2 \cap U \subseteq \ldots \subseteq V_{n-1} \cap U \subseteq V_n \cap U = U. \qquad (3.1)$$

Note that terms of this chain can be equal; that is, (3.1) is in general not a composition series. However, if we remove any repetition, so that each module occurs precisely once, then we get a composition series for U: Consider the factors

$(V_i \cap U)/(V_{i-1} \cap U)$. Using that $V_{i-1} \subset V_i$ and applying the isomorphism theorem (see Theorem 2.24) we obtain

$$(V_i \cap U)/(V_{i-1} \cap U) = (V_i \cap U)/(V_{i-1} \cap (V_i \cap U))$$

$$\cong (V_{i-1} + (V_i \cap U))/V_{i-1} \subseteq V_i/V_{i-1}.$$

Since V_i/V_{i-1} is simple the factor modules $(V_i \cap U)/(V_{i-1} \cap U)$ occurring in (3.1) are either zero or simple.   □

In general, a module can have many composition series, even infinitely many different composition series; see Exercise 3.11. The Jordan–Hölder Theorem shows that any two composition series of a module have the same length and the same factors up to isomorphism and up to order.

Theorem 3.11 (Jordan–Hölder Theorem). *Let A be a K-algebra. Suppose an A-module V has two composition series*

$$0 = V_0 \subset V_1 \subset V_2 \subset \ldots \subset V_{n-1} \subset V_n = V \tag{I}$$

$$0 = W_0 \subset W_1 \subset W_2 \subset \ldots \subset W_{m-1} \subset W_m = V. \tag{II}$$

Then $n = m$, and there is a permutation σ of $\{1, 2, \ldots, n\}$ such that $V_i/V_{i-1} \cong W_{\sigma(i)}/W_{\sigma(i)-1}$ for each $i = 1, \ldots, n$.

Before starting with the proof of this theorem, we give a definition, and we will also deal with a special setting as a preparation.

Definition 3.12.

(a) The composition series (I) and (II) in Theorem 3.11 are called *equivalent*.
(b) If an A-module V has a composition series, the simple factor modules V_i/V_{i-1} are called the *composition factors* of V. By the Jordan–Hölder theorem, they only depend on V, and not on the composition series.

For the proof of the Jordan–Hölder Theorem, which will be given below, we need to compare two given composition series. We consider now the case when V_{n-1} is different from W_{m-1}. Observe that this shows why the simple quotients of two composition series can come in different orders.

Lemma 3.13. *With the notation as in the Jordan–Hölder Theorem, suppose $V_{n-1} \neq W_{m-1}$ and consider the intersection $D := V_{n-1} \cap W_{m-1}$. Then the following holds:*

$$V_{n-1}/D \cong V/W_{m-1} \quad \text{and} \quad W_{m-1}/D \cong V/V_{n-1},$$

and these factor modules are simple A-modules.

Proof. We first observe that $V_{n-1} + W_{m-1} = V$. In fact, we have

$$V_{n-1} \subseteq V_{n-1} + W_{m-1} \subseteq V.$$

If the first inclusion was an equality then $W_{m-1} \subseteq V_{n-1} \subset V$; but both V_{n-1} and W_{m-1} are maximal submodules of V (since V/V_{n-1} and V/W_{m-1} are simple, see Lemma 3.4). Thus $V_{n-1} = W_{m-1}$, a contradiction. Since V_{n-1} is a maximal submodule of V, we conclude that $V_{n-1} + W_{m-1} = V$.

Now we apply the isomorphism theorem (Theorem 2.24), and get

$$V/W_{m-1} = (V_{n-1} + W_{m-1})/W_{m-1} \cong V_{n-1}/(V_{n-1} \cap W_{m-1}) = V_{n-1}/D.$$

Similarly one shows that $V/V_{n-1} \cong W_{m-1}/D$. $\square$

Proof (of the Jordan–Hölder Theorem). We proceed by induction on n (the length of the composition series (I)).

The zero module is the only module with a composition series of length $n = 0$ (see Example 3.7), and the statement of the theorem clearly holds in this case.

Let $n = 1$. Then V is simple, so $W_1 = V$ (since there is no non-zero submodule except V), and $m = 1$. Clearly, the factor modules are the same in both series.

Now suppose $n > 1$. The inductive hypothesis is that the theorem holds for modules which have a composition series of length $\leq n - 1$.

Assume first that $V_{n-1} = W_{m-1} =: U$, say. Then the module U inherits a composition series of length $n - 1$, from (I). By the inductive hypothesis, any two composition series of U have length $n - 1$. So the composition series of U inherited from (II) also has length $n - 1$ and therefore $m - 1 = n - 1$ and $m = n$. Moreover, by the inductive hypothesis, there is a permutation σ of $\{1, \ldots, n - 1\}$ such that $V_i/V_{i-1} \cong W_{\sigma(i)}/W_{\sigma(i)-1}$. We also have $V_n/V_{n-1} = V/V_{n-1} = W_n/W_{n-1}$. So if we view σ as a permutation of $\{1, \ldots, n\}$ fixing n then we have the required permutation.

Now assume $V_{n-1} \neq W_{m-1}$. We define $D := V_{n-1} \cap W_{m-1}$ as in Lemma 3.13. Take a composition series of D (it exists by Proposition 3.10), say

$$0 = D_0 \subset D_1 \subset \ldots \subset D_t = D.$$

Then V has composition series

(III) $0 = D_0 \subset D_1 \subset \ldots \subset D_t = D \subset V_{n-1} \subset V$

(IV) $0 = D_0 \subset D_1 \subset \ldots \subset D_t = D \subset W_{m-1} \subset V$

since, by Lemma 3.13, the quotients V_{n-1}/D and W_{m-1}/D are simple. Moreover, by Lemma 3.13, the composition series (III) and (IV) are equivalent since only the two top factors are interchanged, up to isomorphism.

Next, we claim that $m = n$. The module V_{n-1} inherits a composition series of length $n - 1$ from (I). So by the inductive hypothesis, all composition series of

V_{n-1} have length $n-1$. But the composition series which is inherited from (III) has length $t+1$ and hence $n-1 = t+1$. Similarly, the module W_{m-1} inherits from (IV) a composition series of length $t+1 = n-1$, so by the inductive hypothesis all composition series of W_{m-1} have length $n-1$. In particular, the composition series inherited from (II) does, and therefore $m-1 = n-1$ and $m = n$.

Now we show that the composition series (I) and (III) are equivalent. By the inductive hypothesis, the composition series of V_{n-1} inherited from (I) and (III) are equivalent, that is, there is a permutation of $n-1$ letters, γ say, such that

$$D_i/D_{i-1} \cong V_{\gamma(i)}/V_{\gamma(i)-1}, \quad (i \neq n-1), \text{ and } V_{n-1}/D \cong V_{\gamma(n-1)}/V_{\gamma(n-1)-1}.$$

We view γ as a permutation of n letters (fixing n), and then also

$$V/V_{n-1} = V_n/V_{n-1} \cong V_{\gamma(n)}/V_{\gamma(n)-1},$$

which proves that (I) and (III) are equivalent.

Similarly one shows that (II) and (IV) are equivalent. We have already seen that (III) and (IV) are equivalent as well. Therefore, it follows that (I) and (II) are also equivalent.   □

Example 3.14. Let K be a field.

(1) Let $A = M_n(K)$ and $V = A$ as an A-module. We consider the A-submodules C_i of A (where $1 \leq i \leq n$) consisting of matrices which are zero outside the i-th column. Define $V_i := C_1 \oplus C_2 \oplus \ldots \oplus C_i$ for $0 \leq i \leq n$. Then we have a chain of submodules

$$0 = V_0 \subset V_1 \subset \ldots \subset V_{n-1} \subset V_n = V.$$

This is a composition series, as we have seen in Example 3.7.

On the other hand, the A-module V also has submodules

$$W_j := C_n \oplus C_{n-1} \oplus \ldots \oplus C_{n-j+1},$$

for any $0 \leq j \leq n$, and this gives us a series of submodules

$$0 = W_0 \subset W_1 \subset \ldots \subset W_{n-1} \subset W_n = V.$$

This also is a composition series, since $W_j/W_{j-1} \cong C_{n-j+1} \cong K^n$ which is a simple A-module. As predicted by the Jordan–Hölder theorem, both composition series have the same length n. Since all composition factors are isomorphic as A-modules, we do not need to worry about a permutation.

(2) Let $A = T_n(K)$ be the upper triangular matrix algebra and $V = K^n$ the natural A-module. By Exercise 2.14, V has $T_n(K)$-submodules V_i for $0 \leq i \leq n$ consisting of those vectors in K^n with non-zero entries only in the first i

coordinates; in particular, each V_i has dimension i, with $V_0 = 0$ and $V_n = K^n$. This gives a series of $T_n(K)$-submodules

$$0 = V_0 \subset V_1 \subset \ldots \subset V_{n-1} \subset V_n = K^n.$$

Each factor module V_i/V_{i-1} is 1-dimensional and hence simple, so this is a composition series. Actually, this is the only composition series of the $T_n(K)$-module K^n, since by Exercise 2.14 the V_i are the only $T_n(K)$-submodules of K^n.

Moreover, also by Exercise 2.14, the factor modules V_i/V_{i-1} are pair-wise non-isomorphic. We will see in Example 3.28 that these n simple $T_n(K)$-modules of dimension 1 are in fact all simple $T_n(K)$-modules, up to isomorphism.

(3) This example also shows that a module can have non-isomorphic composition factors. Let $A := K \times K$, the direct product of K-algebras, and view A as an A-module. Let $S_1 := \{(x, 0) \mid x \in K\}$ and $S_2 := \{(0, y) \mid y \in K\}$. Then each S_i is an A-submodule of A. It is 1-dimensional and therefore it is simple. We have a series

$$0 \subset S_1 \subset A$$

and by the isomorphism theorem, $A/S_1 \cong S_2$. This is therefore a composition series. We claim that S_1 and S_2 are not isomorphic: Let $\phi : S_1 \to S_2$ be an A-module homomorphism, then

$$\phi(a(x, 0)) = a\phi((x, 0)) \in S_2$$

for each $a \in A$. We take $a = (1, 0)$, then $a(x, 0) = (x, 0)$ but $a(0, y) = 0$ for all $(0, y) \in S_2$. That is, $\phi = 0$.

3.3 Modules of Finite Length

Because of the Jordan–Hölder theorem we can define the length of a module. This is a very useful natural generalization of the dimension of a vector space.

Definition 3.15. Let A be a K-algebra. For every A-module V the *length* $\ell(V)$ is defined as the length of a composition series of V (see Definition 3.6), if it exists; otherwise we set $\ell(V) = \infty$. An A-module V is said to be of *finite length* if $\ell(V)$ is finite, that is, when V has a composition series.

Note that the length of a module is well-defined because of the Jordan–Hölder theorem which in particular says that all composition series of a module have the same length.

Example 3.16. Let K be a field.

(1) Let $A = K$, the 1-dimensional K-algebra. Then A-modules are just K-vector spaces and simple A-modules are precisely the 1-dimensional K-vector spaces. In particular, the length of a composition series is just the dimension, that is, for every K-vector space V we have $\ell(V) = \dim_K V$.

(2) Let A be a K-algebra and V an A-module. By Example 3.7 we have that $\ell(V) = 0$ if and only if $V = 0$, the zero module, and that $\ell(V) = 1$ if and only if V is a simple A-module. In addition, we have seen there that for $A = M_n(K)$, the natural module $V = K^n$ has $\ell(V) = 1$, and that for the A-module A we have $\ell(A) = n$. Roughly speaking, the length gives a measure of how far a module is away from being simple.

We now collect some fundamental properties of the length of modules. We first prove a result analogous to Proposition 3.10, but now also including factor modules. It generalizes properties from linear algebra: Let V be a finite-dimensional vector space over K and U a subspace, then U and V/U also are finite-dimensional and $\dim_K V = \dim_K U + \dim_K V/U$. Furthermore, $\dim_K U < \dim_K V$ if $U \neq V$.

Proposition 3.17. *Let A be a K-algebra and let V be an A-module which has a composition series. Then for every A-submodule $U \subseteq V$ the following holds.*

(a) The factor module V/U has a composition series.

(b) There exists a composition series of V in which U is one of the terms. Moreover, for the lengths we have $\ell(V) = \ell(U) + \ell(V/U)$.

(c) We have $\ell(U) \leq \ell(V)$. If $U \neq V$ then $\ell(U) < \ell(V)$.

Proof. (a) Let $0 = V_0 \subset V_1 \subset \ldots \subset V_{n-1} \subset V_n = V$ be a composition series for V. We wish to relate this to a sequence of submodules of V/U. In general, U need not be related to any of the modules V_i, but we have a series of submodules of V/U (using the submodule correspondence) of the form

$$0 = (V_0 + U)/U \subseteq (V_1 + U)/U \subseteq \ldots \subseteq (V_{n-1} + U)/U \subseteq (V_n + U)/U = V/U.$$
$$(3.2)$$

Note that in this series terms can be equal. Using the isomorphism theorem (Theorem 2.24) we analyze the factor modules

$$((V_i + U)/U)/((V_{i-1} + U)/U) \cong (V_i + U)/(V_{i-1} + U)$$
$$= (V_{i-1} + U + V_i)/(V_{i-1} + U)$$
$$\cong V_i/((V_{i-1} + U) \cap V_i)$$

where the equality in the second step holds since $V_{i-1} \subset V_i$. We also have $V_{i-1} \subseteq V_{i-1} + U$ and therefore $V_{i-1} \subseteq (V_{i-1} + U) \cap V_i \subseteq V_i$. But V_{i-1} is a maximal submodule of V_i and therefore the factor module $V_i/(V_{i-1} + U) \cap V_i$ is either zero or is simple. We omit terms where the factor in the series (3.2) is zero, and we obtain a composition series for V/U, as required.

(b) By assumption V has a composition series. Then by Proposition 3.10 and by part (a) the modules U and V/U have composition series. Take a composition series for U,

$$0 = U_0 \subset U_1 \subset \ldots \subset U_{t-1} \subset U_t = U.$$

By the submodule correspondence (see Theorem 2.28) every submodule of V/U has the form V_i/U for some submodule $V_i \subseteq V$ containing U, hence a composition series of V/U is of the form

$$0 = V_0/U \subset V_1/U \subset \ldots \subset V_{r-1}/U \subset V_r/U = V/U.$$

We have that $U_t = U = V_0$; and by combining the two series we obtain

$$0 = U_0 \subset \ldots \subset U_t \subset V_1 \subset \ldots \subset V_r = V.$$

In this series all factor modules are simple. This is clear for U_i/U_{i-1}; and furthermore, by the isomorphism theorem (Theorem 2.24) $V_j/V_{j-1} \cong (V_j/U)/(V_{j-1}/U)$ is simple. Therefore, we have constructed a composition series for V in which $U = U_t$ appears as one of the terms.

For the lengths we get $\ell(V) = t + r = \ell(U) + \ell(V/U)$, as claimed.

(c) By part (b) we have $\ell(U) = \ell(V) - \ell(V/U) \leq \ell(V)$. Moreover, if $U \neq V$ then V/U is non-zero; so $\ell(V/U) > 0$ and $\ell(U) = \ell(V) - \ell(V/U) < \ell(V)$. □

3.4 Finding All Simple Modules

The Jordan–Hölder theorem shows that every module which has a composition series can be built from simple modules. Therefore, it is a fundamental problem of representation theory to understand what the simple modules of a given algebra are.

Recall from Example 2.25 the following notion. Let A be a K-algebra and V an A-module. Then for every $v \in V$ we set $\text{Ann}_A(v) = \{a \in A \mid av = 0\}$, and call this the *annihilator* of v in A. We have seen in Example 2.25 that for every $v \in V$ there is an isomorphism of A-modules $A/\text{Ann}_A(v) \cong Av$. In the context of simple modules this takes the following form, which we restate here for convenience.

Lemma 3.18. *Let A be a K-algebra and S a simple A-module. Then for every non-zero $s \in S$ we have that $S \cong A/\text{Ann}_A(s)$ as A-modules.*

Proof. As in Example 2.25 we consider the A-module homomorphism $\psi : A \to S$, $\psi(a) = as$. Since S is simple and s non-zero, this map is surjective by Lemma 3.3, and by definition the kernel is $\text{Ann}_A(s)$. So the isomorphism theorem yields $A/\text{Ann}_A(s) \cong \text{im}(\psi) = As = S$. □

This implies in particular that if an algebra has a composition series, then it can only have finitely many simple modules:

Theorem 3.19. *Let A be a K-algebra which has a composition series as an A-module. Then every simple A-module occurs as a composition factor of A. In particular, there are only finitely many simple A-modules, up to isomorphism.*

Proof. By Lemma 3.18 we know that if S is a simple A-module then $S \cong A/I$ for some A-submodule I of A. By Proposition 3.17 there is some composition series of A in which I is one of the terms. Since A/I is simple there are no further A-submodules between I and A (see Lemma 3.4). This means that I can only appear as the penultimate entry in this composition series, and $S \cong A/I$, so it is a composition factor of A.   □

For finite-dimensional algebras we have an interesting consequence.

Corollary 3.20. *Let A be a finite-dimensional K-algebra. Then every simple A-module is finite-dimensional.*

Proof. Suppose S is a simple A-module, then by Lemma 3.18, we know that S is isomorphic to a factor module of A. Hence if A is finite-dimensional, so is S.   □

Remark 3.21.

(a) In Theorem 3.19 the assumption that A has a composition series as an A-module is essential. For instance, consider the polynomial algebra $A = K[X]$ when K is infinite. There are infinitely many simple A-modules which are pairwise non-isomorphic. In fact, take a one-dimensional vector space $V = \text{span}\{v\}$ and make it into a $K[X]$-module V_λ by setting $X \cdot v = \lambda v$ for $\lambda \in K$. For $\lambda \neq \mu$ the modules V_λ and V_μ are not isomorphic, see Example 2.23; however they are 1-dimensional and hence simple. In particular, we can conclude from Theorem 3.19 that $A = K[X]$ cannot have a composition series as an A-module.

(b) In Corollary 3.20 the assumption on A is essential. Infinite-dimensional algebras can have simple modules of infinite dimension. For instance, let Q be the *two-loop quiver* with one vertex and two loops,

and let $A = KQ$ be the path algebra. Exercise 3.6 constructs for each $n \in \mathbb{N}$ a simple A-module of dimension n and even an infinite-dimensional simple A-module.

Example 3.22. Let $A = M_n(K)$, the algebra of $n \times n$-matrices over K. We have seen a composition series of A in Example 3.7, in which every composition factor is isomorphic to the natural module K^n. So by Theorem 3.19 the algebra $M_n(K)$ has precisely one simple module, up to isomorphism, namely the natural module K^n of dimension n.

3.4.1 Simple Modules for Factor Algebras of Polynomial Algebras

We will now determine the simple modules for an algebra A of the form $K[X]/I$ where I is a non-zero ideal with $I \neq K[X]$; hence $I = (f)$ where f is a polynomial of positive degree. Note that this does not require us to know a composition series of A, in fact we could have done this already earlier, after Lemma 3.4.

Proposition 3.23. *Let* $A = K[X]/(f)$ *with* $f \in K[X]$ *of positive degree.*

(a) *The simple A-modules are up to isomorphism precisely the A-modules* $K[X]/(h)$ *where h is an irreducible polynomial dividing f.*
(b) *Write* $f = f_1^{a_1} \ldots f_r^{a_r}$, *with* $a_i \in \mathbb{N}$, *as a product of irreducible polynomials* $f_i \in K[X]$ *which are pairwise coprime. Then A has precisely r simple modules, up to isomorphism, namely* $K[X]/(f_1), \ldots, K[X]/(f_r)$.

Proof. (a) First, let $h \in K[X]$ be an irreducible polynomial dividing f. Then $K[X]/(h)$ is an A-module, by Exercise 2.23, with A-action given by

$$(g_1 + (f))(g_2 + (h)) = g_1 g_2 + (h).$$

Since h is irreducible, the ideal (h) is maximal, and hence $K[X]/(h)$ is a simple A-module, by Lemma 3.4.

Conversely, let S be any simple A-module. By Lemmas 3.18 and 3.4 we know that S is isomorphic to A/U where U is a maximal submodule of A. By the submodule correspondence, see Theorem 2.28, we know $U = W/(f)$ where W is an ideal of $K[X]$ containing (f), that is, $W = (h)$ where $h \in K[X]$ and h divides f. Applying the isomorphism theorem yields

$$A/U = (K[X]/(f))/(W/(f)) \cong K[X]/W.$$

Isomorphisms preserve simple modules (see Exercise 3.3), so with A/U the module $K[X]/W$ is also simple. This means that $W = (h)$ is a maximal ideal of $K[X]$ and then h is an irreducible polynomial.

(b) By part (a), every simple A-module is isomorphic to one of $K[X]/(f_1), \ldots, K[X]/(f_r)$ (use that $K[X]$ has the unique factorization property, hence $f_1, \ldots, f_r$ are the unique irreducible divisors of f, up to multiplication by units). On the other hand, these A-modules are pairwise non-isomorphic: suppose $\psi : K[X]/(f_i) \to K[X]/(f_j)$ is an A-module homomorphism, we show that for $i \neq j$ it is not injective. Write $\psi(1 + (f_i)) = g + (f_j)$ and consider the coset $f_j + (f_i)$ Since f_i and f_j are irreducible and coprime, this coset is not the zero element in $K[X]/(f_i)$. But it is in the kernel of ψ, since

$$\psi(f_j + (f_i)) = \psi((f_j + (f_i))(1 + (f_i))) = (f_j + (f_i))\psi(1 + (f_i))$$

$$= (f_j + (f_i))(g + (f_j)) = f_j g + (f_j),$$

which is the zero element in $K[X]/(f_j)$. In particular, ψ is not an isomorphism. $\square$

Remark 3.24. We can use this to find a composition series of the algebra $A = K[X]/(f)$ as an A-module: Let $f = f_1 f_2 \ldots f_t$ with $f_i \in K[X]$ irreducible, we allow repetitions (that is, the f_i are not necessarily pairwise coprime). This gives a series of submodules of A

$$0 \subset I_1/(f) \subset I_2/(f) \subset \ldots \subset I_{t-1}/(f) \subset A$$

where I_j is the ideal of $K[X]$ generated by $f_{j+1} \ldots f_t$. Then $I_1/(f) \cong K[X]/(f_1)$ (see Exercise 2.23) and

$$(I_j/(f))/(I_{j-1}/(f)) \cong I_j/I_{j-1} \cong K[X]/(f_j),$$

so that all factor modules are simple. Hence we have found a composition series of A.

Of course, one would also get other composition series by changing the order of the irreducible factors. Note that the factorisation of a polynomial into irreducible factors depends on the field K.

Example 3.25.

(1) Over the complex numbers, every polynomial $f \in \mathbb{C}[X]$ of positive degree splits into linear factors. Hence every simple $\mathbb{C}[X]/(f)$-module is one-dimensional.

   The same works more generally for $K[X]/(f)$ when K is algebraically closed. (Recall that a field K is algebraically closed if every non-constant polynomial in $K[X]$ is a product of linear factors.)

   We will see later, in Corollary 3.38, that this is a special case of a more general result about commutative algebras over algebraically closed fields.

(2) As an explicit example, let $G = \langle g \rangle$ be a cyclic group of order n, and let $A = \mathbb{C}G$ be the group algebra over $\mathbb{C}$. Then A is isomorphic to the factor algebra $\mathbb{C}[X]/(X^n - 1)$, see Example 1.27. The polynomial $X^n - 1$ has n distinct roots in $\mathbb{C}$, namely $e^{2k\pi i/n}$ where $0 \leq k \leq n - 1$, so it splits into linear factors of the form

$$X^n - 1 = \prod_{k=0}^{n-1}(X - e^{2k\pi i/n}).$$

   According to Proposition 3.23 the algebra $\mathbb{C}[X]/(X^n - 1)$ has precisely n simple modules, up to isomorphism, each 1-dimensional, namely $S_k := \mathbb{C}[X]/(X - e^{2k\pi i/n})$ for $k = 0, 1, \ldots, n - 1$. The structure of the module S_k is completely determined by the action of the coset of X in $\mathbb{C}[X]/(X^n - 1)$, which is clearly given by multiplication by $e^{2k\pi i/n}$.

(3) Now we consider the situation over the real numbers. We first observe that every irreducible polynomial $g \in \mathbb{R}[X]$ has degree 1 or 2. In fact, if $g \in \mathbb{R}[X]$

has a non-real root $z \in \mathbb{C}$, then the complex-conjugate $\bar{z}$ is also a root: write $g = \sum_i a_i X^i$, where $a_i \in \mathbb{R}$, then

$$0 = \bar{0} = \overline{g(z)} = \overline{\sum_i a_i z^i} = \sum_i \bar{a}_i \bar{z}^i = \sum_i a_i \bar{z}^i = g(\bar{z}).$$

Then the product of the two linear factors

$$(X - z)(X - \bar{z}) = X^2 - (z + \bar{z})X + z\bar{z}$$

is a polynomial with real coefficients, and a divisor of g.

This has the following immediate consequence for the simple modules of algebras of the form $\mathbb{R}[X]/(f)$: *Every simple $\mathbb{R}[X]/(f)$-module has dimension 1 or 2.*

As an example of a two-dimensional module consider the algebra $A = \mathbb{R}[X]/(X^2 + 1)$, which is a simple A-module since $X^2 + 1$ is irreducible in $\mathbb{R}[X]$.

(4) Over the rational numbers, simple $\mathbb{Q}[X]/(f)$-modules can have arbitrarily large dimensions. For example, consider the algebra $\mathbb{Q}[X]/(X^p - 1)$, where p is a prime number. We have the factorisation

$$X^p - 1 = (X - 1)(X^{p-1} + X^{p-2} + \ldots + X + 1).$$

Since p is prime, the second factor is an irreducible polynomial in $\mathbb{Q}[X]$ (this follows by applying the Eisenstein criterion from basic algebra). So Proposition 3.23 shows that $\mathbb{Q}[X]/(X^{p-1} + \ldots + X + 1)$ is a simple $\mathbb{Q}[X]/(X^p - 1)$-module, it has dimension $p - 1$.

3.4.2   Simple Modules for Path Algebras

In this section we completely describe the simple modules for finite-dimensional path algebras of quivers.

Let Q be a quiver without oriented cycles, then for any field K, the path algebra $A = KQ$ is finite-dimensional (see Exercise 1.2). We label the vertices of Q by $Q_0 = \{1, \ldots, n\}$. Recall that for every vertex $i \in Q_0$ there is a trivial path e_i of length 0. We consider the A-module Ae_i generated by e_i; as a vector space this module is spanned by the paths which start at i. The A-module Ae_i has an A-submodule $J_i := Ae_i^{\geq 1}$ spanned by all paths of positive length starting at vertex i.

Hence we get n one-dimensional (hence simple) A-modules as factor modules of the form

$$S_i := Ae_i/J_i = \mathrm{span}\{e_i + J_i\},$$

for $i = 1, \ldots, n$. The A-action on these simple modules is given by

$$e_i(e_i + J_i) = e_i + J_i \text{ and } p(e_i + J_i) = 0 \text{ for all } p \in \mathcal{P} \setminus \{e_i\},$$

where $\mathcal{P}$ denotes the set of paths in Q. The A-modules $S_1, \ldots, S_n$ are pairwise non-isomorphic. In fact, let $\varphi : S_i \to S_j$ be an A-module homomorphism for some $i \neq j$. Then there exists a scalar $\lambda \in K$ such that $\varphi(e_i + J_i) = \lambda e_j + J_j$. Hence we get

$$\varphi(e_i + J_i) = \varphi(e_i^2 + J_i) = \varphi(e_i(e_i + J_i)) = e_i(\lambda e_j + J_j) = \lambda(e_i e_j + J_j) = 0,$$

since $e_i e_j = 0$ for $i \neq j$. In particular, φ is not an isomorphism.

We now show that this gives all simple KQ-modules, up to isomorphism.

Theorem 3.26. *Let K be a field and let Q be a quiver without oriented cycles. Assume the vertices of Q are denoted $Q_0 = \{1, \ldots, n\}$. Then the finite-dimensional path algebra $A = KQ$ has precisely the simple modules (up to isomorphism) given by $S_1, \ldots, S_n$, where $S_i = Ae_i/J_i$ for $i \in Q_0$. In particular, all simple A-modules are one-dimensional, and they are labelled by the vertices of Q.*

To identify simple modules, we first determine the maximal submodules of Ae_i. This will then be used in the proof of Theorem 3.26.

Lemma 3.27. *With the above notation the following hold:*

(a) For each vertex $i \in Q_0$, the space $e_i Ae_i$ is 1-dimensional, and is spanned by e_i.
(b) The only maximal submodule of Ae_i is $J_i = Ae_i^{\geq 1}$, and $e_i J_i = 0$.

Proof. (a) The elements in $e_i Ae_i$ are linear combinations of paths from vertex i to vertex i. The path of length zero, that is e_i, is the only path from i to i since Q has no oriented cycle.

(b) First, J_i is a maximal submodule of Ae_i, since the factor module is 1-dimensional, hence simple. Furthermore, there are no paths of positive length from i to i and therefore $e_i J_i = 0$. Next, let U be any submodule of Ae_i with $U \neq Ae_i$, we must show that U is contained in J_i. If not, then U contains an element $u = ce_i + u'$ where $0 \neq c \in K$ and $u' \in J_i$. Then also $e_i u \in U$ but $e_i u = c(e_i^2) + e_i u' = ce_i + 0$ (because $e_i J_i = 0$) and $e_i \in U$. It follows that $U = Ae_i$, a contradiction. Hence if U is maximal then $U = J_i$. $\square$

Proof of Theorem 3.26. Each S_i is 1-dimensional, hence is simple. We have already seen that for $i \neq j$, the modules S_i and S_j are not isomorphic.

Now let S be any simple A-module and take $0 \neq s \in S$. Then

$$s = 1_A s = e_1 s + e_2 s + \ldots + e_n s$$

and there is some i such that $e_i s \neq 0$. By Lemma 3.3 we know that $S = Ae_i s$. We have the A-module homomorphism

$$\psi : Ae_i \to S, \quad \psi(ae_i) = ae_i s.$$

It is surjective and hence, by the isomorphism theorem, we have

$$S \cong Ae_i / \ker(\psi).$$

Since S is simple, the kernel of ψ is a maximal submodule of Ae_i, by the submodule correspondence theorem. The only maximal submodule of Ae_i is J_i, by Lemma 3.27. Hence S is isomorphic to S_i.   □

Example 3.28. We have seen in Remark 1.25 (see also Exercise 1.18) that the algebra $T_n(K)$ of upper triangular matrices is isomorphic to the path algebra of the quiver Q

$$1 \longleftarrow 2 \longleftarrow \ldots \longleftarrow n-1 \longleftarrow n$$

Theorem 3.26 shows that KQ, and hence $T_n(K)$, has precisely n simple modules, up to isomorphism. However, we have already seen n pairwise non-isomorphic simple $T_n(K)$-modules in Example 3.14. Thus, these are all simple $T_n(K)$-modules, up to isomorphism.

3.4.3 Simple Modules for Direct Products

In this section we will describe the simple modules for direct products $A = A_1 \times \ldots \times A_r$ of algebras. We will show that the simple A-modules are precisely the simple A_i-modules, viewed as A-modules by letting the other factors act as zero. We have seen a special case in Example 3.14.

Let $A = A_1 \times \ldots \times A_r$. The algebra A contains $\varepsilon_i := (0, \ldots, 0, 1_{A_i}, 0, \ldots, 0)$ for $1 \leq i \leq r$, and ε_i commutes with all elements of A. Moreover, $\varepsilon_i \varepsilon_j = 0$ for $i \neq j$ and also $\varepsilon_i^2 = \varepsilon_i$; and we have

$$1_A = \varepsilon_1 + \ldots + \varepsilon_r.$$

Each A_i is isomorphic to a factor algebra of A, via the projection map

$$\pi_i : A \to A_i, \quad \pi_i(a_1, \ldots, a_n) = a_i.$$

This is convenient for computing with inflations of modules. Indeed, if M is an A_i-module then if we view it as an A-module by the usual inflation (see Remark 2.38), the formula for the action of A is

$$(a_1, \ldots, a_r) \cdot m := a_i m \quad \text{for } (a_1, \ldots, a_r) \in A, m \in M.$$

Proposition 3.29. *Let K be a field and let $A = A_1 \times \ldots \times A_r$ be a direct product of K-algebras. Then for $i \in \{1, \ldots, r\}$, any simple A_i-module S becomes a simple A-module by setting*

$$(a_1, \ldots, a_r) \cdot s = a_i s \text{ for all } (a_1, \ldots, a_r) \in A \text{ and } s \in S.$$

Proof. By the above, we only have to show that S is also simple as an A-module. This is a special case of Lemma 3.5.   □

We will now show that every simple A-module is of the form as in Proposition 3.29. For this we will more generally describe A-modules and we use the elements ε_i and their properties.

Lemma 3.30. *Let K be a field and $A = A_1 \times \ldots \times A_r$ a direct product of K-algebras. Moreover, let $\varepsilon_i := (0, \ldots, 0, 1_{A_i}, 0, \ldots, 0)$ for $1 \leq i \leq r$. Then for every A-module M the following holds.*

(a) Let $M_i := \varepsilon_i M$, then M_i is an A-submodule of M, and $M = M_1 \oplus \ldots \oplus M_r$, the direct sum of these submodules.

(b) If M is a simple A-module then there is precisely one $i \in \{1, \ldots, r\}$ such that $M_i \neq 0$ and this M_i is a simple A-module.

Proof. (a) We have $M_i = \{\varepsilon_i m \mid m \in M\}$. Since ε_i commutes with all elements of A, we see that if $a \in A$ then $a(\varepsilon_i m) = \varepsilon_i am$, therefore each M_i is an A-submodule of M.

To prove M is a direct sum, we first see that $M = M_1 + \ldots + M_r$. In fact, for every $m \in M$ we have

$$m = 1_A m = (\sum_{j=1}^{r} \varepsilon_j) m = \sum_{j=1}^{r} \varepsilon_j m \in M_1 + \ldots + M_r.$$

Secondly, we have to check that $M_i \cap (\sum_{j \neq i} M_j) = 0$ for each $i \in \{1, \ldots, r\}$. To this end, suppose $x := \varepsilon_i m = \sum_{j \neq i} \varepsilon_j m_j \in M_i \cap (\sum_{j \neq i} M_j)$. Since $\varepsilon_i \varepsilon_i = \varepsilon_i$ and $\varepsilon_i \varepsilon_j = 0$ for $j \neq i$ we then have

$$x = \varepsilon_i x = \varepsilon_i (\sum_{j \neq i} \varepsilon_j m_j) = \sum_{j \neq i} \varepsilon_i \varepsilon_j m_j = 0.$$

This shows that $M = M_1 \oplus \ldots \oplus M_r$, as claimed.

(b) By part (a) we have the direct sum decomposition $M = M_1 \oplus \ldots \oplus M_r$, where the M_i are A-submodules of M. If M is a simple A-module, then precisely one of these submodules M_i must be non-zero. In particular, $M \cong M_i$ and M_i is a simple A-module.   □

We can now completely describe all simple modules for a direct product.

Corollary 3.31. *Let K be a field and $A = A_1 \times \ldots \times A_r$ a direct product of K-algebras. Then the simple A-modules are precisely the inflations of the simple A_i-modules, for $i = 1, \ldots, r$.*

Proof. We have seen in Proposition 3.29 that the inflation of any simple A_i-module is a simple A-module.

Conversely, if M is a simple A-module then by Lemma 3.30, $M \cong M_i$ for precisely one $i \in \{1, \ldots, r\}$. The A-action on $M_i = \varepsilon_i M$ is given by the A_i-action and the zero action for all factors A_j with $j \neq i$, since $\varepsilon_j \varepsilon_i = 0$ for $j \neq i$. Note that M_i is also simple as an A_i-module. In other words, M is the inflation of the simple A_i-module M_i. $\qquad\square$

Example 3.32. For a field K let $A = M_{n_1}(K) \times \ldots \times M_{n_r}(K)$ for some natural numbers n_i. By Example 3.22, the matrix algebra $M_{n_i}(K)$ has only one simple module, which is the natural module K^{n_i} (up to isomorphism). Then by Corollary 3.31, the algebra A has precisely r simple modules, given by the inflations of the natural $M_{n_i}(K)$-modules. In particular, the r simple modules of A have dimensions $n_1, \ldots, n_r$.

3.5 Schur's Lemma and Applications

The Jordan–Hölder Theorem shows that simple modules are the 'building blocks' for arbitrary finite-dimensional modules. So it is important to understand simple modules. The first question one might ask is, given two simple modules, how can we find out whether or not they are isomorphic? This is answered by Schur's lemma, which we will now present. Although it is elementary, Schur's lemma has many important applications.

Theorem 3.33 (Schur's Lemma). *Let A be a K-algebra where K is a field. Suppose S and T are simple A-modules and $\phi : S \longrightarrow T$ is an A-module homomorphism. Then the following holds.*

(a) Either $\phi = 0$, or ϕ is an isomorphism. In particular, for every simple A-module S the endomorphism algebra $\mathrm{End}_A(S)$ is a division algebra.

(b) Suppose $S = T$, and S is finite-dimensional, and let K be algebraically closed. Then $\phi = \lambda\,\mathrm{id}_S$ for some scalar $\lambda \in K$.

Proof. (a) Suppose ϕ is non-zero. The kernel $\ker(\phi)$ is an A-submodule of S and $\ker(\phi) \neq S$ since $\phi \neq 0$. But S is simple, so $\ker(\phi) = 0$ and ϕ is injective.

Similarly, the image $\mathrm{im}(\phi)$ is an A-submodule of T, and T is simple. Since $\phi \neq 0$, we know $\mathrm{im}(\phi) \neq 0$ and therefore $\mathrm{im}(\phi) = T$. So ϕ is also surjective, and we have proved that ϕ is an isomorphism.

The second statement is just a reformulation of the first one, using the definition of a division algebra (see Definition 1.7).

(b) Since K is algebraically closed, the K-linear map ϕ on the finite-dimensional vector space S has an eigenvalue, say $\lambda \in K$. That is, there is some non-zero vector $v \in S$ such that $\phi(v) = \lambda v$. The map $\lambda \operatorname{id}_S$ is also an A-module homomorphism, and so is $\phi - \lambda \operatorname{id}_S$. The kernel of $\phi - \lambda \operatorname{id}_S$ is an A-submodule and is non-zero (it contains v). Since S is simple, it follows that $\ker(\phi - \lambda \operatorname{id}_S) = S$, so that we have $\phi = \lambda \operatorname{id}_S$. □

Remark 3.34. The two assumptions in part (b) of Schur's lemma are both needed for the endomorphism algebra to be 1-dimensional.

To see that one needs that K is algebraically closed, consider $K = \mathbb{R}$ and $A = \mathbb{H}$, the quaternions, from Example 1.8. We have seen that $\mathbb{H}$ is a division algebra over $\mathbb{R}$, and hence it is a simple $\mathbb{H}$-module (see Example 3.2). But right multiplication by any fixed element in $\mathbb{H}$ is an $\mathbb{H}$-module endomorphism, and for example multiplication by $i \in \mathbb{H}$ is not of the form $\lambda \operatorname{id}_{\mathbb{H}}$ for any $\lambda \in \mathbb{R}$.

Even if K is algebraically closed, part (b) of Schur's lemma can fail for a simple module S which is not finite-dimensional. For instance, let $K = \mathbb{C}$ and $A = \mathbb{C}(X)$, the field of rational functions. Since it is a field, it is a division algebra over $\mathbb{C}$, hence $\mathbb{C}(X)$ is a simple $\mathbb{C}(X)$-module. For example, left multiplication by X is a $\mathbb{C}(X)$-module endomorphism which is not of the form $\lambda \operatorname{id}_{\mathbb{C}(X)}$ for any $\lambda \in \mathbb{C}$.

One important application of Schur's lemma is that elements in the centre of a K-algebra A act as scalars on finite-dimensional simple A-modules when K is algebraically closed.

Definition 3.35. Let K be a field and let A be a K-algebra. The *centre of A* is defined to be

$$Z(A) := \{z \in A \mid za = az \text{ for all } a \in A\}.$$

Exercise 3.2. Let A be a K-algebra. Show that the centre $Z(A)$ is a subalgebra of A.

Example 3.36.

(1) By definition, a K-algebra A is commutative if and only if $Z(A) = A$. So in some sense, the size of the centre provides a 'measure' of how far an algebra is from being commutative.

(2) For any $n \in \mathbb{N}$ the centre of the matrix algebra $M_n(K)$ has dimension 1, it is spanned by the identity matrix. The proof of this is Exercise 3.16.

Lemma 3.37. *Let K be an algebraically closed field, and let A be a K-algebra. Suppose that S is a finite-dimensional simple A-module. Then for every $z \in Z(A)$ there is some scalar $\lambda_z \in K$ such that $zs = \lambda_z s$ for all $s \in S$.*

Proof. We consider the map $\rho : S \to S$ defined by $\rho(s) = zs$. One checks that it is a K-linear map. Moreover, it is an A-module homomorphism: using that z commutes with every element $a \in A$ we have

$$\rho(as) = z(as) = (za)s = (az)s = a(zs) = a\rho(s).$$

The assumptions allow us to apply part (b) of Schur's lemma, giving some $\lambda_z \in K$ such that $\rho = \lambda_z \, \mathrm{id}_S$, that is, $zs = \lambda_z s$ for all $s \in S$. $\qquad \square$

Corollary 3.38. *Let K be an algebraically closed field and let A be a commutative algebra over K. Then every finite-dimensional simple A-module S is 1-dimensional.*

Proof. Since A is commutative we have $A = Z(A)$, so by Lemma 3.37, every $a \in A$ acts by scalar multiplication on S. For every $0 \neq s \in S$ this implies that $\mathrm{span}\{s\}$ is a (1-dimensional) A-submodule of S. But S is simple, so $S = \mathrm{span}\{s\}$ and S is 1-dimensional. $\qquad \square$

Remark 3.39. Both assumptions in Corollary 3.38 are needed.

(1) If the field is not algebraically closed, simple modules of a commutative algebra need not be 1-dimensional. For example, let $A = \mathbb{R}[X]/(X^2 + 1)$, a 2-dimensional commutative $\mathbb{R}$-algebra. The polynomial $X^2 + 1$ is irreducible over $\mathbb{R}$, and hence by Proposition 3.23, we know that A is simple as an A-module, so A has a 2-dimensional simple module.

(2) The assumption that S is finite-dimensional is needed. As an example, consider the commutative $\mathbb{C}$-algebra $A = \mathbb{C}(X)$ as an A-module, as in Remark 3.34. This is a simple module, but clearly not 1-dimensional.

We will see more applications of Schur's lemma later. In particular, it will be crucial for the proof of the Artin–Wedderburn structure theorem for semisimple algebras.

EXERCISES

3.3. Let A be a K-algebra. Suppose V and W are A-modules and $\phi : V \to W$ is an A-module isomorphism.

(a) Show that V is simple if and only if W is simple.
(b) Suppose $0 = V_0 \subset V_1 \subset \ldots \subset V_n = V$ is a composition series of V. Show that then

$$0 = \phi(0) \subset \phi(V_1) \subset \ldots \subset \phi(V_n) = W$$

is a composition series of W.

3.4. Find a composition series for A as an A-module, where A is the 3-subspace algebra

$$A := \left\{ \begin{pmatrix} a_1 & b_1 & b_2 & b_3 \\ 0 & a_2 & 0 & 0 \\ 0 & 0 & a_3 & 0 \\ 0 & 0 & 0 & a_4 \end{pmatrix} \mid a_i, b_j \in K \right\} \subseteq M_4(K)$$

introduced in Example 1.16.

3.5. Let A be the ring $A = \begin{pmatrix} \mathbb{C} & \mathbb{C} \\ 0 & \mathbb{R} \end{pmatrix}$, that is, A consists of all upper triangular matrices in $M_2(\mathbb{C})$ with $(2, 2)$-entry in $\mathbb{R}$.

(a) Show that A is an algebra over $\mathbb{R}$ (but not over $\mathbb{C}$). What is its dimension over $\mathbb{R}$?

(b) Consider A as an A-module. Check that $\begin{pmatrix} \mathbb{C} & 0 \\ 0 & 0 \end{pmatrix}$ and $\begin{pmatrix} 0 & \mathbb{C} \\ 0 & 0 \end{pmatrix}$ are A-submodules of A. Show that they are simple A-modules and that they are isomorphic.

(c) Find a composition series of A as an A-module.

(d) Determine all simple A-modules, up to isomorphism, and their dimensions over $\mathbb{R}$.

3.6. Let Q be the quiver with one vertex and two loops denoted x and y. For any field K consider the path algebra KQ. Note that for any choice of $n \times n$-matrices X, Y over K, taking $x, y \in KQ$ to X, Y in $M_n(K)$ extends to an algebra homomorphism, hence a representation of KQ, that is, one gets a KQ-module.

(a) Let $V_3 := K^3$ be the 3-dimensional KQ-module on which x and y act via the matrices

$$X = \begin{pmatrix} 0 & 1 & 0 \\ 0 & 0 & 1 \\ 0 & 0 & 0 \end{pmatrix} \text{ and } Y = \begin{pmatrix} 0 & 0 & 0 \\ 1 & 0 & 0 \\ 0 & 1 & 0 \end{pmatrix}.$$

Show that V_3 is a simple KQ-module.

(b) For every $n \in \mathbb{N}$ find a simple KQ-module of dimension n.

(c) Construct an infinite-dimensional simple KQ-module.

3.7. Let V be a 2-dimensional vector space over a field K, and let A be a subalgebra of $\mathrm{End}_K(V)$. Recall that V is then an A-module (by applying linear transformations to vectors). Show that V is *not* simple as an A-module if and only if there is some $0 \neq v \in V$ which is an eigenvector for all $\alpha \in A$. More generally, show that this also holds for a 2-dimensional A-module V, by considering the representation $\psi : A \to \mathrm{End}_K(V)$.

3.8. We consider subalgebras A of $M_2(K)$ and the natural A-module $V = K^2$. Show that in each of the following situations the A-module V is simple, and determine the endomorphism algebra $\mathrm{End}_K(V)$.

(a) $K = \mathbb{R}$, $A = \left\{ \begin{pmatrix} a & b \\ -b & a \end{pmatrix} \mid a, b \in \mathbb{R} \right\}$.

(b) $K = \mathbb{Z}_2$, $A = \left\{ \begin{pmatrix} a & b \\ b & a+b \end{pmatrix} \mid a, b \in \mathbb{Z}_2 \right\}$.

3.9. (a) Find all simple $\mathbb{R}[X]/(X^4 - 1)$-modules, up to isomorphim.

(b) Find all simple $\mathbb{Q}[X]/(X^3 - 2)$-modules, up to isomorphism.

3.10. Consider the $\mathbb{R}$-algebra $A = \mathbb{R}[X]/(f)$ for a non-constant polynomial $f \in \mathbb{R}[X]$. Show that for each simple A-module S the endomorphism algebra $\operatorname{End}_{\mathbb{R}}(S)$ is isomorphic to $\mathbb{R}$ or to $\mathbb{C}$.

3.11. Let $A = M_2(\mathbb{R})$, and V be the A-module $V = A$. The following will show that V has infinitely many different composition series.

(a) Let $\varepsilon \in A \setminus \{0, 1\}$ with $\varepsilon^2 = \varepsilon$. Show that then $A\varepsilon = \{a\varepsilon \mid a \in A\}$ is an A-submodule of A, and that $0 \subset A\varepsilon \subset A$ is a composition series of A. (You may apply the Jordan–Hölder Theorem.)

(b) For $\lambda \in \mathbb{R}$ let $\varepsilon_\lambda := \begin{pmatrix} 1 & \lambda \\ 0 & 0 \end{pmatrix}$. Show that for $\lambda \neq \mu$ the A-modules $A\varepsilon_\lambda$ and $A\varepsilon_\mu$ are different. Hence deduce that V has infinitely many different composition series.

3.12. (a) Let $A = K[X]/(X^n)$. Show that A as an A-module has a unique composition series.

(b) Find all composition series of A as an A-module where A is the K-algebra $K[X]/(X^3 - X^2)$.

(c) Let $f \in K[X]$ be a non-constant polynomial and let $f = f_1^{a_1} \ldots f_r^{a_r}$ be the factorisation into irreducible polynomials in $K[X]$ (where the different f_i are pairwise coprime). Determine the number of different composition series of A as an A-module where A is the K-algebra $K[X]/(f)$.

3.13. Consider the group algebra $A = \mathbb{C}D_4$ where D_4 is the group of symmetries of the square. Let $V = \mathbb{C}^2$, which is an A-module if we take the representation as in Example 3.2. Show that V is a simple $\mathbb{C}D_4$-module.

3.14. For any natural number $n \geq 3$ let D_n be the dihedral group of order $2n$, that is, the symmetry group of a regular n-gon. This group is by definition a subgroup of $GL(\mathbb{R}^2)$, generated by the rotation r by an angle $2\pi/n$ and a reflection s. The elements satisfy $r^n = \operatorname{id}$, $s^2 = \operatorname{id}$ and $s^{-1}rs = r^{-1}$. (For $n = 5$ this group appeared in Exercise 2.20.) Let $\mathbb{C}D_n$ be the group algebra over the complex numbers.

(a) Prove that every simple $\mathbb{C}D_n$-module has dimension at most 2. (Hint: If v is an eigenvector of the action of r, then sv is also an eigenvector for the action of r.)

(b) Find all 1-dimensional $\mathbb{C}D_n$-modules, up to isomorphism. (Hint: Use the identities for r, s; the answers will be different depending on whether n is even or odd.)

3.15. Find all simple modules (up to isomorphism) and their dimensions for the following K-algebras:

(i) $K[X]/(X^2) \times K[X]/(X^3)$.

(ii) $M_2(K) \times M_3(K)$.

(iii) $K[X]/(X^2 - 1) \times K$.

3.16. Let D be a division algebra over K. Let A be the K-algebra $M_n(D)$ of all $n \times n$-matrices with entries in D. Find the centre $Z(A)$ of A.

3.17. Suppose A is a finite-dimensional algebra over a *finite* field K, and S is a (finite-dimensional) simple A-module. Let $D := \text{End}_A(S)$. Show that then D must be a field. More generally, let D be a finite-dimensional division algebra over a finite field K. Then D must be commutative, hence is a field.

3.18. Let A be a K-algebra and M an A-module of finite length $\ell(M)$. Show that $\ell(M)$ is the maximal length of a chain

$$M_0 \subset M_1 \subset \ldots \subset M_{r-1} \subset M_r = M$$

of A-submodules of M with $M_i \neq M_{i+1}$ for all i.

3.19. Let A be a K-algebra and M an A-module of finite length. Show that for all A-submodules U and V of M we have

$$\ell(U + V) = \ell(U) + \ell(V) - \ell(U \cap V).$$

3.20. Assume A is a K-algebra, and M is an A-module which is a direct sum of submodules, $M = M_1 \oplus M_2 \oplus \ldots \oplus M_n$. Consider the sequence of submodules

$$0 \subset M_1 \subset M_1 \oplus M_2 \subset \ldots \subset M_1 \oplus M_2 \oplus \ldots \oplus M_{n-1} \subset M.$$

(a) Apply the isomorphism theorem and show that

$$(M_1 \oplus \ldots \oplus M_j)/(M_1 \oplus \ldots \oplus M_{j-1}) \cong M_j$$

as A-modules.

(b) Explain briefly how to construct a composition series for M if one is given a composition series of M_j for each j.

3.21. Let Q be a quiver without oriented cycles, so the path algebra $A = KQ$ is finite-dimensional. Let $Q_0 = \{1, 2, \ldots, n\}$.

(a) For a vertex i of Q, let r be the maximal length of a path in Q with starting vertex i. Recall Ae_i has a sequence of submodules

$$0 \subset Ae_i^{\geq r} \subset Ae_i^{\geq r-1} \subset \ldots \subset Ae_i^{\geq 2} \subset Ae_i^{\geq 1} \subset Ae_i.$$

Show that $Ae_i^{\geq t}/Ae_i^{\geq t+1}$ for $t \leq r$ is a direct sum of simple modules (spanned by the cosets of paths starting at i of length t). Hence describe a composition series of Ae_i.

(b) Recall that $A = Ae_1 \oplus \ldots \oplus Ae_n$ as an A-module (see Exercise 2.6). Apply the previous exercise and part (a) and describe a composition series for A as an A-module.

Chapter 4
Semisimple Modules and Semisimple Algebras

In the previous chapter we have seen that simple modules are the 'building blocks' for arbitrary finite-dimensional modules. One would like to understand how modules are built up from simple modules. In this chapter we study modules which are direct sums of simple modules, this leads to the theory of semisimple modules. If an algebra A, viewed as an A-module, is a direct sum of simple modules, then surprisingly, every A-module is a direct sum of simple modules. In this case, A is called a semisimple algebra. We will see later that semisimple algebras can be described completely, this is the famous Artin–Wedderburn theorem. Semisimple algebras (and hence semisimple modules) occur in many places in mathematics; for example, as we will see in Chap. 6, many group algebras of finite groups are semisimple.

In this chapter, as an exception, we deal with arbitrary direct sums of modules, as introduced in Definition 2.15. The results have the same formulation, independent of whether we take finite or arbitrary direct sums, and this is an opportunity to understand a result which does not have finiteness assumptions. The only new tool necessary is Zorn's lemma.

4.1 Semisimple Modules

This section deals with modules which can be expressed as a direct sum of simple submodules. Recall Definition 2.15 for the definition of direct sums.

We assume throughout that K is a field.

Definition 4.1. Let A be a K-algebra. An A-module $V \neq 0$ is called *semisimple* if V is the direct sum of simple submodules, that is, there exist simple submodules S_i,

© Springer International Publishing AG, part of Springer Nature 2018
K. Erdmann, T. Holm, *Algebras and Representation Theory*, Springer
Undergraduate Mathematics Series, https://doi.org/10.1007/978-3-319-91998-0_4

for $i \in I$ an index set, such that

$$V = \bigoplus_{i \in I} S_i.$$

Example 4.2.

(1) Every simple module is semisimple, by definition.
(2) Consider the field K as a 1-dimensional algebra $A = K$. Then A-modules are the same as K-vector spaces and submodules are the same as K-subspaces. Recall from linear algebra that every vector space V has a basis. Take a basis $\{b_i \mid i \in I\}$ of V where I is some index set which may or may not be finite, and set $S_i := \mathrm{span}\{b_i\}$. Then S_i is a simple A-submodule of V since it is 1-dimensional, and we have $V = \bigoplus_{i \in I} S_i$, since every element of V has a unique expression as a (finite) linear combination of the basis vectors. This shows that when the algebra is the field K then every non-zero K-module is semisimple.
(3) Let $A = M_n(K)$ and consider $V = A$ as an A-module. We know from Exercise 2.5 that $V = C_1 \oplus C_2 \oplus \ldots \oplus C_n$, where C_i is the space of matrices which are zero outside the i-th column. We have also seen in Exercise 3.1 that each C_i is isomorphic to K^n and hence is a simple A-module. So $A = M_n(K)$ is a semisimple A-module.
(4) Consider again the matrix algebra $M_n(K)$, and the natural module $V = K^n$. As we have just observed, V is a simple $M_n(K)$-module, hence also a semisimple $M_n(K)$-module.

    However, we can also consider $V = K^n$ as a module for the algebra of upper triangular matrices $A = T_n(K)$. Then by Exercise 2.14 the A-submodules of K^n are given by the subspaces V_i for $i = 0, 1, \ldots, n$, where $V_i = \{(x_1, \ldots, x_i, 0, \ldots, 0)^t \mid x_i \in K\}$. Hence the A-submodules of V form a chain

$$0 = V_0 \subset V_1 \subset \ldots \subset V_{n-1} \subset V_n = K^n.$$

    In particular, $V_i \cap V_j \neq 0$ for every $i, j \neq 0$ and hence V cannot be the direct sum of A-submodules if $n \geq 2$. Thus, for $n \geq 2$, the natural $T_n(K)$-module K^n is not semisimple.
(5) As a similar example, consider the algebra $A = K[X]/(X^t)$ for $t \geq 2$, as an A-module. By Theorem 2.10, this module is of the form V_α where V has dimension t and α is the linear map which comes from multiplication with the coset of X. We see $\alpha^t = 0$ and $\alpha^{t-1} \neq 0$. We have seen in Proposition 3.23 that A has a unique simple module, which is isomorphic to $K[X]/(X)$. This is a 1-dimensional space, spanned by an eigenvector for the coset of X with eigenvalue 0. Suppose A is semisimple as an A-module, then it is a direct sum of simple modules, each is spanned by a vector which is mapped to zero by the coset of X, that is, by α. Then the matrix for α is the zero matrix and it follows that $t = 1$. Hence for $t > 1$, A is not semisimple as an A-module.

Given some A-module V, how can we decide whether or not it is semisimple? The following result provides several equivalent criteria, and each of them has its advantages.

Theorem 4.3. *Let A be a K-algebra and let V be a non-zero A-module. Then the following statements are equivalent.*

(1) For every A-submodule U of V there exists an A-submodule C of V such that $V = U \oplus C$.

(2) V is a direct sum of simple submodules (that is, V is semisimple).

(3) V is a sum of simple submodules, that is, there exist simple A-submodules S_i, $i \in I$, such that $V = \sum_{i \in I} S_i$.

The module C in (1) such that $V = U \oplus C$ is called a *complement* to U in V. Condition (1) shows that every submodule U is also isomorphic to a factor module (namely V/C) and every factor module, V/U, is also isomorphic to a submodule (namely C).

The implication (2) $\Rightarrow$ (3) is obvious. So it suffices to show the implications (1) $\Rightarrow$ (2) and (3) $\Rightarrow$ (1). We will first prove these when V is finite-dimensional (or when V has a composition series) and then we will introduce Zorn's lemma, and give a general proof.

Proof when V is finite-dimensional. (1) $\Rightarrow$ (2). Assume every submodule of V has a complement. We want to show that V is a direct sum of simple submodules.

Let $\mathcal{M}$ be the set of submodules of V which can be expressed as a direct sum of simple submodules. We assume for the moment that V is finite-dimensional, and so is every submodule of V. By Lemma 3.9 every submodule of V has a composition series, hence every non-zero submodule of V has a simple submodule.

In particular, the set $\mathcal{M}$ is non-empty. Choose U in $\mathcal{M}$ of largest possible dimension. We claim that then $U = V$.

We assume (1) holds, so there is a submodule C of V such that $V = U \oplus C$. Suppose (for a contradiction) that $U \neq V$, then C is non-zero. Then the module C has a simple submodule, S say. The intersection of U and S is zero, since $U \cap S \subseteq U \cap C = 0$. Hence $U + S = U \oplus S$. Since U is a direct sum of simple modules, $U \oplus S$ is also a direct sum of simple modules. But $\dim_K U < \dim_K (U \oplus S)$ (recall a simple module is non-zero by definition), which contradicts the choice of U. Therefore we must have $C = 0$ and $V = U \in \mathcal{M}$ is a direct sum of simple submodules.

(3) $\Rightarrow$ (1). Assume V is the sum of simple submodules. We claim that then every submodule of V has a complement.

Let U be a submodule of V, we may assume $U \neq V$. Let $\mathcal{M}$ be the set of submodules W of V such that $U \cap W = 0$; clearly $\mathcal{M}$ is non-empty since it contains the zero submodule. Again, any such W has dimension at most $\dim_K V$. Take C in $\mathcal{M}$ of largest possible dimension. We claim that $U \oplus C = V$.

Since $U \cap C = 0$, we have $U + C = U \oplus C$. Suppose that $U \oplus C$ is a proper submodule of V. Then there must be a simple submodule, S say, of V which is not contained in $U \oplus C$ (indeed, if all simple submodules of V were contained in $U \oplus C$

then since V is a sum of simple submodules, we would have that $V \subseteq U \oplus C$). Then the intersection of S with $U \oplus C$ is zero (if not, the intersection is a non-zero submodule of S and since S is simple, it is equal to S). Then we have a submodule of V of the form $U \oplus (C \oplus S)$ and $\dim_K C < \dim_K (C \oplus S)$ which contradicts the choice of C. Therefore we must have $U \oplus C = V$.   □

Note that if we only assume V has a composition series then the proof is identical, by replacing 'dimension' by 'length' (of a composition series).

If V is not finite-dimensional, then we apply Zorn's lemma. This is a more general statement about partially ordered sets. Let $\mathcal{P}$ be a non-empty set and $\leq$ some partial order on $\mathcal{P}$. A chain $(U_i)_{i \in I}$ in $\mathcal{P}$ is a linearly ordered subset. An upper bound of such a chain is an element $U \in \mathcal{P}$ such that $U_i \leq U$ for all $i \in I$. Zorn's lemma states that if every chain in $\mathcal{P}$ has an upper bound then $\mathcal{P}$ has at least one maximal element. For a discussion of this, we refer to the book by Cameron in this series.[1]

Perhaps one of the most important applications of Zorn's lemma in representation theory is that a non-zero cyclic module always has a maximal submodule.

Definition 4.4. Let R be a ring. An R-module M is called *cyclic* if there exists an element $m \in M$ such that $M = Rm = \{rm \mid r \in R\}$, that is, the module M is generated by a single element.

Lemma 4.5. *Assume R is a ring and $M = Rm$, a non-zero cyclic R-module. Then M has a maximal submodule, and hence has a simple factor module.*

Proof. Let $\mathcal{M}$ be the set of submodules of M which do not contain m, this set is partially ordered by inclusion. Then $\mathcal{M}$ is not empty (the zero submodule belongs to $\mathcal{M}$). Take any chain $(U_i)_{i \in I}$ in $\mathcal{M}$, and let $U := \cup_{i \in I} U_i$, this is a submodule of M and it does not contain m. So $U \in \mathcal{M}$, and $U_i \subseteq U$, that is, U is an upper bound in $\mathcal{M}$ for the chain. By Zorn's lemma, the set $\mathcal{M}$ has a maximal element, N say. We claim that N is a maximal submodule of M: Indeed, let $N \subseteq W \subseteq M$, where W is a submodule of M. If $m \in W$ then $M = Rm \subseteq W$ and $W = M$. On the other hand, if $m \notin W$ then W belongs to $\mathcal{M}$, and by maximality of N it follows that $N = W$. Then M/N is a simple R-module, by the submodule correspondence (see Theorem 2.28).   □

We return to the proof of Theorem 4.3 in the general case. One ingredient in the proof of $(1) \Rightarrow (2)$ is that assuming (1) holds for V then every non-zero submodule of V must have a simple submodule. We can prove this for general V as follows.

Lemma 4.6. *Let A be a K-algebra. Assume V is an A-module such that every submodule U of V has a complement. Then every non-zero submodule of V has a simple submodule.*

---

[1]P. J. Cameron, *Sets, Logic and Categories*. Springer Undergraduate Mathematics Series. Springer-Verlag London, Ltd., London, 1999. x+180 pp.

Proof. It is enough to show that if Am is a non-zero cyclic submodule of V then Am has a simple submodule. By Lemma 4.5, the module Am has a maximal submodule, U say, and then Am/U is a simple A-module. By the assumption, there is a submodule C of V such that $V = U \oplus C$. Then we have

$$Am = Am \cap V = Am \cap (U \oplus C) = U \oplus (Am \cap C),$$

where the last equality holds since U is contained in Am. It follows now by the isomorphism theorem that $Am/U \cong Am \cap C$, which is simple and is also a submodule of Am.   □

Proof of Theorem 4.3 in general. (1) $\Rightarrow$ (2). Consider families of simple submodules of V whose sum is a direct sum. We set

$$\mathcal{M} := \{(S_i)_{i \in I} \mid S_i \subseteq V \text{ simple}, \sum_{i \in I} S_i = \bigoplus_{i \in I} S_i\}.$$

By assumption in Theorem 4.3, $V \neq 0$. We assume that (1) holds, then by Lemma 4.6, V has a simple submodule, so $\mathcal{M}$ is non-empty. We consider the 'refinement order' on $\mathcal{M}$, that is, we define

$$(S_i)_{i \in I} \leq (T_j)_{j \in J}$$

if every simple module S_i appears in the family $(T_j)_{j \in J}$. This is a partial order. To apply Zorn's lemma, we must show that any chain in $\mathcal{M}$ has an upper bound in $\mathcal{M}$. We can assume that the index sets of the sequences in the chain are also totally ordered by inclusion. Let $\tilde{I}$ denote the union of the index sets of the families in the chain. Then the family $(S_i)_{i \in \tilde{I}}$ is an upper bound of the chain in $\mathcal{M}$: Suppose (for a contradiction) that $(S_i)_{i \in \tilde{I}}$ does not lie in $\mathcal{M}$, that is, $\sum_{i \in \tilde{I}} S_i$ is not a direct sum. Then for some $k \in \tilde{I}$ we have $S_k \cap \sum_{i \neq k} S_i \neq 0$. This means that there exists a non-zero element $s \in S_k$ which can be expressed as a finite(!) sum $s = s_{i_1} + \ldots + s_{i_r}$ with $s_{i_j} \in S_{i_j}$ for some $i_1, \ldots, i_r \in \tilde{I}$. Since $\tilde{I}$ is a union of index sets, the finitely many indices $k, i_1, \ldots, i_r$ must appear in some index set I' which is an index set for some term of the chain in $\mathcal{M}$. But then $\sum_{i \in I'} S_i \neq \bigoplus_{i \in I'} S_i$, contradicting the assumption that $(S_i)_{i \in I'} \in \mathcal{M}$. So we have shown that every chain in the partially ordered set $\mathcal{M}$ has an upper bound in $\mathcal{M}$. Now Zorn's lemma implies that $\mathcal{M}$ has a maximal element $(S_j)_{j \in J}$. In particular, $U := \sum_{j \in J} S_j = \bigoplus_{j \in J} S_j$. Now we continue as in the first version of the proof: By (1) there is a submodule C of V such that $V = U \oplus C$. If C is non-zero then by Lemma 4.6, it contains a simple submodule S. Since $U \cap C = 0$, we have $U \cap S = 0$ and hence $U + S = U \oplus S$. This means that the family $(S_j)_{j \in J} \cup \{S\} \in \mathcal{M}$, contradicting the maximality of the family $(S_j)_{j \in J}$. Therefore, $C = 0$ and $V = U = \bigoplus_{j \in J} S_j$ is a direct sum of simple submodules, that is, (2) holds.

(3) $\Rightarrow$ (1). Let $U \subseteq V$ be a submodule of V. Consider the set

$$\mathcal{S} := \{W \mid W \subseteq V \text{ an } A\text{-submodule such that } U \cap W = 0\}.$$

Then $\mathcal{S}$ is non-empty (the zero module is in $\mathcal{S}$), it is partially ordered by inclusion and for each chain in $\mathcal{S}$ the union of the submodules is an upper bound in $\mathcal{S}$. So Zorn's lemma gives a maximal element $C \in \mathcal{S}$. Now the rest of the proof is completely analogous to the above proof for the finite-dimensional case.   □

The important Theorem 4.3 has many consequences; we collect a few which will be used later.

Corollary 4.7. *Let A be a K-algebra.*

(a) *Let $\varphi : S \to V$ be an A-module homomorphism, where S is a simple A-module. Then $\varphi = 0$ or the image of φ is a simple A-module isomorphic to S.*
(b) *Let $\varphi : V \to W$ be an isomorphism of A-modules. Then V is semisimple if and only if W is semisimple.*
(c) *All non-zero submodules and all non-zero factor modules of semisimple A-modules are again semisimple.*
(d) *Let $(V_i)_{i \in I}$ be a family of non-zero A-modules. Then the direct sum $\bigoplus_{i \in I} V_i$ (see Definition 2.17) is a semisimple A-module if and only if all the modules V_i, $i \in I$, are semisimple A-modules.*

Proof. (a) The kernel $\ker(\varphi)$ is an A-submodule of S. Since S is simple, there are only two possibilities: if $\ker(\varphi) = S$, then $\varphi = 0$; otherwise $\ker(\varphi) = 0$, then by the isomorphism theorem, $\operatorname{im}(\varphi) \cong S/\ker(\varphi) \cong S$ and is simple.
Towards (b) and (c), we prove:
(∗) If $\varphi : V \to W$ is a non-zero surjective A-module homomorphism, and if V is semisimple, then so is W.
So assume that V is semisimple. By Theorem 4.3, V is a sum of simple submodules, $V = \sum_{i \in I} S_i$ say. Then

$$W = \varphi(V) = \varphi\left(\sum_{i \in I} S_i\right) = \sum_{i \in I} \varphi(S_i).$$

By part (a), each $\varphi(S_i)$ is either zero, or is a simple A-module. We can ignore the ones which are zero, and get that W is a sum of simple A-modules and hence is semisimple, using again Theorem 4.3.
Part (b) follows now, by applying (∗) to φ and also to the inverse isomorphism φ^{-1}.
(c) Suppose that V is a semisimple A-module, and $U \subseteq V$ an A-submodule. We start by dealing with the factor module V/U, we must show that if V/U is non-zero then it is semisimple. Let π be the canonical A-module homomorphism

$$\pi : V \to V/U , \ \pi(v) = v + U.$$

If V/U is non-zero then π is non-zero, and by (∗) we get that V/U is semisimple.
Next, assume U is non-zero, we must show that then U is semisimple. By Theorem 4.3 we know that there exists a complement to U, that is, an A-submodule

$C \subseteq V$ such that $V = U \oplus C$. This implies that $U \cong V/C$. But the non-zero factor module V/C is semisimple by the first part of (c), and then by (b) we deduce that U is also semisimple.

(d) Write $V := \bigoplus_{i \in I} V_i$ and consider the inclusion maps $\iota_i : V_i \to V$. These are injective A-module homomorphisms; in particular, $V_i \cong \mathrm{im}(\iota_i) \subseteq V$ are A-submodules.

Suppose that V is semisimple. Then by parts (b) and (c) each V_i is semisimple, as it is isomorphic to the non-zero submodule $\mathrm{im}(\iota_i)$ of the semisimple module V.

Conversely, suppose that each V_i, $i \in I$, is a semisimple A-module. By Theorem 4.3 we can write V_i as a sum of simple submodules, say $V_i = \sum_{j \in J_i} S_{ij}$ (for some index sets J_i). On the other hand we have that $V = \sum_{i \in I} \iota_i(V_i)$, since every element of the direct sum has only finitely many non-zero entries, see Definition 2.17. Combining these, we obtain that

$$V = \sum_{i \in I} \iota_i(V_i) = \sum_{i \in I} \iota_i \left(\sum_{j \in J_i} S_{ij} \right) = \sum_{i \in I} \sum_{j \in J_i} \iota_i(S_{ij})$$

and V is a sum of simple A-submodules (the $\iota_i(S_{ij})$ are simple by part (a)). Hence V is semisimple by Theorem 4.3. $\qquad\square$

4.2 Semisimple Algebras

In Example 4.2 we have seen that for the 1-dimensional algebra $A = K$, every non-zero A-module is semisimple. We would like to describe algebras for which all non-zero modules are semisimple. If A is such an algebra, then in particular A viewed as an A-module is semisimple. Surprisingly, the converse holds, as we will see soon: If A as an A-module is semisimple, then all non-zero A-modules are semisimple. Therefore, we make the following definition.

Definition 4.8. A K-algebra A is called *semisimple* if A is semisimple as an A-module.

We have already seen some semisimple algebras.

Example 4.9. Every matrix algebra $M_n(K)$ is a semisimple algebra, see Example 4.2.

Remark 4.10. By definition, a semisimple algebra A is a direct sum $A = \bigoplus_{i \in I} S_i$ of simple A-submodules. Luckily, in this situation the index set I must be finite. Indeed, the identity element can be expressed as a finite sum $1_A = \sum_{i \in I} s_i$ with $s_i \in S_i$. This means that there is a finite subset $\{i_1, \ldots, i_k\} \subseteq I$ such that $1_A \in \sum_{j=1}^{k} S_{i_j}$. Then $A = A1_A \subseteq \sum_{j=1}^{k} S_{i_j}$, that is, $A = \bigoplus_{j=1}^{k} S_{i_j}$ is a direct sum of finitely many simple A-submodules. In particular, a semisimple algebra has finite length as an A-module, and every simple A-module is isomorphic to one of the modules $S_{i_1}, \ldots, S_{i_k}$ which appear in the direct sum decomposition of A (see Theorem 3.19).

When A is a semisimple algebra, then we can understand arbitrary non-zero A-modules; they are just direct sums of simple modules, as we will now show.

Theorem 4.11. *Let A be a K-algebra. Then the following assertions are equivalent.*

(i) A is a semisimple algebra.
(ii) Every non-zero A-module is semisimple.

Proof. The implication (ii) $\Rightarrow$ (i) follows by Definition 4.8.

Conversely, suppose that A is semisimple as an A-module. Take an arbitrary non-zero A-module V. We have to show that V is a semisimple A-module. As a K-vector space V has a basis, say $\{v_i \mid i \in I\}$. With the same index set I, we take the A-module

$$\bigoplus_{i \in I} A := \{(a_i)_{i \in I} \mid a_i \in A, \text{ only finitely many } a_i \text{ are non-zero}\}$$

the direct sum of copies of A (see Definition 2.17). We consider the map

$$\psi : \bigoplus_{i \in I} A \to V , \quad (a_i)_{i \in I} \mapsto \sum_{i \in I} a_i v_i .$$

One checks that ψ is an A-module homomorphism. Since the v_i form a K-basis of V, the map ψ is surjective. By the isomorphism theorem,

$$\left(\bigoplus_{i \in I} A \right) / \ker(\psi) \cong \operatorname{im}(\psi) = V.$$

By assumption, A is a semisimple A-module, and hence $\bigoplus_{i \in I} A$ is also a semisimple A-module by Corollary 4.7. But then Corollary 4.7 implies that the non-zero factor module $(\bigoplus_{i \in I} A)/\ker(\psi)$ is also semisimple. Finally, Corollary 4.7 gives that $V \cong (\bigoplus_{i \in I} A)/\ker(\psi)$ is a semisimple A-module. $\qquad\square$

This theorem has many consequences, in particular we can use it to show that factor algebras of semisimple algebras are semisimple, and hence also that an algebra isomorphic to a semisimple algebra is again semisimple:

Corollary 4.12. *Let A and B be K-algebras. Then the following holds:*

(a) Let $\varphi : A \to B$ be a surjective algebra homomorphism. If A is a semisimple algebra then so is B.
(b) If A and B are isomorphic then A is semisimple if and only if B is semisimple.
(c) Every factor algebra of a semisimple algebra is semisimple.

Proof. (a) Let $\varphi : A \to B$ be a surjective algebra homomorphism. By Theorem 4.11 it suffices to show that every non-zero B-module is semisimple. Suppose $M \neq 0$ is a B-module, then we can view it as an A-module, with action

$$a \cdot m = \varphi(a)m \ (\text{for } m \in m, a \in A)$$

by Example 2.4. Since A is semisimple, the A-module M can be written as $M = \sum_{i \in I} S_i$ where the S_i are simple A-modules. We are done if we show that each S_i is a B-submodule of M and that S_i is simple as a B-module.

First, S_i is a non-zero subspace of M. Let $b \in B$ and $v \in S_i$, we must show that $bv \in S_i$. Since φ is surjective we have $b = \varphi(a)$ for some $a \in A$, and then

$$a \cdot v = \varphi(a)v = bv$$

and by assumption $a \cdot v \in S_i$. Hence S_i is a B-submodule of M.

Now let U be a non-zero B-submodule of S_i, then with the above formula for the action of A, we see that U is a non-zero A-submodule of S_i. Since S_i is simple as an A-module, it follows that $U = S_i$. Hence S_i is simple as a B-module.

(b) Suppose $\varphi : A \to B$ is an isomorphism. Then (b) follows by applying (a) to φ and to the inverse isomorphism φ^{-1}.

(c) Assume A is semisimple. Let $I \subset A$ be a two-sided ideal with $I \neq A$, and consider the factor algebra A/I. We have the canonical surjective algebra homomorphism $\varphi : A \to A/I$. By (a) the algebra A/I is semisimple. $\square$

Example 4.13.

(1) The algebra $A = T_n(K)$ of upper triangular matrices is not semisimple when $n \geq 2$ (for $n = 1$ we have $A = K$, which is semisimple). In fact, the natural A-module K^n is not a semisimple module, see Example 4.2. Therefore Theorem 4.11 implies that $T_n(K)$ is not a semisimple algebra.

(2) The algebra $A = K[X]/(X^t)$ is not a semisimple algebra for $t \geq 2$. Indeed, we have seen in Example 4.2 that A is not semisimple as an A-module.

It follows that also the polynomial algebra $K[X]$ is not a semisimple algebra: indeed, apply Corollary 4.12 to the surjective algebra homomorphism $K[X] \to K[X]/(X^t)$, $f \mapsto f + (X^t)$.

We can answer precisely which factor algebras of $K[X]$ are semisimple. We recall from basic algebra that the polynomial ring $K[X]$ over a field K is a unique factorization domain, that is, every polynomial can be (uniquely) factored as a product of irreducible polynomials.

Proposition 4.14. *Let K be a field, $f \in K[X]$ a non-constant polynomial and $f = f_1^{a_1} \cdot \ldots \cdot f_r^{a_r}$ its factorization into irreducible polynomials, where $a_i \in \mathbb{N}$ and $f_1, \ldots, f_r$ are pairwise coprime. Then the following statements are equivalent.*

(i) $K[X]/(f)$ is a semisimple algebra.

(ii) We have $a_i = 1$ for all $i = 1, \ldots, r$, that is, $f = f_1 \cdot \ldots \cdot f_r$ is a product of pairwise coprime irreducible polynomials.

Proof. We consider the algebra $A = K[X]/(f)$ as an A-module. Recall that the A-submodules of A are given precisely by $(g)/(f)$ where the polynomial g divides f. (i) $\Rightarrow$ (ii). We assume A is semisimple. Assume for a contradiction that (say) $a_1 \geq 2$. We will find a non-zero A-module which is not semisimple (and this contradicts Theorem 4.11). Consider $M := K[X]/(f_1^2)$. Then M is a $K[X]$-module which is annihilated by f, and hence is an A-module (see for example Lemma 2.37). The A-submodules of M are of the form $(g)/(f_1^2)$ for polynomials g dividing f_1^2. Since f_1 is irreducible, M has only one non-trivial A-submodule, namely $(f_1)/(f_1^2)$. In particular, M cannot be the direct sum of simple A-submodules, that is, M is not semisimple as an A-module.

(ii) $\Rightarrow$ (i). Assume f is a product of pairwise coprime irreducible polynomials. We want to show that A is a semisimple algebra. It is enough to show that every A-submodule of A has a complement (see Theorem 4.3). Every submodule of A is of the form $(g)/(f)$ where g divides f. Write $f = gh$ then by our assumption, the polynomials g and h are coprime. One shows now, with basic algebra, that $A = (g)/(f) \oplus (h)/(f)$, which is a direct sum of A-modules; see the worked Exercise 4.3.   □

Example 4.15.

(1) By Proposition 4.14 we see that $A = K[X]/(X^t)$ is a semisimple algebra if and only if $t = 1$ (see Example 4.13).

(2) In Sect. 1.4 we have seen that up to isomorphism there are precisely three 2-dimensional $\mathbb{R}$-algebras, namely $\mathbb{R}[X]/(X^2)$, $\mathbb{R}[X]/(X^2 - 1)$ and $\mathbb{R}[X]/(X^2 + 1) \cong \mathbb{C}$. Using Proposition 4.14 we can now see which of these are semisimple. The algebra $\mathbb{R}[X]/(X^2)$ is not semisimple, as we have just seen. The other two algebras are semisimple because $X^2 - 1 = (X - 1)(X + 1)$, the product of two coprime polynomials and $X^2 + 1$ is irreducible in $\mathbb{R}[X]$.

(3) Let p be a prime number and denote by $\mathbb{Z}_p$ the field with p elements. Consider the algebra $A = \mathbb{Z}_p/(X^p - 1)$. We have $X^p - 1 = (X - 1)^p$ in $\mathbb{Z}_p[X]$, hence by Proposition 4.14 the algebra A is not semisimple.

Remark 4.16. Note that subalgebras of semisimple algebras are not necessarily semisimple. For example, the algebra of upper triangular matrices $T_n(K)$ for $n \geq 2$ is not semisimple, by Example 4.13; but it is a subalgebra of the semisimple algebra $M_n(K)$, see Example 4.9.

More generally, in Exercise 1.29 we have seen that every finite-dimensional algebra is isomorphic to a subalgebra of a matrix algebra $M_n(K)$, hence to a subalgebra of a semisimple algebra, and we have seen many algebras which are not semisimple.

On the other hand, the situation is different for factor algebras: we have already seen in Corollary 4.12 that every factor algebra of a semisimple algebra is again semisimple.

If $B = A/I$, where I is an ideal of A with $I \neq A$, then in Lemma 2.37 we have seen that a B-module can be viewed as an A-module on which I acts as zero, and conversely any A-module V with $IV = 0$ (that is, I acts as zero on V), can be viewed as a B-module. The actions are related by the formula

$$(a + I)v = av \quad (a \in A, \; v \in V).$$

The following shows that with this correspondence, semisimple modules correspond to semisimple modules.

Theorem 4.17. *Let A be a K-algebra, $I \subset A$ a two-sided ideal of A with $I \neq A$, and let $B = A/I$ the factor algebra. The following are equivalent for any B-module V.*

(i) V is a semisimple B-module.
(ii) V is a semisimple A-module with $IV = 0$.

Proof. First, suppose that (i) holds. By Theorem 4.3, $V = \sum_{j \in J} S_j$, the sum of simple B-submodules of V. By Lemma 2.37, we can also view the S_j as A-modules with $IS_j = 0$. Moreover, they are also simple as A-modules, by Lemma 3.5. This shows that V is a sum of simple A-modules, and therefore it is a semisimple A-module, by Theorem 4.3.

Conversely, suppose that (ii) holds, assume V is a semisimple A-module with $IV = 0$. By Theorem 4.3 we know $V = \sum_{j \in J} S_j$, a sum of simple A-submodules of V. Then $IS_j \subseteq IV = 0$, and Lemma 2.37 says that we can view the S_j as B-modules. One checks that these are also simple as B-modules (with the same reasoning as in Lemma 3.5). So as a B-module, $V = \sum_{j \in J} S_j$, a sum of simple B-modules, and hence is a semisimple B-module by Theorem 4.3. $\qquad\square$

Corollary 4.18. *Let $A_1, \ldots, A_r$ be finitely many K-algebras. Then the direct product $A_1 \times \ldots \times A_r$ is a semisimple algebra if and only if each A_i for $i = 1, \ldots, r$ is a semisimple algebra.*

Proof. Set $A = A_1 \times \ldots \times A_r$. Suppose first that A is semisimple. For any $i \in \{1, \ldots, r\}$, the projection $\pi_i : A \to A_i$ is a surjective algebra homomorphism. By Corollary 4.12 each A_i is a semisimple algebra.

Conversely, suppose that all algebras $A_1, \ldots, A_r$ are semisimple. We want to use Theorem 4.11, that is, we have to show that every non-zero A-module is semisimple. Let $M \neq 0$ be an A-module. We use Lemma 3.30, which gives that $M = M_1 \oplus M_2 \oplus \ldots \oplus M_r$, where $M_i = \varepsilon_i M$, with $\varepsilon_i = (0, \ldots, 0, 1_{A_i}, 0, \ldots, 0)$, and M_i is an A-submodule of M. Then M_i is also an A_i-module, since the kernel of π_i annihilates M_i (using Lemma 2.37). We can assume that $M_i \neq 0$; otherwise we can ignore this summand in $M = M_1 \oplus M_2 \oplus \ldots \oplus M_r$. Then by assumption and Theorem 4.11, M_i is semisimple as a module for A_i, and then by Theorem 4.17 it is also semisimple as an A-module. Now part (d) of Corollary 4.7 shows that M is semisimple as an A-module. $\qquad\square$

Example 4.19. Let K be a field. We have already seen that matrix algebras $M_n(K)$ are semisimple (see Example 4.9). Corollary 4.18 now shows that arbitrary finite direct products

$$M_{n_1}(K) \times \ldots \times M_{n_r}(K)$$

are semisimple algebras.

4.3 The Jacobson Radical

In this section we give an alternative characterisation of semisimplicity of algebras. We will introduce the Jacobson radical $J(A)$ of an algebra A. We will see that this is an ideal which measures how far away A is from being semisimple.

Example 4.20. Let $A = \mathbb{R}[X]/(f)$ where $f = (X - 1)^a(X + 1)^b$ for $a, b \geq 1$. By Proposition 3.23 this algebra has two simple modules (up to isomorphism), which we can take as $S_1 := A/M_1$ and $S_2 = A/M_2$, where M_1, M_2 are the maximal left ideals of A given by $M_1 = (X - 1)/(f)$ and $M_2 = (X + 1)/(f)$. Since $X - 1$ and $X + 1$ are coprime we observe that

$$M_1 \cap M_2 = ((X - 1)(X + 1))/(f).$$

We know that A is semisimple if and only if $a = b = 1$ (see Proposition 4.14). This is the same as $M_1 \cap M_2 = 0$. This motivates the definition of the Jacobson radical below.

Exercise 4.1. Assume A is a semisimple algebra, say $A = S_1 \oplus \ldots \oplus S_n$, where the S_i are simple A-submodules of A (see Remark 4.10). Then $S_i \cong A/M_i$, where M_i is a maximal left ideal of A (see Lemma 3.18). Show that the intersection $M_1 \cap \ldots \cap M_n$ is zero. (Hint: Let a be in the intersection, show that $aS_i = 0$ for $1 \leq i \leq n$.)

Definition 4.21. Let A be a K-algebra. The *Jacobson radical* $J(A)$ of A is defined to be the intersection of all maximal left ideals of A. In other words, $J(A)$ is the intersection of all maximal A-submodules of A.

Example 4.22.

(1) Assume K is an infinite field, then the polynomial algebra $A = K[X]$ has Jacobson radical $J(A) = 0$: In fact, for each $\lambda \in K$ the left ideal generated by $X - \lambda$ is maximal (the factor module $K[X]/(X - \lambda)$ has dimension 1; then use the submodule correspondence, Theorem 2.28). Then $J(A) \subseteq \bigcap_{\lambda \in K} (X - \lambda)$. We claim this is zero. Suppose $0 \neq f$ is in the intersection, then f is a polynomial of degree n (say), and $X - \lambda$ divides f for each $\lambda \in K$, but there are infinitely many such factors, a contradiction. It follows that $J(A) = 0$.

In general, $J(K[X]) = 0$ for arbitrary fields K. One can adapt the above proof by using that $K[X]$ has infinitely many irreducible polynomials. The proof is Exercise 4.11.

(2) Consider a factor algebra of a polynomial algebra, $A = K[X]/(f)$, where f is a
non-constant polynomial, and write $f = f_1^{a_1} \ldots f_r^{a_r}$, where the f_i are pairwise
coprime irreducible polynomials in $K[X]$. We have seen before that the left
ideals of A are of the form $(g)/(f)$ where g is a divisor of f. The maximal left
ideals are those where g is irreducible. One deduces that

$$J(A) = \bigcap_{i=1}^{r} (f_i)/(f) = (\prod_{i=1}^{r} f_i)/(f).$$

In particular, $J(A) = 0$ if and only if $f = f_1 \cdot \ldots \cdot f_r$, that is, f is the product
of pairwise coprime irreducible polynomials.

The following theorem collects some of the basic properties of the Jacobson
radical.

Recall that for any left ideals I, J of an algebra A, the product is defined as

$$IJ = \text{span}\{xy \mid x \in I, y \in J\}$$

and this is also a left ideal of A. In particular, for any left ideal I of A we define
powers inductively by setting $I^0 = A$, $I^1 = I$ and $I^k = I^{k-1}I$ for all $k \geq 2$. Thus
for every left ideal I we get a chain of left ideals of the form

$$A \supseteq I \supseteq I^2 \supseteq I^3 \supseteq \ldots$$

The ideal I is called *nilpotent* if there is some $r \geq 1$ such that $I^r = 0$.

We also need the *annihilator* of a simple A-module S, this is defined as

$$\text{Ann}_A(S) := \{a \in A \mid as = 0 \text{ for every } s \in S\}.$$

This is contained in $\text{Ann}_A(s)$ whenever $S = As$ for $s \in S$, and since
$S \cong A/\text{Ann}_A(s)$ (see Lemma 3.18) we know that $\text{Ann}_A(s)$ is a maximal left
ideal of A.

Exercise 4.2. Let A be a K-algebra. Show that for any A-module M we have

$$\text{Ann}_A(M) := \{a \in A \mid am = 0 \text{ for every } m \in M\}$$

is a two-sided ideal of A.

Theorem 4.23. *Let K be a field and A a K-algebra which has a composition series
as an A-module (that is, A has finite length as an A-module). Then the following
holds for the Jacobson radical $J(A)$.*

(a) $J(A)$ is the intersection of finitely many maximal left ideals.
(b) We have that

$$J(A) = \bigcap_{s \text{ simple}} \text{Ann}_A(S),$$

that is, $J(A)$ consists of those $a \in A$ such that $aS = 0$ for every simple A-module S.

(c) *$J(A)$ is a two-sided ideal of A.*

(d) *$J(A)$ is a nilpotent ideal, we have $J(A)^n = 0$ where n is the length of a composition series of A as an A-module.*

(e) *The factor algebra $A/J(A)$ is a semisimple algebra.*

(f) *Let $I \subseteq A$ be a two-sided ideal with $I \neq A$ such that the factor algebra A/I is semisimple. Then $J(A) \subseteq I$.*

(g) *A is a semisimple algebra if and only $J(A) = 0$.*

Remark 4.24. The example of a polynomial algebra $K[X]$ shows that the assumption of finite length in the theorem is needed. We have seen in Example 4.22 that $J(K[X]) = 0$. However, $K[X]$ is not semisimple, see Example 4.13. So, for instance, part (g) of Theorem 4.23 is not valid for $K[X]$.

Proof. (a) Suppose that $M_1, \ldots, M_r$ are finitely many maximal left ideals of A. Hence we have that $J(A) \subseteq M_1 \cap \ldots \cap M_r$. If we have equality then we are done. Otherwise there exists another maximal left ideal M_{r+1} such that

$$M_1 \cap \ldots \cap M_r \supsetneq M_1 \cap \ldots \cap M_r \cap M_{r+1}$$

and this is a proper inclusion. Repeating the argument gives a sequence of left ideals of A,

$$A \supsetneq M_1 \supsetneq M_1 \cap M_2 \supsetneq M_1 \cap M_2 \cap M_3 \supsetneq \ldots$$

Each quotient is non-zero, so the process must stop at the latest after n steps where n is the length of a composition series of A, see Exercise 3.18. This means that $J(A)$ is the intersection of finitely many maximal left ideals.

(b) We first prove that the intersection of the annihilators of simple modules is contained in $J(A)$. Take an element $a \in A$ such that $aS = 0$ for all simple A-modules S. We want to show that a belongs to every maximal left ideal of A. Suppose M is a maximal left ideal, then A/M is a simple A-module. Therefore by assumption $a(A/M) = 0$, so that $a + M = a(1_A + M) = 0$ and hence $a \in M$. Since M is arbitrary, this shows that a is in the intersection of all maximal left ideals, that is, $a \in J(A)$.

Assume (for a contradiction) that the inclusion is not an equality, then there is a simple A-module S such that $J(A)S \neq 0$; let $s \in S$ with $J(A)s \neq 0$. Then $J(A)s$ is an A-submodule of S (since $J(A)$ is a left ideal), and it is non-zero. Because S is simple we get that $J(A)s = S$. In particular, there exists an $x \in J(A)$ such that $xs = s$, that is, $x - 1_A \in \mathrm{Ann}_A(s)$. Now, $\mathrm{Ann}_A(s)$ is a maximal left ideal (since $A/\mathrm{Ann}_A(s) \cong S$, see Lemma 3.18). Hence $J(A) \subseteq \mathrm{Ann}_A(s)$. Therefore we have $x \in \mathrm{Ann}_A(s)$ and $x - 1_A \in \mathrm{Ann}_A(s)$ and it follows that $1_A \in \mathrm{Ann}_A(s)$, a contradiction since $s \neq 0$.

(c) This follows directly from part (b), together with Exercise 4.2.

(d) Take a composition series of A as an A-module, say

$$0 = V_0 \subset V_1 \subset \ldots \subset V_{n-1} \subset V_n = A.$$

We will show that $J(A)^n = 0$. For each i with $1 \leq i \leq n$, the factor module V_i/V_{i-1} is a simple A-module. By part (b) it is therefore annihilated by $J(A)$, and this implies that $J(A)V_i \subseteq V_{i-1}$ for all $i = 1, \ldots, n$. Hence

$$J(A)V_1 = 0, \quad J(A)^2 V_2 \subseteq J(A)V_1 = 0$$

and inductively we see that $J(A)^r V_r = 0$ for all r. In particular, $J(A)^n A = 0$, and this implies $J(A)^n = 0$, as required.

(e) By Definition 4.8 we have to show that $A/J(A)$ is semisimple as an $A/J(A)$-module. According to Theorem 4.17 this is the same as showing that $A/J(A)$ is semisimple as an A-module. From part (a) we know that $J(A) = \bigcap_{i=1}^{r} M_i$ for finitely many maximal left ideals M_i; moreover, we can assume that for each i we have $\bigcap_{j \neq i} M_j \not\subseteq M_i$ (otherwise we may remove M_i from the intersection $M_1 \cap \ldots \cap M_r$). We then consider the following map

$$\Phi : A/J(A) \to A/M_1 \oplus \ldots \oplus A/M_r, \quad x + J(A) \mapsto (x + M_1, \ldots, x + M_r).$$

This map is well-defined since $J(A) \subseteq M_i$ for all i, and it is injective since $J(A) = \bigcap_{i=1}^{r} M_i$. Moreover, it is an A-module homomorphism since the action on the direct sum is componentwise. It remains to show that Φ is surjective, and hence an isomorphism; then the claim in part (e) follows since each A/M_i is a simple A-module. To prove that Φ is surjective, it suffices to show that for each i the element $(0, \ldots, 0, 1_A + M_i, 0, \ldots, 0)$ is in the image of Φ. Fix some i. By our assumption we have that M_i is a proper subset of $M_i + (\bigcap_{j \neq i} M_j)$. Since M_i is maximal this implies that $M_i + (\bigcap_{j \neq i} M_j) = A$. So there exist $m_i \in M_i$ and $y \in \bigcap_{j \neq i} M_j$ such that $1_A = m_i + y$. Therefore $\Phi(y) = (0, \ldots, 0, 1_A + M_i, 0, \ldots, 0)$, as desired.

(f) We have by assumption, $A/I = S_1 \oplus \ldots \oplus S_r$ with finitely many simple A/I-modules S_i, see Remark 4.10. The S_i can also be viewed as simple A-modules (see the proof of Theorem 4.17). From part (b) we get $J(A)S_i = 0$, which implies

$$J(A)(A/I) = J(A)(S_1 \oplus \ldots \oplus S_r) = 0,$$

that is, $J(A) = J(A)A \subseteq I$.

(g) If A is semisimple then $J(A) = 0$ by part (f), taking $I = 0$. Conversely, if $J(A) = 0$ then A is semisimple by part (e). $\qquad \square$

Remark 4.25. We now obtain an alternative proof of Proposition 4.14 which characterizes which algebras $A = K[X]/(f)$ are semisimple. Let $f \in K[X]$ be a non-constant polynomial and write $f = f_1^{a_1} \ldots f_r^{a_r}$ with pairwise coprime irreducible polynomials $f_1, \ldots, f_r \in K[X]$. We have seen in Example 4.22 that

$J(A) = 0$ if and only if $a_i = 1$ for all $i = 1, \ldots, r$. By Theorem 4.23 this is exactly the condition for the algebra $A = K[X]/(f)$ to be semisimple.

We will now describe the Jacobson radical for a finite-dimensional path algebra. Let $A = KQ$ where Q is a quiver, recall A is finite-dimensional if and only if Q does not have an oriented cycle of positive length, see Exercise 1.2.

Proposition 4.26. *Let K be a field, and let Q be a quiver without oriented cycles, so the path algebra $A = KQ$ is finite-dimensional. Then the Jacobson radical $J(A)$ is the subspace of A spanned by all paths in Q of positive length.*

Note that this result does not generalize to infinite-dimensional path algebras. For example, consider the path algebra of the one loop quiver, which is isomorphic to the polynomial algebra $K[X]$. Its Jacobson radical is therefore zero but the subspace generated by the paths of positive length is non-zero (even infinite-dimensional).

Proof. We denote the vertices of Q by $\{1, \ldots, n\}$. We apply part (b) of Theorem 4.23. The simple A-modules are precisely the modules $S_i := Ae_i/J_i$ for $1 \le i \le n$, and J_i is spanned by all paths in Q of positive length starting at i, see Theorem 3.26. Therefore we see directly that

$$\mathrm{Ann}_A(S_i) = J_i \oplus (\bigoplus_{j \neq i} Ae_j).$$

Taking the intersection of all these, we get precisely $J(A) = \bigoplus_{i=1}^{n} J_i$ which is the span of all paths in Q of length ≥ 1.   □

Corollary 4.27. *Let KQ be a finite-dimensional path algebra. Then KQ is a semisimple algebra if and only if Q has no arrows, that is, Q is a union of vertices. In particular, the semisimple path algebras KQ are isomorphic to direct products $K \times \ldots \times K$ of copies of the field K.*

Proof. By Theorem 4.23, KQ is semisimple if and only if $J(KQ) = 0$. Then the first statement directly follows from Proposition 4.26. The second statement is easily verified by mapping each vertex of Q to one of the factors of the direct product.   □

EXERCISES

4.3. Let $f \in K[X]$ be the product of two non-constant coprime polynomials, $f = gh$. Show that then $K[X]/(f) = (g)/(f) \oplus (h)/(f)$.

4.4. Let $A = T_n(K)$, the K-algebra of upper triangular matrices. Let $R \subseteq A$ be the submodule of all matrices which are zero outside the first row; similarly, let $C \subseteq A$ be the submodule of all matrices which are zero outside the n-th column. For each of R and C, determine whether or not it is a semisimple A-module.

4.5. Let $G = S_3$ be the symmetric group of order 6. Consider the 3-dimensional $K S_3$-module $V = \text{span}\{v_1, v_2, v_3\}$ on which S_3 acts by permuting the basis vectors, that is, $\sigma \cdot v_i = v_{\sigma(i)}$ for all $\sigma \in S_3$ and where the action is extended to arbitrary linear combinations in $K S_3$.

   (a) Verify that $U := \text{span}\{v_1 + v_2 + v_3\}$ is a $K S_3$-submodule of V.
   (b) Assume the characteristic of K is not equal to 3. Show that V is a semisimple $K S_3$-module by expressing V as a direct sum of two simple $K S_3$-submodules.
   (c) Suppose that K has characteristic 3. Show that then V is not a semisimple $K S_3$-module.

4.6. Let A be a K-algebra. Let I be a left ideal of A which is nilpotent, that is, there exists an $r \in \mathbb{N}$ such that $I^r = 0$. Show that I is contained in the Jacobson radical $J(A)$.

4.7. Which of the following subalgebras of $M_3(K)$ are semisimple? (Each asterisk stands for an arbitrary element from K.)

$$A_1 = \begin{pmatrix} * & 0 & 0 \\ 0 & * & 0 \\ 0 & 0 & * \end{pmatrix} \quad A_2 = \begin{pmatrix} * & * & * \\ 0 & * & 0 \\ 0 & 0 & * \end{pmatrix} \quad A_3 = \begin{pmatrix} * & 0 & * \\ 0 & * & 0 \\ 0 & 0 & * \end{pmatrix} \quad A_4 = \begin{pmatrix} * & 0 & * \\ 0 & * & 0 \\ * & 0 & * \end{pmatrix}$$

4.8. Which of the following K-algebras $K[X]/(f)$ are semisimple?

   (i) $K[X]/(X^3 - X^2 + X - 1)$ for $K = \mathbb{C}, \mathbb{R}$ and $\mathbb{Q}$,
   (ii) $\mathbb{C}[X]/(X^3 + X^2 - X - 1)$,
   (iii) $\mathbb{R}[X]/(X^3 - 3X^2 + 4)$,
   (iv) $\mathbb{Q}[X]/(X^2 - 1)$,
   (v) $\mathbb{Q}[X]/(X^4 - X^2 - 2)$,
   (vi) $\mathbb{Z}_2[X]/(X^4 + X + 1)$.

4.9. (a) Let $f \in \mathbb{R}[X]$ be a non-constant polynomial. Show that the $\mathbb{R}$-algebra $\mathbb{R}[X]/(f)$ is semisimple if and only if the $\mathbb{C}$-algebra $\mathbb{C}[X]/(f)$ is semisimple.
   (b) Let $f \in \mathbb{Q}[X]$ be a non-constant polynomial. Either prove the following, or find a counterexample: the $\mathbb{Q}$-algebra $\mathbb{Q}[X]/(f)$ is semisimple if and only if the $\mathbb{R}$-algebra $\mathbb{R}[X]/(f)$ is semisimple.

4.10. Suppose A is an algebra and N is some A-module. We define a subquotient of N to be a module Y/X where X, Y are submodules of N such that $0 \subseteq X \subseteq Y \subseteq N$.

   Suppose N has composition length 3, and assume that every subquotient of N which has composition length 2 is semisimple. Show that then N must be semisimple. (Hint: Choose a simple submodule X of N and show that there are submodules $U_1 \neq U_2$ of N, both containing X, of composition length 2. Then show that $U_1 + U_2$ is the direct sum of three simple modules.)

4.11. (a) Show that there exist infinitely many irreducible polynomials in $K[X]$.
(Hint: try a variation of Euclid's famous proof that there are infinitely many prime numbers.)

(b) Deduce that the Jacobson radical of $K[X]$ is zero.

4.12. For each of the following subalgebras of $M_3(K)$, find the Jacobson radical.

$$A_1 = \begin{pmatrix} * & 0 & 0 \\ 0 & * & 0 \\ 0 & 0 & * \end{pmatrix} \qquad A_2 = \left\{ \begin{pmatrix} x & y & z \\ 0 & x & 0 \\ 0 & 0 & x \end{pmatrix} \mid x, y, z \in K \right\}$$

$$A_3 = \left\{ \begin{pmatrix} x & y & 0 \\ 0 & z & 0 \\ 0 & 0 & x \end{pmatrix} \mid x, y, z \in K \right\} \qquad A_4 = \begin{pmatrix} * & * & 0 \\ 0 & * & 0 \\ 0 & * & * \end{pmatrix}$$

4.13. Let $\varphi : A \to B$ be a surjective K-algebra homomorphism. Prove the following statements.

(a) For the Jacobson radicals we have $\varphi(J(A)) \subseteq J(B)$.

(b) If $\ker(\varphi) \subseteq J(A)$ then $\varphi(J(A)) = J(B)$.

(c) For all K-algebras A we have $J(A/J(A)) = 0$.

4.14. Assume A is a commutative K-algebra of dimension n. Show that if A has n pairwise non-isomorphic simple modules then A must be semisimple. (Hint: Consider a map Φ as in the proof of Theorem 4.23 (e), from A to $A/M_1 \oplus \ldots \oplus A/M_n$, where the A/M_i are the distinct simple modules. Show that Φ must be an isomorphism.)

4.15. Which of the following commutative algebras over $\mathbb{C}$ are semisimple? Note that the algebras in (i) have dimension 2, and the others have dimension 4.

(i) $\mathbb{C}[X]/(X^2 - X)$, $\mathbb{C}[X]/(X^2)$, $\mathbb{C}[X]/(X^2 - 1)$,

(ii) $\mathbb{C}[X_1]/(X_1^2 - X_1) \times \mathbb{C}[X_2]/(X_2^2 - X_2)$,

(iii) $\mathbb{C}[X_1, X_2]/(X_1^2 - X_1, X_2^2 - X_2)$,

(iv) $\mathbb{C}[X_1]/(X_1^2) \times \mathbb{C}[X_2]/(X_2^2)$,

(v) $\mathbb{C}[X_1, X_2]/(X_1^2, X_2^2)$.

Chapter 5
The Structure of Semisimple Algebras: The Artin–Wedderburn Theorem

In this chapter we will prove the fundamental Artin–Wedderburn Theorem, which completely classifies semisimple K-algebras. We have seen in Example 4.19 that finite direct products $M_{n_1}(K) \times \ldots \times M_{n_r}(K)$ of matrix algebras are semisimple K-algebras. When the field K is algebraically closed the Artin–Wedderburn theorem shows that in fact every semisimple algebra is isomorphic to such an algebra. In general, a semisimple algebra is isomorphic to a direct product of matrix algebras but where the matrix coefficients in the matrix blocks are elements of some division algebras over K.

Our starting point is a direct sum decomposition, as an A-module, of the algebra A. The first observation is how to obtain information on the algebra just from such a direct sum decomposition. This is in fact more general, in the following lemma the algebra need not be semisimple.

Lemma 5.1. *Assume a K-algebra A as an A-module is the direct sum of non-zero submodules,*

$$A = M_1 \oplus M_2 \oplus \ldots \oplus M_r.$$

Write the identity element of A as $1_A = \varepsilon_1 + \varepsilon_2 + \ldots + \varepsilon_r$ with $\varepsilon_i \in M_i$. Then

(a) $\varepsilon_i \varepsilon_j = 0$ for $i \neq j$ and $\varepsilon_i^2 = \varepsilon_i$;
(b) $M_i = A\varepsilon_i$ and $\varepsilon_i \neq 0$.

Proof. (a) We have

$$\varepsilon_i = \varepsilon_i 1_A = \varepsilon_i \varepsilon_1 + \varepsilon_i \varepsilon_2 + \ldots + \varepsilon_i \varepsilon_r$$

and therefore

$$\varepsilon_i - \varepsilon_i^2 = \varepsilon_i \varepsilon_1 + \ldots + \varepsilon_i \varepsilon_{i-1} + \varepsilon_i \varepsilon_{i+1} + \ldots + \varepsilon_i \varepsilon_r.$$

© Springer International Publishing AG, part of Springer Nature 2018
K. Erdmann, T. Holm, *Algebras and Representation Theory*, Springer
Undergraduate Mathematics Series, https://doi.org/10.1007/978-3-319-91998-0_5

The left-hand side belongs to M_i, and the right-hand side belongs to $\sum_{j \neq i} M_j$. The sum $A = M_1 \oplus M_2 \oplus \ldots \oplus M_r$ is direct, therefore $M_i \cap \sum_{j \neq i} M_j = 0$. So $\varepsilon_i^2 = \varepsilon_i$, which proves part of (a). Moreover, this implies

$$0 = \varepsilon_i \varepsilon_1 + \ldots + \varepsilon_i \varepsilon_{i-1} + \varepsilon_i \varepsilon_{i+1} + \ldots + \varepsilon_i \varepsilon_r,$$

where the summands $\varepsilon_i \varepsilon_j$ are in M_j for each $j \neq i$. Since we have a direct sum, each of these summands must be zero, and this completes the proof of (a).

(b) We show now that $A\varepsilon_i = M_i$. Since $\varepsilon_i \in M_i$ and M_i is an A-module, it follows that $A\varepsilon_i \subseteq M_i$. For the converse, take some $m \in M_i$, then

$$m = m 1_A = m\varepsilon_1 + \ldots + m\varepsilon_i + \ldots + m\varepsilon_r.$$

Now $m - m\varepsilon_i \in M_i \cap \sum_{j \neq i} M_j$, which is zero, and therefore $m = m\varepsilon_i \in A\varepsilon_i$. We assume $M_i \neq 0$ and therefore $\varepsilon_i \neq 0$. $\square$

Elements $\varepsilon_i \in A$ with $\varepsilon_i^2 = \varepsilon_i$ are called idempotents, the properties in (a) are referred to as an orthogonal idempotent decomposition of the identity. (Such a decomposition for a path algebra has already appeared in Exercise 2.6.)

5.1 A Special Case

In this section we classify finite-dimensional commutative semisimple K-algebras where K is algebraically closed. This particularly nice result is a special case of the Artin–Wedderburn theorem.

Proposition 5.2. *Let K be an algebraically closed field. Suppose A is a finite-dimensional commutative K-algebra. Then A is semisimple if and only if A is isomorphic to the direct product of copies of K, that is, as algebras we have $A \cong K \times K \times \ldots \times K$.*

Proof. The direct product of copies of K is semisimple, even for arbitrary fields, see Example 4.19.

Conversely, assume A is a finite-dimensional semisimple commutative K-algebra. By Remark 4.10, as an A-module, A is the direct sum of finitely many simple submodules,

$$A = S_1 \oplus S_2 \oplus \ldots \oplus S_r.$$

We apply Lemma 5.1 with $M_i = S_i$, so we get an orthogonal idempotent decomposition of the identity element of A, as $1_A = \varepsilon_1 + \varepsilon_2 + \ldots + \varepsilon_r$, and $S_i = A\varepsilon_i$.

Since A is finite-dimensional, every simple A-module is finite-dimensional by Corollary 3.20. Moreover, since K is algebraically closed and A commutative, Corollary 3.38 implies that the simple A-module S_i is 1-dimensional, hence has a basis ε_i.

We will now construct a map $\psi : A \to K \times K \times \ldots \times K$ (with r factors in the direct product) and show that it is an algebra isomorphism. Take an arbitrary element $a \in A$. Then $a\varepsilon_i \in S_i$ for $i = 1, \ldots, r$. Since ε_i is a basis for S_i, there exist unique $\alpha_i \in K$ such that $a\varepsilon_i = \alpha_i \varepsilon_i$. It follows that we have

$$a = a1_A = a\varepsilon_1 + a\varepsilon_2 + \ldots + a\varepsilon_r = \alpha_1\varepsilon_1 + \alpha_2\varepsilon_2 + \ldots + \alpha_r\varepsilon_r.$$

Define a map $\psi : A \to K \times K \times \ldots \times K$ by setting

$$\psi(a) := (\alpha_1, \alpha_2, \ldots, \alpha_r).$$

We now show that ψ is an algebra isomorphism. From the definition one sees that ψ is K-linear. It is also surjective, since $\psi(\varepsilon_i) = (0, \ldots, 0, 1, 0, \ldots, 0)$ for each i. Moreover, it is injective: if $\psi(a) = 0$, so that all α_i are zero, then $a = 0$, by definition. It only remains to show that the map ψ is an algebra homomorphism. For any $a, b \in A$ suppose that $\psi(a) = (\alpha_1, \alpha_2, \ldots, \alpha_r)$ and $\psi(b) = (\beta_1, \beta_2, \ldots, \beta_r)$; then we have

$$(ab)1_A = a(b1_A) = a(\beta_1\varepsilon_1 + \beta_2\varepsilon_2 + \ldots + \beta_r\varepsilon_r)$$
$$= a\beta_1\varepsilon_1 + a\beta_2\varepsilon_2 + \ldots + a\beta_r\varepsilon_r = \beta_1(a\varepsilon_1) + \beta_2(a\varepsilon_2) + \ldots + \beta_r(a\varepsilon_r)$$
$$= \beta_1\alpha_1\varepsilon_1 + \beta_2\alpha_2\varepsilon_2 + \ldots + \beta_r\alpha_r\varepsilon_r = \alpha_1\beta_1\varepsilon_1 + \alpha_2\beta_2\varepsilon_2 + \ldots + \alpha_r\beta_r\varepsilon_r,$$

where the fourth equality uses axiom (Alg) from Definition 1.1 and the last equality holds since the α_i and β_i are in K and hence commute. This implies that

$$\psi(a)\psi(b) = (\alpha_1, \alpha_2, \ldots, \alpha_r)(\beta_1, \beta_2, \ldots, \beta_r) = (\alpha_1\beta_1, \alpha_2\beta_2, \ldots, \alpha_r\beta_r) = \psi(ab).$$

Finally, it follows from the definition that $\psi(1_A) = (1, 1, \ldots, 1) = 1_{K \times \ldots \times K}$. This proves that $\psi : A \to K \times K \times \ldots \times K$ is an isomorphism of algebras. $\square$

Remark 5.3.

(1) Proposition 5.2 need not hold if K is not algebraically closed. For example, consider the commutative $\mathbb{R}$-algebra $A = \mathbb{R}[X]/(X^2 + 1)$. Since $X^2 + 1$ is irreducible in $\mathbb{R}[X]$, we know from Proposition 3.23 that A as an A-module is simple, and it is a semisimple algebra, by Proposition 4.14. However, $A \not\cong \mathbb{R} \times \mathbb{R}$ since $A \cong \mathbb{C}$ is a field, whereas $\mathbb{R} \times \mathbb{R}$ contains non-zero zero divisors.

(2) Proposition 5.2 does not hold for infinite-dimensional algebras, even if the field K is algebraically closed. We have seen in Remark 3.34 that $\mathbb{C}(X)$ is a simple $\mathbb{C}(X)$-module. In particular, $\mathbb{C}(X)$ is a semisimple $\mathbb{C}$-algebra. As in (1), the field $\mathbb{C}(X)$ cannot be isomorphic to a product of copies of $\mathbb{C}$.

5.2   Towards the Artin–Wedderburn Theorem

We want to classify semisimple algebras. The input for this is more general: the first ingredient is to relate any algebra A to its algebra of A-module endomorphisms. The second ingredient is the observation that one can view the endomorphisms of a direct sum of A-modules as an algebra of matrices, where the entries are homomorphisms between the direct summands. We will discuss these now.

Let A be a K-algebra. For any A-module V we denote by $\mathrm{End}_A(V)$ the K-algebra of A-module homomorphisms from V to V (see Exercise 2.7). Recall from Definition 1.6 the definition of the opposite algebra: For any K-algebra B, the opposite algebra B^{op} has the same K-vector space structure as B and has multiplication $*$ defined by $b * b' := b'b$ for any $b, b' \in B$. The following result compares an algebra with its endomorphism algebra.

Lemma 5.4. *Let K be a field and let A be a K-algebra. Then A is isomorphic to $\mathrm{End}_A(A)^{op}$ as K-algebras.*

Proof. For any $a \in A$ we consider the right multiplication map,

$$r_a : A \to A \, , \ r_a(x) = xa \ \text{for all } x \in A.$$

One sees that this is an A-module homomorphism, so that $r_a \in \mathrm{End}_A(A)$ for every $a \in A$.

Conversely, we claim that every element in $\mathrm{End}_A(A)$ is of this form, that is, we have $\mathrm{End}_A(A) = \{r_a \,|\, a \in A\}$. In fact, let $\varphi \in \mathrm{End}_A(A)$ and let $a := \varphi(1_A)$; then for every $x \in A$ we have

$$\varphi(x) = \varphi(x 1_A) = x\varphi(1_A) = xa = r_a(x),$$

that is $\varphi = r_a$, proving the claim. We define therefore a map

$$\psi : A \to \mathrm{End}_A(A)^{op} \, , \ \psi(a) = r_a.$$

Then ψ is surjective, as we have just seen. It is also injective: If $\psi(a) = \psi(a')$ then for all $x \in A$ we have $xa = xa'$, and taking $x = 1_A$ shows $a = a'$.

We will now complete the proof of the lemma, by showing that ψ is a homomorphism of K-algebras. First, the map ψ is K-linear: For every $\lambda, \mu \in K$ and $a, b, x \in A$ we have

$$r_{\lambda a + \mu b}(x) = x(\lambda a + \mu b) = \lambda(xa) + \mu(xb) = \lambda r_a(x) + \mu r_b(x) = (\lambda r_a + \mu r_b)(x)$$

(where the second equality uses axiom (Alg) from Definition 1.1). Therefore,

$$\psi(\lambda a + \mu b) = r_{\lambda a + \mu b} = \lambda r_a + \mu r_b = \lambda \psi(a) + \mu \psi(b).$$

Moreover, it is clear that $\psi(1_A) = \mathrm{id}_A$. Finally, we show that ψ preserves the multiplication. For every $a, b, x \in A$ we have

$$r_{ab}(x) = x(ab) = (xa)b = r_b(r_a(x)) = (r_a * r_b)(x)$$

and hence

$$\psi(ab) = r_{ab} = \psi(a) * \psi(b).$$

Note that here we used the multiplication in the opposite algebra.   □

As promised, we will now analyse endomorphisms of direct sums of A-modules. In analogy to the definition of matrix algebras in linear algebra, if we start with a direct sum of A-modules, we can define a matrix algebra where the entries are homomorphisms between the direct summands. If U, W are A-modules, then we write $\mathrm{Hom}_A(U, W)$ for the vector space of all A-module homomorphisms $U \to W$, a subspace of the K-linear maps from U to W.

Lemma 5.5. *Let A be a K-algebra. Given finitely many A-modules $U_1, \ldots, U_r$, we consider $r \times r$-matrices whose (i, j)-entry is an A-module homomorphism from U_j to U_i,*

$$\Lambda := \left\{ \begin{pmatrix} \varphi_{11} & \cdots & \varphi_{1r} \\ \vdots & & \vdots \\ \varphi_{r1} & \cdots & \varphi_{rr} \end{pmatrix} \,\middle|\, \varphi_{ij} \in \mathrm{Hom}_A(U_j, U_i) \right\}.$$

Then Λ becomes a K-algebra with respect to matrix addition and matrix multiplication, where the product of two matrix entries is composition of maps.

Proof. It is clear that matrix addition and scalar multiplication turn Λ into a K-vector space (where homomorphisms are added pointwise as usual). Furthermore, matrix multiplication induces a multiplication on Λ. To see this, consider the product of two elements $\varphi = (\varphi_{ij})$ and $\psi = (\psi_{ij})$ from Λ. The product $\varphi\psi$ has as the (i, j)-entry the homomorphism $\sum_{\ell=1}^{r} \varphi_{i\ell} \circ \psi_{\ell j}$, which is indeed an element from $\mathrm{Hom}_A(U_j, U_i)$, as needed. The identity element in Λ is the diagonal matrix with diagonal entries $\mathrm{id}_{U_1}, \ldots, \mathrm{id}_{U_r}$. All axioms follow from the usual rules for matrix addition and matrix multiplication.   □

In linear algebra one identifies the algebra of endomorphisms of an n-dimensional K-vector space with the algebra $M_n(K)$ of $n \times n$-matrices over K. In analogy, we can identify the algebra of endomorphisms of a direct sum of A-modules with the matrix algebra as introduced in Lemma 5.5.

Lemma 5.6. *Let A be a K-algebra, suppose $U_1, \ldots, U_r$ are A-modules and $V := U_1 \oplus \ldots \oplus U_r$, their direct sum. Then the algebra Λ from Lemma 5.5 is isomorphic as a K-algebra to the endomorphism algebra $\mathrm{End}_A(V)$.*

Proof. For every $i \in \{1, \ldots, r\}$ we consider the projections $\pi_i : V \to U_i$ and the embeddings $\kappa_i : U_i \to V$. These are A-module homomorphisms and they satisfy $\sum_{i=1}^{r} \kappa_i \circ \pi_i = \mathrm{id}_V$.

Now let $\gamma \in \mathrm{End}_A(V)$ be an arbitrary element. Then for every $i, j \in \{1, \ldots, r\}$ we define

$$\gamma_{ij} := \pi_i \circ \gamma \circ \kappa_j \in \mathrm{Hom}_A(U_j, U_i).$$

This leads us to define the following map

$$\Phi : \mathrm{End}_A(V) \to \Lambda , \quad \gamma \mapsto \begin{pmatrix} \gamma_{11} & \cdots & \gamma_{1r} \\ \vdots & & \vdots \\ \gamma_{r1} & \cdots & \gamma_{rr} \end{pmatrix} \in \Lambda.$$

We are going to show that Φ is an isomorphism of K-algebras, thus proving the lemma. Let $\beta, \gamma \in \mathrm{End}_A(V)$. For every pair of scalars $a, b \in K$ the (i, j)-entry in $\Phi(a\beta + b\gamma)$ is

$$\pi_i \circ (a\beta + b\gamma) \circ \kappa_j = a(\pi_i \circ \beta \circ \kappa_j) + b(\pi_i \circ \gamma \circ \kappa_j),$$

which is equal to the sum of the (i, j)-entries of the matrices $a\Phi(\beta)$ and $b\Phi(\gamma)$. Thus, Φ is a K-linear map. Furthermore, it is clear from the definition that

$$\Phi(1_{\mathrm{End}_A(V)}) = \Phi(\mathrm{id}_V) = 1_\Lambda,$$

the diagonal matrix with identity maps on the diagonal, since $\pi_i \circ \kappa_j = 0$ for $i \neq j$ and $\pi_i \circ \kappa_i = \mathrm{id}_{U_i}$ for all i. Next we show that Φ is multiplicative. The (i, j)-entry in the product $\Phi(\beta)\Phi(\gamma)$ is given by

$$\sum_{\ell=1}^{r} \beta_{i\ell} \circ \gamma_{\ell j} = \sum_{\ell=1}^{r} \pi_i \circ \beta \circ \kappa_\ell \circ \pi_\ell \circ \gamma \circ \kappa_j$$

$$= \pi_i \circ \beta \circ (\sum_{\ell=1}^{r} \kappa_\ell \circ \pi_\ell) \circ \gamma \circ \kappa_j$$

$$= \pi_i \circ (\beta \circ \mathrm{id}_V \circ \gamma) \circ \kappa_j = (\beta \circ \gamma)_{ij}.$$

Thus, $\Phi(\beta \circ \gamma) = \Phi(\beta)\Phi(\gamma)$.

It remains to show that Φ is bijective. For injectivity suppose that $\Phi(\gamma) = 0$, that is, $\gamma_{ij} = 0$ for all i, j. Then we get

$$\gamma = \mathrm{id}_V \circ \gamma \circ \mathrm{id}_V = (\sum_{i=1}^{r} \kappa_i \circ \pi_i) \circ \gamma \circ (\sum_{j=1}^{r} \kappa_j \circ \pi_j)$$

$$= \sum_{i=1}^{r} \sum_{j=1}^{r} \kappa_i \circ (\pi_i \circ \gamma \circ \kappa_j) \circ \pi_j = \sum_{i=1}^{r} \sum_{j=1}^{r} \kappa_i \circ \gamma_{ij} \circ \pi_j = 0.$$

To show that Φ is surjective, let $\lambda \in \Lambda$ be an arbitrary element, with (i, j)-entry λ_{ij}. We have to find a preimage under Φ. To this end, we define

$$\gamma := \sum_{k=1}^{r} \sum_{\ell=1}^{r} \kappa_k \circ \lambda_{k\ell} \circ \pi_\ell \in \mathrm{End}_A(V).$$

Using that $\pi_i \circ \kappa_k = 0$ for $i \neq k$ and $\pi_i \circ \kappa_i = \mathrm{id}_{U_i}$ we then obtain for every $i, j \in \{1, \ldots, r\}$ that

$$\gamma_{ij} = \pi_i \circ \gamma \circ \kappa_j = \pi_i \circ \left(\sum_{k=1}^{r} \sum_{\ell=1}^{r} \kappa_k \circ \lambda_{k\ell} \circ \pi_\ell \right) \circ \kappa_j$$

$$= \sum_{k=1}^{r} \sum_{\ell=1}^{r} \pi_i \circ \kappa_k \circ \lambda_{k\ell} \circ \pi_\ell \circ \kappa_j = \lambda_{ij}.$$

Thus, $\Phi(\gamma) = \lambda$ and Φ is surjective. $\qquad\square$

Theorem 5.7. *Let A be a K-algebra and let $V = S_1 \oplus \ldots \oplus S_t$, where $S_1, \ldots, S_t$ are simple A-modules. Then there exist positive integers r and $n_1, \ldots, n_r$, and division algebras $D_1, \ldots, D_r$ over K such that*

$$\mathrm{End}_A(V) \cong M_{n_1}(D_1) \times \ldots \times M_{n_r}(D_r).$$

Proof. The crucial input is Schur's lemma (see Theorem 3.33) which we recall now. If S_i and S_j are simple A-modules then $\mathrm{Hom}_A(S_i, S_j) = 0$ if S_i and S_j are not isomorphic, and otherwise we have that $\mathrm{End}_A(S_i)$ is a division algebra over K.

We label the summands of the module V so that isomorphic ones are grouped together, explicitly we take

$$S_1 \cong \ldots \cong S_{n_1}, \ S_{n_1+1} \cong \ldots \cong S_{n_1+n_2}, \ldots, S_{n_1+\ldots+n_{r-1}+1} \cong \ldots \cong S_{n_1+\ldots+n_r} = S_t$$

and there are no other isomorphisms. That is, we have r different isomorphism types amongst the S_i, and they come with multiplicities $n_1, \ldots, n_r$. Define

$$D_1 := \mathrm{End}_A(S_1), \ D_2 := \mathrm{End}_A(S_{n_1+1}), \ \ldots, \ D_r := \mathrm{End}_A(S_{n_1+\ldots+n_{r-1}+1});$$

these are division algebras, by Schur's lemma. Then Lemma 5.6 and Schur's lemma show that the endomorphism algebra of V can be written as a matrix algebra, with block matrices

$$\mathrm{End}_A(V) \cong \Lambda = (\mathrm{Hom}_A(S_j, S_i))_{i,j} \cong \begin{pmatrix} M_{n_1}(D_1) & & 0 \\ & \ddots & \\ 0 & & M_{n_r}(D_r) \end{pmatrix}$$

$$\cong M_{n_1}(D_1) \times \ldots \times M_{n_r}(D_r).$$

$\square$

We will now consider the algebras in Theorem 5.7 in more detail. In particular, we want to show that they are semisimple K-algebras. This is part of the Artin–Wedderburn theorem, which will come in the next section.

Example 4.19 is a special case, and we have seen that for every field K the algebras $M_{n_1}(K) \times \ldots \times M_{n_r}(K)$ are semisimple. The proof for the algebras in Theorem 5.7 is essentially the same.

Lemma 5.8.

(a) *Let D be a division algebra over K. Then for every $n \in \mathbb{N}$ the matrix algebra $M_n(D)$ is a semisimple K-algebra. Moreover, the opposite algebra $M_n(D)^{op}$ is isomorphic to $M_n(D^{op})$, as a K-algebra.*

(b) *Let $D_1, \ldots, D_r$ be division algebras over K. Then for any $n_1, \ldots, n_r \in \mathbb{N}$ the direct product $M_{n_1}(D_1) \times \ldots \times M_{n_r}(D_r)$ is a semisimple K-algebra.*

Proof. (a) Let $A = M_n(D)$, and let D^n be the natural A-module. We claim that this is a simple module. The proof of this is exactly the same as the proof for $D = K$ in Example 2.14, since there we only used that non-zero elements have inverses (and not that elements commute). As for the case $D = K$, we see that A as an A-module is the direct sum $A = C_1 \oplus C_2 \oplus \ldots \oplus C_n$, where C_i consists of the matrices in A which are zero outside the i-th column. As for the case $D = K$, each C_i is isomorphic to the natural module D^n, as an A-module. This shows that A is a direct sum of simple submodules and hence is a semisimple algebra.

We show now that the opposite algebra $M_n(D)^{op}$ is isomorphic to $M_n(D^{op})$. Note that both algebras have the same underlying K-vector space. Let τ be the map which takes an $n \times n$-matrix to its transpose. That is, if $a = (a_{ij}) \in M_n(D)$ then $\tau(a)$ is the matrix with (s, t)-entry equal to a_{ts}. Then τ defines a K-linear isomorphism on the vector space $M_n(D)$. We show that τ is an algebra isomorphism $M_n(D)^{op} \to M_n(D^{op})$. The identity elements in both algebras are the identity matrices, and τ takes the identity to the identity. It remains to show that for $a, b \in M_n(D)^{op}$ we have $\tau(a * b)$ is equal to $\tau(a)\tau(b)$.

(i) We have $\tau(a * b) = \tau(ba)$, and this has (s, t)-entry equal to the (t, s)-entry of the matrix product ba, which is

$$\sum_{j=1}^{n} b_{tj} a_{js}.$$

(ii) Now we write $\tau(a) = (\hat{a}_{ij})$, where $\hat{a}_{ij} = a_{ji}$, and similarly let $\tau(b) = (\hat{b}_{ij})$. We compute $\tau(a)\tau(b)$ in $M_n(D^{op})$. This has (s, t)-entry equal to

$$\sum_{j=1}^n \hat{a}_{sj} * \hat{b}_{jt} = \sum_{j=1}^n a_{js} * b_{tj} = \sum_{j=1}^n b_{tj} a_{js}$$

(where in the first step we removed the $\hat{}$ and in the second step we removed the $*$.) This holds for all s, t, hence $\tau(a * b) = \tau(a)\tau(b)$.

(b) By part (a) we know that $M_n(D)$ is a semisimple algebra. Now part (b) follows directly using Corollary 4.18, which shows that finite direct products of semisimple algebras are semisimple. $\qquad\square$

5.3 The Artin–Wedderburn Theorem

We have seen that any K-algebra $M_{n_1}(D_1) \times \ldots \times M_{n_r}(D_r)$ is semisimple, where $D_1 \ldots, D_r$ are division algebras over K. The Artin–Wedderburn theorem shows that up to isomorphism, every semisimple K-algebra is of this form.

Theorem 5.9 (Artin–Wedderburn Theorem). *Let K be a field and A a semisimple K-algebra. Then there exist positive integers r and $n_1, \ldots, n_r$, and division algebras $D_1, \ldots, D_r$ over K such that*

$$A \cong M_{n_1}(D_1) \times \ldots \times M_{n_r}(D_r).$$

Conversely, each K-algebra of the form $M_{n_1}(D_1) \times \ldots \times M_{n_r}(D_r)$ is semisimple.

We will refer to this direct product as the *Artin–Wedderburn decomposition* of the semisimple algebra A.

Proof. The last statement has been proved in Lemma 5.8.

Suppose that A is a semisimple K-algebra. By Remark 4.10, A as an A-module is a finite direct sum $A = S_1 \oplus \ldots \oplus S_t$ with simple A-submodules $S_1, \ldots, S_t$. Now Theorem 5.7 implies that there exist positive integers r and $n_1, \ldots, n_r$ and division algebras $\widetilde{D}_1, \ldots, \widetilde{D}_r$ over K such that

$$\mathrm{End}_A(A) \cong M_{n_1}(\widetilde{D}_1) \times \ldots \times M_{n_r}(\widetilde{D}_r).$$

We can now deduce the structure of the algebra A, namely

$$
\begin{aligned}
A &\cong \mathrm{End}_A(A)^{op} &&\text{(by Lemma 5.4)}\\
&\cong (M_{n_1}(\widetilde{D}_1) \times \ldots \times M_{n_r}(\widetilde{D}_r))^{op}\\
&= M_{n_1}(\widetilde{D}_1)^{op} \times \ldots \times M_{n_r}(\widetilde{D}_r)^{op}\\
&= M_{n_1}(\widetilde{D}_1^{op}) \times \ldots \times M_{n_r}(\widetilde{D}_r^{op}). &&\text{(by Lemma 5.8)}
\end{aligned}
$$

We set $D_i := \tilde{D}_i^{op}$ (note that this is also a division algebra, since reversing the order in the multiplication does not affect whether elements are invertible).   □

Remark 5.10. Note that a matrix algebra $M_n(D)$ is commutative if and only if $n = 1$ and D is a field. Therefore, let A be a commutative semisimple K-algebra. Then the Artin–Wedderburn decomposition of A has the form

$$A \cong M_1(D_1) \times \ldots \times M_1(D_r) \cong D_1 \times \ldots \times D_r,$$

where D_i are fields containing K. From the start, D_i is the endomorphism algebra of a simple A-module. Furthermore, taking D_i as the i-th factor in the above product decomposition, it is a simple A-module, and hence this simple module is identified with its endomorphism algebra.

In the rest of this section we will derive some consequences from the Artin–Wedderburn theorem and we also want to determine the Artin–Wedderburn decomposition for some classes of semisimple algebras explicitly.

We will now see that one can read off the number and the dimensions of simple modules of a semisimple algebra from the Artin–Wedderburn decomposition. This is especially nice when the underlying field is algebraically closed, such as the field of complex numbers.

Corollary 5.11.

(a) *Let $D_1, \ldots, D_r$ be division algebras over K, and let $n_1, \ldots, n_r$ be positive integers. The semisimple K-algebra $M_{n_1}(D_1) \times \ldots \times M_{n_r}(D_r)$ has precisely r simple modules, up to isomorphism. The K-vector space dimensions of these simple modules are $n_1 \dim_K D_1, , \ldots, n_r \dim_K D_r$. (Note that these dimensions of simple modules need not be finite.)*

(b) *Suppose the field K is algebraically closed, and that A is a finite-dimensional semisimple K-algebra. Then there exist positive integers $n_1, \ldots, n_r$ such that*

$$A \cong M_{n_1}(K) \times \ldots \times M_{n_r}(K).$$

Then A has precisely r simple modules, up to isomorphism, of dimensions $n_1, \ldots, n_r$.

Proof. (a) We apply Corollary 3.31 to describe the simple modules: Let $A = A_1 \times \ldots \times A_r$ be a direct product of K-algebras. Then the simple A-modules are given precisely by the simple A_i-modules (where $1 \le i \le r$) such that the factors A_j for $j \neq i$ act as zero. In our situation, $A = M_{n_1}(D_1) \times \ldots \times M_{n_r}(D_r)$ and the simple A-modules are given by the simple $M_{n_i}(D_i)$-modules for $i = 1, \ldots, r$. As observed in the proof of Lemma 5.8, each matrix algebra $M_{n_i}(D_i)$ is a direct sum of submodules, each isomorphic to the natural module $D_i^{n_i}$; then Theorem 3.19 implies that $M_{n_i}(D_i)$ has a unique simple module, namely $D_i^{n_i}$. Therefore, A has precisely r simple modules, up to isomorphism, as claimed. Clearly, the dimensions of these simple A-modules are $\dim_K D_i^{n_i} = n_i \dim_K D_i$.

(b) Since A is finite-dimensional by assumption, every simple A-module is finite-dimensional, see Corollary 3.20. By the assumption on K, Schur's lemma (see Theorem 3.33) shows that $\mathrm{End}_A(S) \cong K$ for every simple A-module S. These are the division algebras appearing in the Artin–Wedderburn decomposition, hence $A \cong M_{n_1}(K) \times \ldots \times M_{n_r}(K)$. The statements on the number and the dimensions of simple modules follow from part (a). □

Remark 5.12. Note that in the proof above we described explicit versions of the simple modules for the algebra $M_{n_1}(D_1) \times \ldots \times M_{n_r}(D_r)$. For each i, take the natural module $D_i^{n_i}$ as a module for the i-th component, and for $j \neq i$ the j-th component acts as zero.

We will now find the Artin–Wedderburn decomposition for semisimple algebras of the form $A = K[X]/(f)$. Recall from Proposition 4.14 that A is semisimple if and only if $f = f_1 \cdot \ldots \cdot f_r$ with pairwise coprime irreducible polynomials $f_1, \ldots, f_r$.

Proposition 5.13. *Let $A = K[X]/(f)$, where $f \in K[X]$ is a non-constant polynomial and $f = f_1 \cdot \ldots \cdot f_r$ is a product of pairwise coprime irreducible polynomials $f_1, \ldots, f_r \in K[X]$. Then the Artin–Wedderburn decomposition of the semisimple algebra A has the form*

$$A \cong M_1(K[X]/(f_1)) \times \ldots \times M_1(K[X]/(f_r)) \cong K[X]/(f_1) \times \ldots \times K[X]/(f_r).$$

Moreover, A has r simple modules, up to isomorphism, of dimensions $\deg(f_1), \ldots, \deg(f_r)$ (where $\deg$ denotes the degree of a polynomial).

Note that this recovers the Chinese Remainder Theorem in this case. Note also that this Artin–Wedderburn decomposition depends on the field K. We will consider the cases $K = \mathbb{R}$ and $K = \mathbb{C}$ explicitly at the end of the section.

Proof. By Proposition 4.23 the simple A-modules are up to isomorphism the modules $K[X]/(f_i)$ for $1 \leq i \leq r$. We may apply Remark 5.10, which shows that A is isomorphic to $D_1 \times \ldots \times D_r$ where D_i are fields. If we label the components so that D_i is the endomorphism algebra of $K[X]/(f_i)$ then by the remark again, this simple module can be identified with D_i, and the claim follows. □

Example 5.14.

(1) Let K be an algebraically closed field, for example $K = \mathbb{C}$. Then the irreducible polynomials in $K[X]$ are precisely the polynomials of degree 1. So the semisimple algebras $A = K[X]/(f)$ appear for polynomials of the form $f = (X - \lambda_1) \cdot \ldots \cdot (X - \lambda_r)$ for pairwise different $\lambda_1, \ldots, \lambda_r \in K$. (Note that we can assume f to be monic since $(f) = (uf)$ for every non-zero $u \in K$.) Then Proposition 5.13 gives the Artin–Wedderburn decomposition

$$A \cong K[X]/(X - \lambda_1) \times \ldots \times K[X]/(X - \lambda_r) \cong K \times \ldots \times K.$$

In particular, A has precisely r simple modules, up to isomorphism, and they are all 1-dimensional.

Note that the Artin–Wedderburn decomposition in this situation also follows from Proposition 5.2; in Exercise 5.11 it is shown how to construct the orthogonal idempotent decomposition of the identity of A, which then gives an explicit Artin–Wedderburn decomposition.

(2) If K is arbitrary and f is a product of pairwise coprime irreducible polynomials of degree 1, then $K[X]/(f)$ also has the Artin–Wedderburn decomposition as in (1). For example, this applies when $K = \mathbb{Z}_p$ and $f(X) = X^p - X$, this has roots precisely the elements in $\mathbb{Z}_p$ and therefore is a product of pairwise coprime linear factors.

(3) Let $K = \mathbb{R}$. The irreducible polynomials in $\mathbb{R}[X]$ have degree 1 or 2 (where a degree 2 polynomial is irreducible if and only if it has no real root). In fact, for any polynomial in $\mathbb{R}[X]$ take a root $z \in \mathbb{C}$; then $(X - z)(X - \overline{z}) \in \mathbb{R}[X]$ is a factor. Thus we consider semisimple algebras $\mathbb{R}[X]/(f)$ where

$$f = f_1 \cdot \ldots \cdot f_r \cdot g_1 \cdot \ldots \cdot g_s$$

with pairwise coprime irreducible polynomials f_i of degree 1 and g_j of degree 2. Set $S_i := \mathbb{R}[X]/(f_i)$ for $i = 1, \ldots, r$ and $T_j := \mathbb{R}[X]/(g_j)$ for $j = 1, \ldots, s$. By Exercise 3.10 we have that $\text{End}_A(S_i) \cong \mathbb{R}$ for $i = 1, \ldots, r$ and $\text{End}_A(T_j) \cong \mathbb{C}$ for $j = 1, \ldots, s$. So, by Remark 5.10, the Artin–Wedderburn decomposition takes the form

$$A \cong \underbrace{\mathbb{R} \times \ldots \times \mathbb{R}}_{r} \times \underbrace{\mathbb{C} \times \ldots \times \mathbb{C}}_{s}.$$

In particular, A has precisely $r + s$ simple modules, up to isomorphism, r of them of dimension 1 and s of dimension 2.

EXERCISES

5.1. Find all semisimple $\mathbb{C}$-algebras of dimension 9, up to isomorphism.

5.2. Find all 4-dimensional semisimple $\mathbb{R}$-algebras, up to isomorphism. You may use without proof that there are no 3-dimensional division algebras over $\mathbb{R}$ and that the quaternions $\mathbb{H}$ form the only 4-dimensional real division algebra, up to isomorphism.

5.3. Find all semisimple $\mathbb{Z}_2$-algebras, up to isomorphism (where $\mathbb{Z}_2$ is the field with two elements).

5.4. Let K be a field and A a semisimple K-algebra.

(a) Show that the centre $Z(A)$ of A is isomorphic to a direct product of fields; in particular, the centre of a semisimple algebra is a commutative, semisimple algebra.

(b) Suppose A is a finite-dimensional semisimple algebra over K. Suppose x is an element in the centre $Z(A)$. Show that if x is nilpotent then $x = 0$.

5.5. Let K be a field and A a K-algebra. We consider the matrix algebra $M_n(A)$ with entries in A; note that this is again a K-algebra.

(a) For $A = M_m(K)$ show that $M_n(M_m(K)) \cong M_{nm}(K)$ as K-algebras.

(b) Let $A = A_1 \times \ldots \times A_r$ be a direct product of K-algebras. Show that then $M_n(A) \cong M_n(A_1) \times \ldots \times M_n(A_r)$ as K-algebras.

(c) Show that if A is a semisimple K-algebra then $M_n(A)$ is also a semisimple K-algebra.

5.6. Let $A = A_1 \times \ldots \times A_r$ be a direct product of K-algebras. Show that all two-sided ideals of A are of the form $I_1 \times \ldots \times I_r$ where I_j is a two-sided ideal of A_j for all $j = 1, \ldots, r$.

5.7. (a) Let D be a division algebra over K. Show that the matrix algebra $M_n(D)$ only has the trivial two-sided ideals 0 and $M_n(D)$.

(b) Let A be a semisimple K-algebra. Determine all two-sided ideals of A and in particular give the number of these ideals.

5.8. Consider the $\mathbb{C}$-algebra $A := \mathbb{C}[X, Y]/(X^2, Y^2, XY)$ (note that A is commutative and 3-dimensional). Find all two-sided ideals of A which have dimension 1 as a $\mathbb{C}$-vector space. Deduce from this (using the previous exercise) that A is not a semisimple algebra.

5.9. Let A be a K-algebra, and $\varepsilon \in A$ be an idempotent, that is, $\varepsilon^2 = \varepsilon$.

(a) Verify that $\varepsilon A \varepsilon = \{\varepsilon a \varepsilon \mid a \in A\}$ is a K-algebra.

(b) Consider the A-module $A\varepsilon$ and show that every A-module homomorphism $A\varepsilon \to A\varepsilon$ is of the form $x \mapsto xb$ for some $b = \varepsilon b \varepsilon \in \varepsilon A \varepsilon$.

(c) By following the strategy of Lemma 5.4, show that $\varepsilon A \varepsilon \cong \mathrm{End}_A(A\varepsilon)^{op}$ as K-algebras.

5.10. Let A be a commutative semisimple K-algebra, where $A = S_1 \oplus \ldots \oplus S_r$ with S_i simple A-modules. By Lemma 5.1 we know $S_i = A\varepsilon_i$, where ε_i are idempotents of A. Deduce from Exercise 5.9 that the endomorphism algebra $\mathrm{End}_A(A\varepsilon_i)$ can be identified with $A\varepsilon_i$.

5.11. Consider the K-algebra $A = K[X]/I$ where $I = (f)$ and $f = \prod_{i=1}^{r}(X - \lambda_i)$ with pairwise distinct $\lambda_i \in K$. For $i = 1, \ldots, r$ we define elements

$$c_i = \prod_{j \neq i}(\lambda_i - \lambda_j) \in K \quad \text{and} \quad \varepsilon_i = (1/c_i)\prod_{j \neq i}(X - \lambda_j) + I \in A.$$

(a) For all i show that $(X + I)\varepsilon_i = \lambda_i \varepsilon_i$ in A, that is, ε_i is an eigenvector for the action of the coset of X.

(b) Deduce from (a) that $\varepsilon_i \varepsilon_j = 0$ for $i \neq j$. Moreover, show that $\varepsilon_i^2 = \varepsilon_i$.

(c) Show that $\varepsilon_1 + \ldots + \varepsilon_r = 1_A$.

5.12. Let $K = \mathbb{Z}_p$ and let $f = X^p - X$. Consider the algebra $A = K[X]/(f)$ as an A-module. Explain how Exercise 5.11 can be applied to express A as a direct sum of 1-dimensional modules. (Hint: The roots of $X^p - X$ are precisely the elements of $\mathbb{Z}_p$, by Lagrange's theorem from elementary group theory. Hence $X^p - X$ factors into p distinct linear factors in $\mathbb{Z}_p[X]$.)

Chapter 6
Semisimple Group Algebras and Maschke's Theorem

The Artin–Wedderburn theorem gives a complete description of the structure of semisimple algebras. We will now investigate when a group algebra of a finite group is semisimple, this is answered by Maschke's theorem. If this is the case, then we will see how the Artin–Wedderburn decomposition of the group algebra gives information about the group, and vice versa.

6.1 Maschke's Theorem

Let G be a finite group and let K be a field. The main idea of the proof of Maschke's theorem is more general: Given any K-linear map between KG-modules, one can construct a KG-module homomorphism, by 'averaging over the group'.

Lemma 6.1. *Let G be a finite group and let K be a field. Suppose M and N are KG-modules, and $f : M \to N$ is a K-linear map. Define*

$$T(f) : M \to N, \quad m \mapsto \sum_{x \in G} x(f(x^{-1}m)).$$

Then $T(f)$ is a KG-module homomorphism.

Proof. We check that $T(f)$ is K-linear. Let $\alpha, \beta \in K$ and $m_1, m_2 \in M$. Using that multiplication by elements in KG is linear, and that f is linear, we get

$$T(f)(\alpha m_1 + \beta m_2) = \sum_{x \in G} x(f(x^{-1}(\alpha m_1 + \beta m_2)))$$

$$= \sum_{x \in G} x(f(x^{-1}\alpha m_1 + x^{-1}\beta m_2))$$

© Springer International Publishing AG, part of Springer Nature 2018
K. Erdmann, T. Holm, *Algebras and Representation Theory*, Springer
Undergraduate Mathematics Series, https://doi.org/10.1007/978-3-319-91998-0_6

$$= \sum_{x \in G} x(\alpha f(x^{-1}m_1) + \beta f(x^{-1}m_2))$$

$$= \alpha\, T(f)(m_1) + \beta\, T(f)(m_2).$$

To see that $T(f)$ is indeed a KG-module homomorphism it suffices to check, by Remark 2.21, that the action of $T(f)$ commutes with the action of elements in the group basis of KG. So take $y \in G$, then we have for all $m \in M$ that

$$T(f)(ym) = \sum_{x \in G} x(f(x^{-1}ym)) = \sum_{x \in G} y(y^{-1}x)(f((y^{-1}x)^{-1}m)).$$

But for a fixed $y \in G$, as x varies through all the elements of G, so does $y^{-1}x$. Hence we get from above that

$$T(f)(ym) = \sum_{\tilde{x} \in G} y\tilde{x}(f(\tilde{x}^{-1}m)) = yT(f)(m),$$

which shows that $T(f)$ is a KG-module homomorphism.   □

Example 6.2. As an illustration of the above lemma, consider a cyclic group G of order 2, generated by an element g, and let $M = N$ be the regular $\mathbb{C}G$-module (that is, $\mathbb{C}G$ with left multiplication). Any $\mathbb{C}$-linear map $f : \mathbb{C}G \to \mathbb{C}G$ is given by a 2×2-matrix $\begin{pmatrix} a & b \\ c & d \end{pmatrix} \in M_2(\mathbb{C})$, with respect to the standard basis $\{1, g\}$ of $\mathbb{C}G$. Then f defines a $\mathbb{C}G$-module homomorphism if and only if this matrix commutes with the matrix describing the action of g with respect to the same basis, which is $\begin{pmatrix} 0 & 1 \\ 1 & 0 \end{pmatrix}$. We compute the map $T(f)$. Since $g = g^{-1}$ in G we have

$$T(f)(1) = f(1) + gf(g^{-1}) = a + cg + g(b + dg) = (a + d) + (b + c)g$$

and similarly

$$T(f)(g) = f(g) + gf(g^{-1}g) = (b + c) + (a + d)g.$$

So the linear map $T(f)$ is, with respect to the standard basis $\{1, g\}$, given by the matrix

$$\begin{pmatrix} a+d & b+c \\ b+c & a+d \end{pmatrix}.$$

One checks that it commutes with $\begin{pmatrix} 0 & 1 \\ 1 & 0 \end{pmatrix}$, hence $T(f)$ is indeed a $\mathbb{C}G$-module homomorphism.

We will now state and prove Maschke's theorem. This is an easy and completely general criterion to decide when a group algebra of a finite group is semisimple.

Theorem 6.3 (Maschke's Theorem).   *Let K be a field and G a finite group. Then the group algebra KG is semisimple if and only if the characteristic of K does not divide the order of G.*

Proof. Assume first that the characteristic of K does not divide the group order $|G|$. By definition the group algebra KG is semisimple if and only if KG is semisimple as a KG-module. So let W be a submodule of KG, then by Theorem 4.3 we must show that W has a complement, that is, there is a KG-submodule C of KG such that $W \oplus C = KG$.

Considered as K-vector spaces there is certainly a K-subspace V such that $W \oplus V = KG$ (this is the standard result from linear algebra that every linearly independent subset can be completed to a basis). Let $f : KG \rightarrow W$ be the projection onto W with kernel V; note that this is just a K-linear map. By assumption, $|G|$ is invertible in K and using Lemma 6.1 we can define

$$\gamma : KG \rightarrow W , \quad \gamma := \frac{1}{|G|} T(f).$$

Note that γ is a KG-module homomorphism by Lemma 6.1. So $C := \ker(\gamma) \subseteq KG$ is a KG-submodule.

We apply Lemma 2.30, with $M = KG$, and with $N = N' = W$, and where $\gamma : KG \rightarrow W$ is the map π of the lemma and $j : W \rightarrow KG$ is the inclusion map. We will show that $\gamma \circ j$ is the identity map on W, hence is an isomorphism. Then Lemma 2.30 shows that $KG = W \oplus C$ is a direct sum of KG-submodules, that is, W has a complement.

Let $w \in W$, then for all $g \in G$ we have $g^{-1}w \in W$ and so $f(g^{-1}w) = g^{-1}w$. Therefore $gf(g^{-1}w) = w$ and

$$(\gamma \circ j)(w) = \gamma(w) = \frac{1}{|G|} \sum_{g \in G} gf(g^{-1}w) = \frac{1}{|G|} \sum_{g \in G} w = w.$$

Hence, $\gamma \circ j$ is the identity map, as required.

For the converse of Maschke's Theorem, suppose KG is a semisimple algebra. We have to show that the characteristic of the field K does not divide the order of G. Consider the element $w := \sum_{g \in G} g \in KG$, and observe that

$$xw = w \quad \text{for all } x \in G. \tag{6.1}$$

Therefore, the 1-dimensional K-subspace $U = \text{span}\{w\}$ is a KG-submodule of KG. We assume that KG is a semisimple algebra, hence there is a KG-submodule C of KG such that $KG = U \oplus C$ (see Theorem 4.3). Write the identity element in the form $1_{KG} = u + c$ with $u \in U$ and $c \in C$. Then u is non-zero, since otherwise we would have $KG = KGc \subseteq C \neq KG$, a contradiction. Now,

$u = \lambda w$ and $0 \neq \lambda \in K$. By (6.1), we deduce that $w^2 = |G|w$, but also $w = w1_{KG} = w(\lambda w) + wc$. We have $wc \in C$ since C is a KG-submodule, and it follows that

$$w - \lambda |G|w \in U \cap C = 0.$$

Now, $w \neq 0$ and therefore $|G|$ must be non-zero in K. $\qquad\qquad\qquad\square$

Note that when the field K has characteristic 0 the condition in Maschke's theorem is satisfied for every finite group G and hence KG is semisimple.

6.2   Some Consequences of Maschke's Theorem

Suppose G is a finite group, then by Maschke's Theorem the group algebra $\mathbb{C}G$ is semisimple. We can therefore apply the Artin–Wedderburn theorem (see Corollary 5.11 (b)) and obtain that

$$\mathbb{C}G \cong M_{n_1}(\mathbb{C}) \times M_{n_2}(\mathbb{C}) \times \ldots \times M_{n_k}(\mathbb{C})$$

for some positive integers $n_1, \ldots, n_k$. This Artin–Wedderburn decomposition of the group algebra $\mathbb{C}G$ gives us new information.

Theorem 6.4. *Let G be a finite group, and let*

$$\mathbb{C}G \cong M_{n_1}(\mathbb{C}) \times M_{n_2}(\mathbb{C}) \times \ldots \times M_{n_k}(\mathbb{C})$$

be the Artin–Wedderburn decomposition of the group algebra $\mathbb{C}G$. Then the following hold:

(a) *The group algebra $\mathbb{C}G$ has precisely k simple modules, up to isomorphism, and the dimensions of these simple modules are $n_1, n_2, \ldots, n_k$.*
(b) *We have $|G| = \sum_{i=1}^{k} n_i^2$.*
(c) *The group G is abelian if and only if all simple $\mathbb{C}G$-modules are of dimension 1.*

Proof. (a) This follows directly from Corollary 5.11.
(b) This follows from comparing $\mathbb{C}$-vector space dimensions on both sides in the Artin–Wedderburn decomposition.
(c) If G is abelian then the group algebra $\mathbb{C}G$ is commutative. Therefore, by Proposition 5.2 we have $\mathbb{C}G \cong \mathbb{C} \times \mathbb{C} \times \ldots \times \mathbb{C}$. By part (a), all simple $\mathbb{C}G$-modules are 1-dimensional.

Conversely, if all simple $\mathbb{C}G$-modules are 1-dimensional, then again by part (a) the Artin–Wedderburn decomposition has the form $\mathbb{C}G \cong \mathbb{C} \times \mathbb{C} \times \ldots \times \mathbb{C}$. Hence $\mathbb{C}G$ is commutative and therefore G is abelian. $\qquad\qquad\square$

Remark 6.5. The statements of Theorem 6.4 need not hold if the field is not algebraically closed. For instance, consider the group algebra $\mathbb{R}C_3$ where C_3 is the cyclic group of order 3. This algebra is isomorphic to the algebra $\mathbb{R}[X]/(X^3 - 1)$, see Example 1.27. In $\mathbb{R}[X]$, we have the factorization $X^3 - 1 = (X-1)(X^2+X+1)$ into irreducible polynomials. Hence, by Example 5.14 the algebra has Artin–Wedderburn decomposition

$$\mathbb{R}C_3 \cong M_1(\mathbb{R}) \times M_1(\mathbb{C}) \cong \mathbb{R} \times \mathbb{C}.$$

Moreover, it has two simple modules, up to isomorphism, of dimensions 1 and 2, by Proposition 3.23. Thus all parts of Theorem 6.4 do not hold in this case.

Exercise 6.1. Consider the group algebra $\mathbb{R}C_4$ of the cyclic group C_4 of order 4 over the real numbers. Determine the Artin–Wedderburn decomposition of this algebra and also the number and the dimensions of the simple modules.

Remark 6.6. We would like to explore examples of Artin–Wedderburn decompositions for groups which are not abelian. We do this using subgroups of small symmetric groups. Recall that S_n is the group of all permutations of $\{1, 2, \ldots, n\}$, it contains $n!$ elements. It has as a normal subgroup the alternating group, A_n, of even permutations, of order $\frac{n!}{2}$, and the factor group S_n/A_n has two elements. A permutation is even if it can be expressed as the product of an even number of 2-cycles. For $n \geq 5$, the group A_n is the only normal subgroup of S_n except the identity subgroup and S_n itself.

Let $n = 4$, then the non-trivial normal subgroups of S_4 are A_4 and also the Klein 4-group. This is the subgroup of S_4 which consists of the identity together with all elements of the form $(a\ b)(c\ d)$ where $\{a, b, c, d\} = \{1, 2, 3, 4\}$. We denote it by V_4. It is contained in A_4 and therefore it is also a normal subgroup of A_4.

Example 6.7. We can sometimes find the dimensions of simple modules from the numerical data obtained from the Artin–Wedderburn decomposition. Let $G = S_3$ be the symmetric group on three letters. Apart from the trivial module there is another 1-dimensional $\mathbb{C}G$-module given by the sign of a permutation, see Exercise 6.2. Moreover, by Theorem 6.4 (c) there must exist a simple $\mathbb{C}G$-module of dimension > 1 since S_3 is not abelian. From Theorem 6.4 (b) we get

$$6 = 1 + 1 + \sum_{i=3}^{k} n_i^2.$$

The only possible solution for this is $6 = 1 + 1 + 2^2$, that is, the group algebra $\mathbb{C}S_3$ has three simple modules, up to isomorphism, of dimensions 1, 1, 2.

Exercise 6.2. Let $G = S_n$, the group of permutations of $\{1, 2, \ldots, n\}$. Recall that every permutation $g \in G$ is either even or odd. Define

$$\sigma(g) = \begin{cases} 1 & g \text{ is even} \\ -1 & g \text{ is odd} \end{cases}$$

Deduce that this defines a representation $\sigma : G \to GL_1(\mathbb{C})$, usually called the sign representation. Describe the corresponding $\mathbb{C}G$-module.

6.3   One-Dimensional Simple Modules and Commutator Groups

In Example 6.7 we found the dimensions of the simple modules for the group algebra $\mathbb{C}S_3$ from the numerical data coming from the Artin–Wedderburn decomposition as in Theorem 6.4, and knowing that the group algebra is not commutative. In general, one needs further information if one wants to find the dimensions of simple modules for a group algebra $\mathbb{C}G$. For instance, take the alternating group A_4. We ask for the integers n_i in Theorem 6.4, that is the sizes of the matrix blocks in the Artin–Wedderburn decomposition of $\mathbb{C}A_4$. That is, we must express 12 as a sum of squares of integers, not all equal to 1 (but at least one summand equal to 1, coming from the trivial module). The possibilities are $12 = 1 + 1 + 1 + 3^2$, or $12 = 1+1+1+1+2^2+2^2$, or also as $12 = 1+1+1+1+1+1+1+1+2^2$. Fortunately, there is further general information on the number of 1-dimensional (simple) $\mathbb{C}G$-modules, which uses the group theoretic description of the largest abelian factor group of G.

We briefly recall a notion from elementary group theory. For a finite group G, the *commutator subgroup* G' is defined as the subgroup of G generated by all elements of the form $[x, y] := xyx^{-1}y^{-1}$ for $x, y \in G$; then we have:

(i) G' is a normal subgroup of G.
(ii) Let N be a normal subgroup of G. Then the factor group G/N is abelian if and only if $G' \subseteq N$. In particular, G/G' is abelian.

Details can be found, for example, in the book by Smith and Tabachnikova in this series.[1]

This allows us to determine the number of one-dimensional simple modules for a group algebra $\mathbb{C}G$, that is, the number of factors $\mathbb{C}$ in the Artin–Wedderburn decomposition.

---

[1] G. Smith, O. Tabachnikova, *Topics in Group Theory*. Springer Undergraduate Mathematics Series. Springer-Verlag London, Ltd., London, 2000. xvi+255 pp.

Corollary 6.8. *Let G be a finite group. Then the number of 1-dimensional simple $\mathbb{C}G$-modules (up to isomorphism) is equal to the order of the factor group G/G', in particular it divides the order of G.*

Proof. Let V be a 1-dimensional $\mathbb{C}G$-module, say $V = \mathrm{span}\{v\}$. We claim that an element $n \in G'$ acts trivially on V. It is enough to prove this for an element $n = [x, y]$ with $x, y \in G$. There exist scalars $\alpha, \beta \in \mathbb{C}$ such that $x \cdot v = \alpha v$ and $y \cdot v = \beta v$. Then

$$[x, y] \cdot v = (xyx^{-1}y^{-1}) \cdot v = \alpha\beta\alpha^{-1}\beta^{-1}v = v$$

and the claim follows.

The bijection in Lemma 2.43 (with $N = G'$) preserves dimensions, so we get a bijection between 1-dimensional representations of $\mathbb{C}(G/G')$ and 1-dimensional $\mathbb{C}G$-modules on which G' acts as identity, that is, with 1-dimensional $\mathbb{C}G$-modules (by what we have seen above).

The group G/G' is abelian, so the 1-dimensional $\mathbb{C}(G/G')$-modules are precisely the simple $\mathbb{C}(G/G')$-modules, by Corollary 3.38 (note that by Corollary 3.20 every simple $\mathbb{C}(G/G')$-module is finite-dimensional). By Theorem 6.4 the number of these is $|G/G'|$. □

Example 6.9. We return to the example at the beginning of this section. Consider the alternating group A_4 of order 12, we determine the commutator subgroup A_4'. As we have mentioned, the Klein 4-group $V_4 = \{\mathrm{id}, (1\,2)(3\,4), (1\,3)(2\,4), (1\,4)(2\,3)\}$ is a normal subgroup in A_4. The factor group has order 3, hence is cyclic and in particular abelian. Thus we obtain $A_4' \subseteq V_4$. On the other hand, every element in V_4 is a commutator in A_4, for example $(1\,2)(3\,4) = [(1\,2\,3), (1\,2\,4)]$. Thus, $A_4' = V_4$ and Corollary 6.8 shows that $\mathbb{C}A_4$ has precisely $|A_4/V_4| = 3$ one-dimensional simple modules. This information is sufficient to determine the number and the dimensions of the simple $\mathbb{C}A_4$-modules. By Theorem 6.4 we know that

$$12 = |A_4| = 1^2 + 1^2 + 1^2 + n_4^2 + \ldots + n_k^2 \text{ with } n_i \geq 2.$$

The only possibility is that $k = 4$ and $n_4 = 3$. Hence, $\mathbb{C}A_4$ has four simple modules (up to isomorphism), of dimensions $1, 1, 1, 3$, and its Artin–Wedderburn decomposition is

$$\mathbb{C}A_4 \cong \mathbb{C} \times \mathbb{C} \times \mathbb{C} \times M_3(\mathbb{C}).$$

Example 6.10. Let G be the dihedral group of order 10, as in Exercise 2.20. Then by Exercise 3.14, the dimension of any simple $\mathbb{C}G$-module is at most 2. The trivial module is 1-dimensional, and by Theorem 6.4 there must be a simple $\mathbb{C}G$-module of dimension 2 since G is not abelian. From Theorem 6.4 we have now

$$10 = a \cdot 1 + b \cdot 4$$

with positive integers a and b. By Corollary 6.8, the number a divides 10, the order of G. The only solution is that $a = 2$ and $b = 2$. So there are two non-isomorphic 2-dimensional simple $\mathbb{C}G$-modules. The Artin–Wedderburn decomposition has the form

$$\mathbb{C}G \cong \mathbb{C} \times \mathbb{C} \times M_2(\mathbb{C}) \times M_2(\mathbb{C}).$$

6.4 Artin–Wedderburn Decomposition and Conjugacy Classes

The number of matrix blocks in the Artin–Wedderburn decomposition of the group algebra $\mathbb{C}G$ for G a finite group also has an interpretation in terms of the group G. Namely it is equal to the number of conjugacy classes of the group. We will see this by showing that both numbers are equal to the dimension of the centre of $\mathbb{C}G$.

Let K be an arbitrary field, and let G be a finite group. Recall that the conjugacy class of an element $x \in G$ is the set $\{gxg^{-1} \mid g \in G\}$. For example, if G is the symmetric group S_n then the conjugacy class of an element $g \in G$ consists precisely of all elements which have the same cycle type. For each conjugacy class C of the group G we consider its 'class sum'

$$\overline{C} := \sum_{g \in C} g \in KG.$$

Proposition 6.11. *Let G be a finite group, and let K be an arbitrary field. Then the class sums $\overline{C} := \sum_{g \in C} g$, as C varies through the conjugacy classes of G, form a K-basis of the centre $Z(KG)$ of the group algebra KG.*

Proof. We begin by showing that each class sum $\overline{C}$ is contained in the centre of KG. It suffices to show that $x\overline{C} = \overline{C}x$ for all $x \in G$. Note that with g also xgx^{-1} varies through all elements of the conjugacy class C. Then we have

$$x\overline{C}x^{-1} = \sum_{g \in C} xgx^{-1} = \sum_{y \in C} y = \overline{C},$$

which is equivalent to $x\overline{C} = \overline{C}x$.

Since each $g \in G$ occurs in precisely one conjugacy class it is clear that the class sums $\overline{C}$ for the different conjugacy classes C are linearly independent over K.

It remains to show that the class sums span the centre. Let $w = \sum_{x \in G} \alpha_x x$ be an element in the centre $Z(KG)$. Then for every $g \in G$ we have

$$w = gwg^{-1} = \sum_{x \in G} \alpha_x gxg^{-1} = \sum_{y \in G} \alpha_{g^{-1}yg} y.$$

The group elements form a basis of KG, we compare coefficients and deduce that

$$\alpha_x = \alpha_{g^{-1}xg} \quad \text{for all } g, x \in G,$$

that is, the coefficients α_x are constant on conjugacy classes. So we can write each element $w \in Z(KG)$ in the form

$$w = \sum_C \alpha_C \overline{C},$$

where the sum is over the different conjugacy classes C of G. Hence the class sums span the centre $Z(KG)$ as a K-vector space. $\square$

We now return to the Artin–Wedderburn decomposition of $\mathbb{C}G$, and relate the number of matrix blocks occurring there to the number of conjugacy classes of G.

Theorem 6.12. *Let G be a finite group and let $\mathbb{C}G \cong M_{n_1}(\mathbb{C}) \times \ldots \times M_{n_k}(\mathbb{C})$ be the Artin–Wedderburn decomposition of the group algebra $\mathbb{C}G$. Then the following are equal:*

(i) The number k of matrix blocks.
(ii) The number of conjugacy classes of G.
(iii) The number of simple $\mathbb{C}G$-modules, up to isomorphism.

Proof. By Theorem 6.4, the numbers in (i) and (iii) are equal. In order to prove the equality of the numbers in (i) and (ii) we consider the centres of the algebras. The centre of $\mathbb{C}G$ has dimension equal to the number of conjugacy classes of G, by Proposition 6.11. On the other hand, the centre of $M_{n_1}(\mathbb{C}) \times \ldots \times M_{n_k}(\mathbb{C})$ is equal to $Z(M_{n_1}(\mathbb{C})) \times \ldots \times Z(M_{n_k}(\mathbb{C}))$ and this has dimension equal to k, the number of matrix blocks, by Exercise 3.16. $\square$

Example 6.13. We consider the symmetric group S_4 on four letters. As we have mentioned, the conjugacy classes of symmetric groups are determined by the cycle type. There are five cycle types for elements of S_4, we have the identity, 2-cycles, 3-cycles, 4-cycles and products of two disjoint 2-cycles. Hence, by Theorem 6.12, there are five matrix blocks in the Artin–Wedderburn decomposition of $\mathbb{C}S_4$, at least one of them has size 1. One can see directly that there is a unique solution for expressing $24 = |S_4|$ as a sum of five squares where at least one of them is equal to 1. Furthermore, we can see from Remark 6.6 and the fact that S_4/V_4 is not abelian, that the commutator subgroup of S_4 is A_4. So $\mathbb{C}S_4$ has precisely two 1-dimensional simple modules (the trivial and the sign module), by Corollary 6.8. From Theorem 6.4 (b) we get

$$24 = |S_4| = 1^2 + 1^2 + n_3^2 + n_4^2 + n_5^2 \quad \text{with } n_i \geq 2.$$

The only possible solution (up to labelling) is $n_3 = 2$ and $n_4 = n_5 = 3$. So $\mathbb{C}S_4$ has five simple modules, of dimensions $1, 1, 2, 3, 3$, and the Artin–Wedderburn decomposition has the form

$$\mathbb{C}S_4 \cong \mathbb{C} \times \mathbb{C} \times M_2(\mathbb{C}) \times M_3(\mathbb{C}) \times M_3(\mathbb{C}).$$

Remark 6.14. The consequences of Maschke's Theorem as discussed also hold if the field K is not $\mathbb{C}$ but is some algebraically closed field whose characteristic does not divide the order of G.

EXERCISES

6.3. Let $G = D_n$ be the dihedral group of order $2n$ (that is, the symmetry group of a regular n-gon). Determine the Artin–Wedderburn decomposition of the group algebra $\mathbb{C}D_n$. (Hint: Apply Exercise 3.14.)

6.4. Let $G = S_3$ be the symmetric group of order 6 and $A = \mathbb{C}G$. We consider the elements $\sigma = (1\,2\,3)$ and $\tau = (1\,2)$ in S_3. We want to show directly that this group algebra is a direct product of three matrix algebras. (We know from Example 6.7 that there should be two blocks of size 1 and one block of size 2.)

 (a) Let $e_\pm := \frac{1}{6}(1 \pm \tau)(1 + \sigma + \sigma^2)$; show that $e_\pm$ are idempotents in the centre of A, and that $e_+ e_- = 0$.

 (b) Let $f = \frac{1}{3}(1 + \omega^{-1}\sigma + \omega\sigma^2)$ where $\omega \in \mathbb{C}$ is a primitive 3rd root of unity. Let $f_1 := \tau f \tau^{-1}$. Show that f and f_1 are orthogonal idempotents, and that

$$f + f_1 = 1_A - e_- - e_+.$$

 (c) Show that $\mathrm{span}\{f, f\tau, \tau f, f_1\}$ is an algebra, isomorphic to $M_2(\mathbb{C})$.

 (d) Apply the above calculations, and show directly that A is isomorphic to a direct product of three matrix algebras.

6.5. Let S_n be the symmetric group on n letters, where $n \geq 2$. It acts on the $\mathbb{C}$-vector space $V := \{(v_1, \ldots, v_n) \in \mathbb{C}^n \mid \sum_{i=1}^n v_i = 0\}$ by permuting the coordinates, that is, $\sigma \cdot (v_1, \ldots, v_n) = (v_{\sigma(1)}, \ldots, v_{\sigma(n)})$ for all $\sigma \in S_n$ (and extension by linearity). Show that V is a simple $\mathbb{C}S_n$-module.

6.6. (a) Let A_n be the alternating group on n letters. Consider the $\mathbb{C}S_n$-module V from Exercise 6.5, and view it as a $\mathbb{C}A_n$-module, by restricting the action. Show that V is also simple as a module for $\mathbb{C}A_n$.

 (b) Show that the group A_5 has five conjugacy classes.

 (c) By applying (a) and (b), and Theorem 6.4, and using the known fact that $A_5' = A_5$, find the dimensions of the simple $\mathbb{C}A_5$-modules and describe the Artin–Wedderburn decomposition of the group algebra $\mathbb{C}A_5$.

6.7. (a) Let G_1, G_2 be two abelian groups of the same order. Explain why $\mathbb{C}G_1$ and $\mathbb{C}G_2$ have the same Artin–Wedderburn decomposition, and hence are isomorphic as $\mathbb{C}$-algebras.

(b) Let G be any non-abelian group of order 8. Show that there is a unique possibility for the Artin–Wedderburn decomposition of $\mathbb{C}G$.

6.8. Let $G = \{\pm 1, \pm i, \pm j, \pm k\}$ be the quaternion group, as defined in Remark 1.9.

(i) Show that the commutator subgroup of G is the cyclic group generated by the element -1.

(ii) Determine the number of simple $\mathbb{C}G$-modules (up to isomorphism) and their dimensions.

(iii) Compare the Artin–Wedderburn decomposition of $\mathbb{C}G$ with that of the group algebra of the dihedral group D_4 of order 8 (that is, the symmetry group of the square). Are the group algebras $\mathbb{C}G$ and $\mathbb{C}D_4$ isomorphic?

6.9. In each of the following cases, does there exist a finite group G such that the Artin–Wedderburn decomposition of the group algebra $\mathbb{C}G$ has the following form?

(i) $M_3(\mathbb{C})$,

(ii) $\mathbb{C} \times M_2(\mathbb{C})$,

(iii) $\mathbb{C} \times \mathbb{C} \times M_2(\mathbb{C})$,

(iv) $\mathbb{C} \times \mathbb{C} \times M_3(\mathbb{C})$.

Chapter 7
Indecomposable Modules

We have seen that for a semisimple algebra, any non-zero module is a direct sum of simple modules (see Theorem 4.11). We investigate now how this generalizes when we consider finite-dimensional modules. If the algebra is not semisimple, one needs to consider indecomposable modules instead of just simple modules, and then one might hope that any finite-dimensional module is a direct sum of indecomposable modules. We will show that this is indeed the case. In addition, we will show that a direct sum decomposition into indecomposable summands is essentially unique; this is known as the Krull–Schmidt Theorem.

In Chap. 3 we have studied simple modules, which are building blocks for arbitrary modules. They might be thought of as analogues of 'elementary particles', and then indecomposable modules could be viewed as analogues of 'molecules'. Throughout this chapter, A is a K-algebra where K is a field.

7.1 Indecomposable Modules

In this section we define indecomposable modules and discuss several examples. In addition, we will show that every finite-dimensional module is a direct sum of indecomposable modules.

Definition 7.1. Let A be a K-algebra, and assume M is a non-zero A-module. Then M is called *indecomposable* if it cannot be written as a direct sum $M = U \oplus V$ for non-zero submodules U and V. Otherwise, M is called *decomposable*.

Remark 7.2.

(1) Every simple module is indecomposable.
(2) Consider the algebra $A = K[X]/(X^t)$ for some $t \geq 2$, recall that the A-modules are of the form V_α with α the linear map of the underlying vector space V

© Springer International Publishing AG, part of Springer Nature 2018
K. Erdmann, T. Holm, *Algebras and Representation Theory*, Springer
Undergraduate Mathematics Series, https://doi.org/10.1007/978-3-319-91998-0_7

given by the action of the coset of X, and note that $\alpha^t = 0$. Let V_α be the 2-dimensional A-module where α has matrix

$$\begin{pmatrix} 0 & 1 \\ 0 & 0 \end{pmatrix}$$

with respect to some basis. It has a unique one-dimensional submodule (spanned by the first basis vector). So it is not simple, and it is indecomposable, since otherwise it would be a direct sum of two 1-dimensional submodules.

(3) Let $A = K[X]/(X^t)$ where $t \geq 2$, and let $M = A$ as an A-module. By the submodule correspondence, every non-zero submodule of M is of the form $(g)/(X^t)$ where g divides X^t but g is not a scalar multiple of X^t. That is, we can take $g = X^r$ for $r < t$. We see that any such submodule contains the element $X^{t-1} + (X^t)$. This means that any two non-zero submodules of M have a non-zero intersection. Therefore M must be indecomposable.

(4) Let A be a semisimple K-algebra. Then an A-module is simple if and only if it is indecomposable. Indeed, by (1) we know that a simple module is indecomposable. For the converse, let M be an indecomposable A-module and let $U \subseteq M$ be a non-zero submodule; we must show that $U = M$. Since A is a semisimple algebra, the module M is semisimple (see Theorem 4.11). So by Theorem 4.3, the submodule U has a complement, that is, $M = U \oplus C$ for some A-submodule C of M. But M is indecomposable, and $U \neq 0$, so $C = 0$ and then $U = M$.

This means that to study indecomposable modules, we should focus on algebras which are not semisimple.

Suppose A is a K-algebra and M an A-module such that $M = U \oplus V$ with U, V non-zero submodules of M. Then in particular, this is a direct sum of K-vector spaces. In linear algebra, having a direct sum decomposition $M = U \oplus V$ of a non-zero vector space M is the same as specifying a projection, ε say, which maps V to zero, and is the identity on U. If U and V are A-submodules of M then ε is an A-module homomorphism, for example by observing that $\varepsilon = \iota \circ \pi$ where $\pi : M \to U$ is the canonical surjection, and $\iota : U \to M$ is the inclusion homomorphism (see Example 2.22). Then $U = \varepsilon(M)$ and $V = (\mathrm{id}_M - \varepsilon)(M)$, and $\varepsilon^2 = \varepsilon$, and $(\mathrm{id}_M - \varepsilon)^2 = \mathrm{id}_M - \varepsilon$.

Recall that an idempotent of an algebra is an element ε in this algebra such that $\varepsilon^2 = \varepsilon$. We see that from the above direct sum decomposition of M as an A-module we get an idempotent ε in the endomorphism algebra $\mathrm{End}_A(M)$.

Lemma 7.3. *Let A be a K-algebra, and let M be a non-zero A-module. Then M is indecomposable if and only if the endomorphism algebra $\mathrm{End}_A(M)$ does not contain any idempotents except 0 and id_M.*

Proof. Assume first that M is indecomposable. Suppose $\varepsilon \in \mathrm{End}_A(M)$ is an idempotent, then since $\mathrm{id}_M = \varepsilon + (\mathrm{id}_M - \varepsilon)$ we have $M = \varepsilon(M) + (\mathrm{id}_M - \varepsilon)(M)$. Moreover, this sum is direct: if $x \in \varepsilon(M) \cap (\mathrm{id}_M - \varepsilon)(M)$ then $\varepsilon(m) = x = (\mathrm{id}_M - \varepsilon)(n)$ with $m, n \in M$, and then

$$x = \varepsilon(m) = \varepsilon^2(m) = \varepsilon(\mathrm{id}_M - \varepsilon)(m) = \varepsilon(m) - \varepsilon^2(m) = 0.$$

Therefore, $M = \varepsilon(M) \oplus (\mathrm{id}_M - \varepsilon)(M)$ is a direct sum of A-submodules. Since M is indecomposable, $\varepsilon(M) = 0$ or $(\mathrm{id}_M - \varepsilon)(M) = 0$. That is, $\varepsilon = 0$ or else $\mathrm{id}_M - \varepsilon = 0$, which means $\varepsilon = \mathrm{id}_M$.

For the converse, if $M = U \oplus V$, where U and V are submodules of M, then as we have seen above, the projection $\varepsilon : M \to M$ with $\varepsilon(u + v) = u$ for $u \in U, v \in V$ is an idempotent in $\mathrm{End}_A(M)$. By assumption, ε is zero or the identity, which means that $U = 0$ or $V = 0$. Hence M is indecomposable. □

Example 7.4. Submodules, or factor modules of indecomposable modules need not be indecomposable. As an example, consider the path algebra $A := KQ$ of the Kronecker quiver as in Example 1.13,

$$1 \underset{b}{\overset{a}{\rightrightarrows}} 2$$

(1) We consider the A-submodule $M := Ae_1 = \mathrm{span}\{e_1, a, b\}$ of A, and $U := \mathrm{span}\{a, b\}$. Each element in the basis $\{e_1, e_2, a, b\}$ of A acts on U by scalars, and hence U and every subspace of U is an A-submodule, and U is the direct sum of non-zero A-submodules

$$U = \mathrm{span}\{a, b\} = \mathrm{span}\{a\} \oplus \mathrm{span}\{b\}.$$

However, M is indecomposable. We will prove this using the criterion of Lemma 7.3. Let $\varepsilon : M \to M$ be an A-module homomorphism with $\varepsilon^2 = \varepsilon$, then we must show that $\varepsilon = \mathrm{id}_M$ or $\varepsilon = 0$. We have $\varepsilon(e_1 M) = e_1 \varepsilon(M) \subseteq e_1 M$, but $e_1 M$ is spanned by e_1, so $\varepsilon(e_1) = \lambda e_1$ for some scalar $\lambda \in K$. Next, note $a = ae_1$ and therefore

$$\varepsilon(a) = \varepsilon(ae_1) = a\varepsilon(e_1) = a(\lambda e_1) = \lambda(ae_1) = \lambda a.$$

Similarly, since $b = be_1$ we have $\varepsilon(b) = \lambda b$. We have proved that $\varepsilon = \lambda \cdot \mathrm{id}_M$. Now, $\varepsilon^2 = \varepsilon$ and therefore $\lambda^2 = \lambda$ and hence $\lambda = 0$ or $\lambda = 1$. That is, ε is the zero map, or the identity map.

(2) Let N be the A-module with basis $\{v_1, v_1', v_2\}$ where $e_1 N$ has basis $\{v_1, v_1'\}$ and $e_2 N$ has basis $\{v_2\}$ and where the action of a and b is defined by

$$av_1 = v_2, \quad av_1' = 0, \quad av_2 = 0 \quad \text{and} \quad bv_1 = 0, \quad bv_1' = v_2, \quad bv_2 = 0.$$

We see that $U := \mathrm{span}\{v_2\}$ is an A-submodule of N. The factor module N/U has basis consisting of the cosets of v_1, v_1', and from the definition, each element in the basis $\{e_1, e_2, a, b\}$ of A acts by scalar multiplication on N/U. As before, this implies that N/U is a direct sum of two 1-dimensional modules. On the other hand, we claim that N is indecomposable; to show this we want to use Lemma 7.3, as in (1). So let $\varepsilon \in \mathrm{End}_A(N)$, then $\varepsilon(v_2) = \varepsilon(e_2 v_2) = e_2 \varepsilon(v_2) \in e_2 N$ and hence $\varepsilon(v_2) = \lambda v_2$, for some $\lambda \in K$. Moreover, $\varepsilon(v_1) = \varepsilon(e_1 v_1) = e_1 \varepsilon(v_1) \in e_1 N$, so $\varepsilon(v_1) = \mu v_1 + \rho v_1'$ for some $\mu, \rho \in K$. Using that $bv_1 = 0$ and $bv_1' = v_2$ this implies that $\rho = 0$. Then $\lambda v_2 = \varepsilon(v_2) = \varepsilon(av_1) = a\varepsilon(v_1) = \mu v_2$, hence $\lambda = \mu$ and $\varepsilon(v_1) = \lambda v_1$. Similarly, one shows $\varepsilon(v_1') = \lambda v_1'$, that is, we get $\varepsilon = \lambda \cdot \mathrm{id}_N$. If $\varepsilon^2 = \varepsilon$ then $\varepsilon = 0$ or $\varepsilon = \mathrm{id}_N$.

We will now show that every non-zero finite-dimensional module can be written as a direct sum of indecomposable modules.

Theorem 7.5. *Let A be a K-algebra, and let M be a non-zero finite-dimensional A-module. Then M can be expressed as a direct sum of finitely many indecomposable A-submodules.*

Proof. We use induction on the vector space dimension $\dim_K M$. If $\dim_K M = 1$, then M is a simple A-module, hence is an indecomposable A-module, and we are done. So let $\dim_K M > 1$. If M is indecomposable, there is nothing to do. Otherwise, $M = U \oplus V$ with non-zero A-submodules U and V of M. Then both U and V have strictly smaller dimension than M. Hence by the inductive hypothesis, each of U and V can be written as a direct sum of finitely many indecomposable A-submodules. Since $M = U \oplus V$, it follows that M can be expressed as a direct sum of finitely many indecomposable A-submodules. $\square$

Remark 7.6. There is a more general version of Theorem 7.5. Recall from Sect. 3.3 that an A-module M is said to be of finite length if M has a composition series; in this case, the length $\ell(M)$ of M is defined as the length of a composition series of M (which is uniquely determined by the Jordan–Hölder theorem). If we replace 'dimension' in the above theorem and its proof by 'length' then everything works the same, noting that proper submodules of a module of finite length have strictly smaller lengths, by Proposition 3.17. We get therefore that any non-zero module of finite length can be expressed as a direct sum of finitely many indecomposable submodules.

There are many modules which cannot be expressed as a finite direct sum of indecomposable modules. For example, let $K = \mathbb{Q}$, and let $\mathbb{R}$ be the set of real numbers. Then $\mathbb{R}$ is a vector space over $\mathbb{Q}$, hence is a $\mathbb{Q}$-module. The indecomposable $\mathbb{Q}$-modules are the 1-dimensional $\mathbb{Q}$-vector spaces. Since $\mathbb{R}$ has infinite dimension over $\mathbb{Q}$, it cannot be a finite direct sum of indecomposable $\mathbb{Q}$-modules. This shows that the condition that the module is finite-dimensional (or has finite length) in Theorem 7.5 cannot be removed.

7.2 Fitting's Lemma and Local Algebras

We would like to have criteria which tell us when a given module is indecomposable. Obviously Definition 7.1 is not so helpful; we would need to inspect all submodules of a given module. One criterion is Lemma 7.3; in this section we will look for further information from linear algebra.

Given a linear transformation of a finite-dimensional vector space, one gets a direct sum decomposition of the vector space, in terms of the kernel and the image of some power of the linear transformation.

Lemma 7.7 (Fitting's Lemma I).   *Let K be a field. Assume V is a finite-dimensional K-vector space, and $\theta : V \to V$ is a linear transformation. Then there is some $n \geq 1$ such that the following hold:*

(i)   For all $k \geq 0$ we have $\ker(\theta^n) = \ker(\theta^{n+k})$ and $\operatorname{im}(\theta^n) = \operatorname{im}(\theta^{n+k})$.
(ii) $V = \ker(\theta^n) \oplus \operatorname{im}(\theta^n)$.

Proof. This is elementary linear algebra, but since it is important, we give the proof. (i) We have inclusions of subspaces

$$\ker(\theta) \subseteq \ker(\theta^2) \subseteq \ker(\theta^3) \subseteq \ldots \subseteq V \tag{7.1}$$

and

$$V \supseteq \operatorname{im}(\theta) \supseteq \operatorname{im}(\theta^2) \supseteq \operatorname{im}(\theta^3) \supseteq \ldots \tag{7.2}$$

Since V is finite-dimensional, the ascending chain (7.1) cannot contain infinitely many strict inequalities, so there exists some $n_1 \geq 1$ such that $\ker(\theta^{n_1}) = \ker(\theta^{n_1+k})$ for all $k \geq 0$. Similarly, for the descending chain (7.2) there is some $n_2 \geq 1$ such that $\operatorname{im}(\theta^{n_2}) = \operatorname{im}(\theta^{n_2+k})$ for all $k \geq 0$. Setting n as the maximum of n_1 and n_2 proves (i).
(ii) We show $\ker(\theta^n) \cap \operatorname{im}(\theta^n) = 0$ and $\ker(\theta^n) + \operatorname{im}(\theta^n) = V$. Let $x \in \ker(\theta^n) \cap \operatorname{im}(\theta^n)$, that is, $\theta^n(x) = 0$ and $x = \theta^n(y)$ for some $y \in V$. We substitute, and then we have that $0 = \theta^n(x) = \theta^{2n}(y)$. This implies $y \in \ker(\theta^{2n}) = \ker(\theta^n)$ by part (i), and thus $x = \theta^n(y) = 0$. Hence $\ker(\theta^n) \cap \operatorname{im}(\theta^n) = 0$. Now, by the rank-nullity theorem we have

$$\dim_K V = \dim_K \ker(\theta^n) + \dim_K \operatorname{im}(\theta^n)$$

$$= \dim_K \ker(\theta^n) + \dim_K \operatorname{im}(\theta^n) - \dim_K(\ker(\theta^n) \cap \operatorname{im}(\theta^n))$$

$$= \dim_K(\ker(\theta^n) + \operatorname{im}(\theta^n)).$$

Hence the sum $\ker(\theta^n) + \operatorname{im}(\theta^n)$ is equal to V since it is a subspace whose dimension is equal to $\dim_K V$. $\qquad\square$

Corollary 7.8. *Let A be a K-algebra. Assume M is a finite-dimensional A-module, and $\theta : M \to M$ is an A-module homomorphism. Then there is some $n \geq 1$ such that $M = \ker(\theta^n) \oplus \operatorname{im}(\theta^n)$, a direct sum of A-submodules of M.*

Proof. This follows directly from Lemma 7.7. Indeed, θ is in particular a linear transformation. The map θ^n is an A-module homomorphism and therefore its kernel and its image are A-submodules of M.   □

Corollary 7.9 (Fitting's Lemma II). *Let A be a K-algebra and M a non-zero finite-dimensional A-module. Then the following statements are equivalent:*

(i) M is an indecomposable A-module.
(ii) Every homomorphism $\theta \in \operatorname{End}_A(M)$ is either an isomorphism, or is nilpotent.

Proof. We first assume that statement (i) holds. By Corollary 7.8, for every $\theta \in \operatorname{End}_A(M)$ we have $M = \ker(\theta^n) \oplus \operatorname{im}(\theta^n)$ as A-modules, for some $n \geq 1$. But M is indecomposable by assumption, so we conclude that $\ker(\theta^n) = 0$ or $\operatorname{im}(\theta^n) = 0$. In the second case we have $\theta^n = 0$, that is, θ is nilpotent. In the first case, θ^n and hence also θ are injective, and moreover, $M = \operatorname{im}(\theta^n)$, therefore θ^n and hence θ are surjective. So θ is an isomorphism.

Conversely, suppose that (ii) holds. To show that M is indecomposable, we apply Lemma 7.3. So let ε be an endomorphism of M such that $\varepsilon^2 = \varepsilon$. By assumption, ε is either nilpotent, or is an isomorphism. In the first case $\varepsilon = 0$ since $\varepsilon = \varepsilon^2 = \ldots = \varepsilon^n$ for all $n \geq 1$. In the second case, $\operatorname{im}(\varepsilon) = M$ and ε is the identity on M: If $m \in M$ then $m = \varepsilon(y)$ for some $y \in M$, hence $\varepsilon(m) = \varepsilon^2(y) = \varepsilon(y) = m$.   □

Remark 7.10. Corollary 7.9 shows that the endomorphism algebra $E := \operatorname{End}_A(M)$ of an indecomposable A-module M has the following property: *if $a \in E$ is not invertible, then $1_E - a \in E$ is invertible.* In fact, if $a \in E$ is not invertible then it is nilpotent, by Corollary 7.9, say $a^n = 0$. Then we have

$$(1_E + a + a^2 + \ldots + a^{n-1})(1_E - a) = 1_E = (1_E - a)(1_E + a + a^2 + \ldots + a^{n-1}),$$

that is, $1_E - a$ is invertible.

Note that in a non-commutative algebra one has to be slightly careful when speaking of invertible elements. More precisely, one should speak of elements which have a left inverse or a right inverse, respectively. If for some element both a left inverse and a right inverse exist, then they coincide (since invertible elements form a group).

Algebras (or more generally rings) with the property in Remark 7.10 appear naturally in many places in mathematics. We now study them in some more detail.

Theorem 7.11. *Assume A is any K-algebra, then the following are equivalent:*

(i) The set N of elements $x \in A$ which do not have a left inverse is a left ideal of A.
(ii) For all $a \in A$, at least one of a and $1_A - a$ has a left inverse in A.

We observe the following: Let N be as in (i). If $x \in N$ and $a \in A$ then ax cannot have a left inverse in A, since otherwise x would have a left inverse. That is, we have $ax \in N$, so that $AN \subseteq N$.

Proof. Assume (ii) holds, we show that then N is a left ideal of A. By the above observation, we only have to show that N is an additive subgroup of A. Clearly $0 \in N$. Now let $x, y \in N$, assume (for a contradiction) that $x - y$ is not in N. Then there is some $a \in A$ such that $a(x - y) = 1_A$, so that

$$(-a)y = 1_A - ax.$$

We know that ax does not have a left inverse, and therefore, using (ii) we deduce that $(-a)y$ has a left inverse. But then y has a left inverse, and $y \notin N$, a contradiction. We have now shown that N is a left ideal of A.

Now assume (i) holds, we prove that this implies (ii). Assume $a \in A$ does not have a left inverse in A. We have to show that then $1_A - a$ has a left inverse in A. If this is false then both a and $1_A - a$ belong to N. By assumption (i), N is closed under addition, therefore $1_A \in N$, which is not true. This contradiction shows that $1_A - a$ must belong to N. $\square$

Definition 7.12. A K-algebra A is called a *local algebra* (or just *local*) if it satisfies the equivalent conditions from Theorem 7.11.

Exercise 7.1. Let A be a local K-algebra. Show that the left ideal N in Theorem 7.11 is a maximal left ideal of A, and that it is the only maximal left ideal of A.

Remark 7.13. Let A be a local K-algebra. By Exercise 7.1 the left ideal N in Theorem 7.11 is then precisely the Jacobson radical as defined and studied in Sect. 4.3 (see Definition 4.21). In particular, if A is finite-dimensional then this unique maximal left ideal is even a two-sided ideal (see Theorem 4.23).

Lemma 7.14.

(a) *Assume A is a local K-algebra. Then the only idempotents in A are 0 and 1_A.*
(b) *Assume A is a finite-dimensional algebra. Then A is local if and only if the only idempotents in A are 0 and 1_A.*

Proof. (a) Let $\varepsilon \in A$ be an idempotent. If ε has no left inverse, then by Theorem 7.11 we know that $1_A - \varepsilon$ has a left inverse, say $a(1_A - \varepsilon) = 1_A$ for some $a \in A$. Then it follows that $\varepsilon = 1_A\varepsilon = a(1_A - \varepsilon)\varepsilon = a\varepsilon - a\varepsilon^2 = 0$. On the other hand, if ε has a left inverse, say $b\varepsilon = 1_A$ for some $b \in A$, then $\varepsilon = 1_A\varepsilon = b\varepsilon^2 = b\varepsilon = 1_A$.

(b) We must show the converse of (a). Assume that 0 and 1_A are the only idempotents in A. We will verify condition (ii) of Theorem 7.11, that is, let $a \in A$, then we show that at least one of a and $1_A - a$ has a left inverse in A. Consider the map $\theta : A \to A$ defined by $\theta(x) := xa$. This is an A-module homomorphism if we view A as a left A-module. By Corollary 7.8 we have $A = \ker(\theta^n) \oplus \mathrm{im}(\theta^n)$ for some $n \geq 1$. So we have a unique expression

$$1_A = \varepsilon_1 + \varepsilon_2 \text{ with } \varepsilon_1 \in \ker(\theta^n), \ \varepsilon_2 \in \mathrm{im}(\theta^n).$$

By Lemma 5.1, the ε_i are idempotents. By assumption, $\varepsilon_1 = 0$ or $\varepsilon_1 = 1_A$. Furthermore, by Lemma 5.1 we have $A = A\varepsilon_1 \oplus A\varepsilon_2$. If $\varepsilon_1 = 0$ then $A = \mathrm{im}(\theta^n) = Aa^n$ and then a has a left inverse in A (since $1_A = ba^n$ for some $b \in A$). Otherwise, $\varepsilon_1 = 1_A$ and then $A = \ker(\theta^n)$, that is, $a^n = 0$. A computation as in Remark 7.10 then shows that $1_A - a$ has a left inverse, namely $1_A + a + a^2 + \ldots + a^{n-1}$.   □

We will now investigate some examples.

Example 7.15.

(1) Let $A = K$, the one-dimensional K-algebra. Then for every $a \in A$, at least one of a or $1_A - a$ has a left inverse in A and hence A is local, by Theorem 7.11 and Definition 7.12. The same argument works to show that every division algebra $A = D$ over the field K (see Definition 1.7) is local.
(2) Let $A = M_n(K)$ where $n \geq 2$. Let $a := E_{11}$, then a and $1_A - a$ do not have a left inverse. Hence A is not local. (We have $M_1(K) = K$, which is local, by (1).)
(3) Consider the factor algebra $A = K[X]/(f)$ for a non-constant polynomial $f \in K[X]$. Then A is a local algebra if and only if $f = g^m$ for some $m \geq 1$, where $g \in K[X]$ is an irreducible polynomial. The proof of this is Exercise 7.6.

Exercise 7.2. Assume $A = KQ$, where Q is a quiver with no oriented cycles. Show that A is local if and only if Q consists just of one vertex.

For finite-dimensional modules, we can now characterize indecomposability in terms of local endomorphism algebras.

Corollary 7.16 (Fitting's Lemma III). *Let A be a K-algebra and let M be a non-zero finite-dimensional A-module. Then M is an indecomposable A-module if and only if the endomorphism algebra $\mathrm{End}_A(M)$ is a local algebra.*

Proof. Let $E := \mathrm{End}_A(M)$. Note that E is finite-dimensional (it is contained in the space of all K-linear maps $M \to M$, which is finite-dimensional by elementary linear algebra). By Lemma 7.3, the module M is indecomposable if and only if the algebra E does not have idempotents other than 0 and id_M. By Lemma 7.14 this is true if and only if E is local.   □

Remark 7.17. All three versions of Fitting's Lemma have a slightly more general version, if one replaces 'finite dimensional' by 'finite length', see Remark 7.6.

The assumption that M is finite-dimensional, or has finite length, cannot be omitted. For instance, consider the polynomial algebra $K[X]$ as a $K[X]$-module. Multiplication by X defines a $K[X]$-module homomorphism $\theta : K[X] \to K[X]$. For every $n \in \mathbb{N}$ we have $\ker(\theta^n) = 0$ and $\mathrm{im}(\theta^n) = (X^n)$, the ideal generated by X^n. But $K[X] \neq \ker(\theta^n) \oplus \mathrm{im}(\theta^n)$. So Lemma 7.7 fails for A. Exercise 7.11 contains further illustrations.

7.3 The Krull–Schmidt Theorem

We have seen in Theorem 7.5 that every non-zero finite-dimensional module can be decomposed into a finite direct sum of indecomposable modules. The fundamental Krull–Schmidt Theorem states that such a decomposition is unique up to isomorphism and up to reordering indecomposable summands. This is one of the most important results in representation theory.

Theorem 7.18 (Krull–Schmidt Theorem). *Let A be a K-algebra, and let M be a non-zero finite-dimensional A-module. Suppose there are two direct sum decompositions*

$$M_1 \oplus \ldots \oplus M_r = M = N_1 \oplus \ldots \oplus N_s$$

of M into indecomposable A-submodules M_i $(1 \leq i \leq r)$ and N_j $(1 \leq j \leq s)$. Then $r = s$, and there is a permutation σ such that $M_i \cong N_{\sigma(i)}$ for all $i = 1, \ldots, r$.

Before starting with the proof, we introduce some notation we will use for the canonical homomorphisms associated to a direct sum decomposition, similar to the notation used in Lemma 5.6. Let $\mu_i : M \to M_i$ be the homomorphism defined by $\mu(m_1 + \ldots + m_r) = m_i$, and let $\iota_i : M_i \to M$ be the inclusion map. Then $\mu_i \circ \iota_i = \mathrm{id}_{M_i}$ and hence if $e_i := \iota_i \circ \mu_i : M \to M$ then e_i is the projection with image M_i and kernel the direct sum of all M_j for $j \neq i$. Then the e_i are orthogonal idempotents in $\mathrm{End}_A(M)$ with $\mathrm{id}_M = \sum_{i=1}^r e_i$.

Similarly let $v_t : M \to N_t$ be the homomorphism defined by $v_t(n_1 + \ldots + n_s) = n_t$, and let $\kappa_t : N_t \to M$ be the inclusion. Then $v_t \circ \kappa_t$ is the identity map of N_t, and if $f_t := \kappa_t \circ v_t : M \to M$ then f_t is the projection with image N_t and kernel the direct sum of all N_j for $j \neq t$. In addition, the f_t are orthogonal idempotents in $\mathrm{End}_A(M)$ whose sum is the identity id_M.

In the proof below we sometimes identify M_i with $\iota_i(M_i)$ and N_t with $\kappa_t(N_t)$, to ease notation.

Proof. We use induction on r, the number of summands in the first decomposition. When $r = 1$ we have $M_1 = M$. This means that M is indecomposable, and we conclude that $s = 1$ and $N_1 = M = M_1$. Assume now that $r > 1$.
(1) We will find an isomorphism between N_1 and some M_i. We have

$$\mathrm{id}_{N_1} = v_1 \circ \kappa_1 = v_1 \circ \mathrm{id}_M \circ \kappa_1 = \sum_{i=1}^r v_1 \circ e_i \circ \kappa_1. \qquad (*)$$

The module N_1 is indecomposable and finite-dimensional, so by Corollary 7.9, every endomorphism of N_1 is either nilpotent, or is an isomorphism. Assume for a contradiction that each summand in the above sum (∗) is nilpotent, and hence does not have a left inverse. We have that $\mathrm{End}_A(N_1)$ is a local algebra, by Corollary 7.16. Hence by Theorem 7.11, the set of elements with no left inverse is closed under addition, so it follows that the sum also has no left inverse. But the sum is the identity of N_1 which has a left inverse, a contradiction.

Hence at least one of the summands in (∗) is an isomorphism, and we may relabel the M_i and assume that ϕ is an isomorphism $N_1 \to M_1$, where

$$\phi := \nu_1 \circ e_1 \circ \kappa_1 = \nu_1 \circ \iota_1 \circ \mu_1 \circ \kappa_1.$$

Now we apply Lemma 2.30 with $M = M_1$, and $N_1 = N = N'$ and $j = \mu_1 \circ \kappa_1$ and $\pi = \nu_1 \circ \iota_1$. We obtain

$$M_1 = \mathrm{im}(\mu_1 \circ \kappa_1) \oplus \ker(\nu_1 \circ \iota_1).$$

Now, since M_1 is indecomposable and $\mu_1 \circ \kappa_1$ is non-zero we have $M_1 = \mathrm{im}(\mu_1 \circ \kappa_1)$ and $\ker(\nu_1 \circ \iota_1) = 0$. Hence the map $\mu_1 \circ \kappa_1 : N_1 \to M_1$ is surjective. It is also injective (since $\phi = \nu_1 \circ \iota_1 \circ \mu_1 \circ \kappa_1$ is injective). This shows that $\mu_1 \circ \kappa_1$ is an isomorphism $N_1 \to M_1$.

(2) As a tool for the inductive step, we construct an A-module isomorphism $\gamma : M \to M$ such that $\gamma(N_1) = M_1$ and $\gamma(N_j) = N_j$ for $2 \leq j \leq s$. Define

$$\gamma := \mathrm{id}_M - f_1 + e_1 \circ f_1.$$

We first show that $\gamma : M \to M$ is an isomorphism. It suffices to show that γ is injective, by dimensions. Let $\gamma(x) = 0$ for some $x \in M$. Using that f_1 is an idempotent we have

$$0 = f_1(0) = (f_1 \circ \gamma)(x) = f_1(x) - f_1^2(x) + (f_1 \circ e_1 \circ f_1)(x) = (f_1 \circ e_1 \circ f_1)(x).$$

By definition, $f_1 \circ e_1 \circ f_1 = \kappa_1 \circ \nu_1 \circ \iota_1 \circ \mu_1 \circ \kappa_1 \circ \nu_1 = \kappa_1 \circ \phi \circ \nu_1$ with the isomorphism $\phi : N_1 \to M_1$ from (1). Since κ_1 and ϕ are injective, it follows from $(f_1 \circ e_1 \circ f_1)(x) = 0$ that $\nu_1(x) = 0$. Then also $f_1(x) = (\kappa_1 \circ \nu_1)(x) = 0$ and this implies $x = \gamma(x) = 0$, as desired.

We now show that $\gamma(N_1) = M_1$ and $\gamma(N_j) = N_j$ for $2 \leq j \leq s$. From (1) we have the isomorphism $\mu_1 \circ \kappa_1 : N_1 \to M_1$, and this is viewed as a homomorphism $e_1 \circ f_1 : M \to M$, by $e_1 \circ f_1 = \iota_1 \circ (\mu_1 \circ \kappa_1) \circ \nu_1$, noting that ν_1 is the identity on N_1 and ι_1 is the identity on M_1. Furthermore, $\mathrm{id}_M - f_1$ maps N_1 to zero, and in total we see that $\gamma(N_1) = M_1$.

Moreover, if $x \in N_j$ and $j \geq 2$ then $x = f_j(x)$ and $f_1 \circ f_j = 0$ and it follows that $\gamma(x) = x$. This proves $\gamma(N_j) = N_j$.

(3) We complete the proof of the Krull–Schmidt Theorem. Note that an isomorphism takes a direct sum decomposition to a direct sum decomposition, see Exercise 7.3. By (2) we have

$$M = \gamma(M) = \gamma(N_1) \oplus \gamma(N_2) \oplus \ldots \oplus \gamma(N_s) = M_1 \oplus N_2 \oplus \ldots \oplus N_s.$$

Using the isomorphism theorem, we have

$$M_2 \oplus \ldots \oplus M_r \cong M/M_1 = (M_1 \oplus N_2 \oplus \ldots \oplus N_s)/M_1 \cong N_2 \oplus \ldots \oplus N_s.$$

To apply the induction hypothesis, we need two direct sum decompositions of the same module. Let $M' := M_2 \oplus \ldots \oplus M_r$. We have obtained an isomorphism $\psi : M' \to N_2 \oplus \ldots \oplus N_s$. Let $N_i' := \psi^{-1}(N_i)$, this is a submodule of M', and we have, again by Exercise 7.3, the direct sum decomposition

$$M' = N_2' \oplus \ldots \oplus N_s'.$$

By the induction hypothesis, $r - 1 = s - 1$ and there is a permutation σ of $\{1, 2, 3, \ldots, r\}$ with $M_i \cong N_{\sigma(i)}' \cong N_{\sigma(i)}$ for $i \geq 2$ (and $\sigma(1) = 1$). This completes the proof.   □

EXERCISES

7.3. Let A be a K-algebra, and M a non-zero A-module. Assume $M = U_1 \oplus \ldots \oplus U_r$ is a direct sum of A-submodules, and assume that $\gamma : M \to M$ is an A-module isomorphism. Show that then $M = \gamma(M) = \gamma(U_1) \oplus \ldots \oplus \gamma(U_r)$.

7.4. Let $T_n(K)$ be the K-algebra of upper triangular $n \times n$-matrices. In the natural $T_n(K)$-module K^n we consider for $0 \leq i \leq n$ the submodules

$$V_i := \{(\lambda_1, \ldots, \lambda_n)^t \mid \lambda_j = 0 \text{ for all } j > i\} = \text{span}\{e_1, \ldots, e_i\},$$

where e_i denotes the i-th standard basis vector. Recall from Exercise 2.14 that $V_0, V_1, \ldots, V_n$ are the only $T_n(K)$-submodules of K^n, and that $V_{i,j} := V_i/V_j$ (for $0 \leq j < i \leq n$) are the $\frac{n(n+1)}{2}$ pairwise non-isomorphic $T_n(K)$-modules.

(a) Determine the endomorphism algebra $\text{End}_{T_n(K)}(V_{i,j})$ for all $0 \leq j < i \leq n$.

(b) Deduce that each $V_{i,j}$ is an indecomposable $T_n(K)$-module.

7.5. Recall that for any K-algebra A and every element $a \in A$ the map $\theta_a : A \to A, b \mapsto ba$, is an A-module homomorphism.

We consider the algebra $A = T_n(K)$ of upper triangular $n \times n$-matrices. Determine for each of the following elements $a \in T_n(K)$ the minimal $d \in \mathbb{N}$ such that $\ker(\theta_a^d) = \ker(\theta_a^{d+j})$ and $\operatorname{im}(\theta_a^d) = \operatorname{im}(\theta_a^{d+j})$ for all $j \in \mathbb{N}$. Moreover, give explicitly the decomposition $T_n(K) = \ker(\theta_a^d) \oplus \operatorname{im}(\theta_a^d)$ (which exists by the Fitting Lemma, see Corollary 7.7):

(i) $a = E_{11}$;
(ii) $a = E_{12} + E_{23} + \ldots + E_{n-1,n}$;
(iii) $a = E_{1n} + E_{2n} + \ldots + E_{nn}$.

7.6. (a) Let $A = K[X]/(f)$ with a non-constant polynomial $f \in K[X]$. Show that A is a local algebra if and only $f = g^m$ for some irreducible polynomial $g \in K[X]$ and some $m \in \mathbb{N}$ (up to multiplication by a non-zero scalar in K).

(b) Which of the following algebras are local?

(i) $\mathbb{Q}[X]/(X^n - 1)$, where $n \geq 2$;
(ii) $\mathbb{Z}_p[X]/(X^p - 1)$ where p is a prime number;
(iii) $K[X]/(X^3 - 6X^2 + 12X - 8)$.

7.7. Which of the following K-algebras A are local?

(i) $A = T_n(K)$, the algebra of upper triangular $n \times n$-matrices;
(ii) $A = \{a = (a_{ij}) \in T_n(K) \mid a_{ii} = a_{jj}$ for all $i, j\}$.

7.8. For a field K let $K[[X]] := \{\sum_{i=0}^{\infty} \lambda_i X^i \mid \lambda_i \in K\}$ be the set of formal power series. On $K[[X]]$ define the following operations:

– addition $(\sum_i \lambda_i X^i) + (\sum_i \mu_i X^i) = \sum_i (\lambda_i + \mu_i) X^i$,
– scalar multiplication $\lambda(\sum_i \lambda_i X^i) = \sum_i \lambda\lambda_i X^i$,
– multiplication $(\sum_i \lambda_i X^i)(\sum_j \mu_j X^j) = \sum_k (\sum_{i+j=k} \lambda_i \mu_j) X^k$.

(a) Verify that $K[[X]]$ with these operations becomes a commutative K-algebra.

(b) Determine the invertible elements in $K[[X]]$.

(c) Show that $K[[X]]$ is a local algebra.

7.9. Let K be a field and A a K-algebra of finite length (as an A-module). Show that if A is a local algebra then A has only one simple module, up to isomorphism.

7.10. For a field K consider the path algebra KQ of the Kronecker quiver

For $\lambda \in K$ we consider a 2-dimensional K-vector space $V_\lambda = \mathrm{span}\{v_1, v_2\}$. This becomes a KQ-module via the representation $\Theta_\lambda : KQ \to \mathrm{End}_K(V_\lambda)$, where

$$\Theta_\lambda(e_1) = \begin{pmatrix} 1 & 0 \\ 0 & 0 \end{pmatrix}, \quad \Theta_\lambda(e_2) = \begin{pmatrix} 0 & 0 \\ 0 & 1 \end{pmatrix}$$

$$\Theta_\lambda(a) = \begin{pmatrix} 0 & 0 \\ 1 & 0 \end{pmatrix}, \quad \Theta_\lambda(b) = \begin{pmatrix} 0 & 0 \\ \lambda & 0 \end{pmatrix}.$$

(a) Show that for $\lambda \neq \mu$ the KQ-modules V_λ and V_μ are not isomorphic.
(b) Show that for all $\lambda \in K$ the KQ-module V_λ is indecomposable.

7.11. Let $K[X]$ be the polynomial algebra.

(a) Show that $K[X]$ is indecomposable as a $K[X]$-module.
(b) Show that the equivalence in the second version of Fitting's Lemma (Corollary 7.9) does not hold, by giving a $K[X]$-module endomorphism of $K[X]$ which is neither invertible nor nilpotent.
(c) Show that the equivalence in the third version of Fitting's Lemma (Corollary 7.16) also does not hold for $K[X]$.

7.12. (a) By applying the Artin–Wedderburn theorem, characterize which semisimple K-algebras are local algebras.
(b) Let G be a finite group such that the group algebra KG is semisimple (that is, the characteristic of K does not divide $|G|$, by Maschke's theorem). Deduce that KG is not a local algebra, except for the group G with one element.

Chapter 8
Representation Type

We have seen in the previous chapter that understanding the finite-dimensional modules of an algebra reduces to understanding all indecomposable modules. However, we will see some examples which show that this may be too ambitious in general: An algebra can have infinitely many indecomposable modules, and classifying them appears not to be feasible. Algebras can roughly be divided into those that have only finitely many indecomposable modules (finite representation type), and those that have infinitely many indecomposable modules (infinite representation type). In this chapter we will introduce the notion of representation type, consider some classes of algebras, and determine whether they have finite or infinite representation type.

8.1 Definition and Examples

We have seen that any finite-dimensional module of a K-algebra A can be expressed as a direct sum of indecomposable modules, and moreover, the indecomposable summands occuring are unique up to isomorphism (see Theorem 7.18). Therefore, for a given algebra A, it is natural to ask how many indecomposable A-modules it has, and if possible, to give a complete description up to isomorphism. The notion for this is the representation type of an algebra, we introduce this now and give several examples, and some reduction methods.

Throughout K is a field.

Definition 8.1. A K-algebra A has *finite representation type* if there are only finitely many finite-dimensional indecomposable A-modules, up to isomorphism. Otherwise, A is said to be of *infinite representation type*.

© Springer International Publishing AG, part of Springer Nature 2018
K. Erdmann, T. Holm, *Algebras and Representation Theory*, Springer
Undergraduate Mathematics Series, https://doi.org/10.1007/978-3-319-91998-0_8

Remark 8.2.

(1) One sometimes alternatively defines finite representation type for an algebra A by requesting that there are only finitely many indecomposable A-modules of finite length, up to isomorphism.

For a finite-dimensional algebra A the two versions are the same: That is, an A-module has finite length if and only if it is finite-dimensional. This follows since in this case every simple A-module is finite-dimensional, see Corollary 3.20.

(2) Isomorphic algebras have the same representation type. To see this, let $\Phi : A \to B$ be an isomorphism of K-algebras. Then every B-module M becomes an A-module by setting $a \cdot m = \Phi(a)m$ and conversely, every A-module becomes a B-module by setting $b \cdot m = \Phi^{-1}(b)m$. This correspondence preserves dimensions of modules and it preserves isomorphisms, and moreover indecomposable A-modules correspond to indecomposable B-modules. We shall use this tacitly from now on.

Example 8.3.

(1) Every semisimple K-algebra has finite representation type.

In fact, suppose A is a semisimple K-algebra. By Remark 7.2, an A-module M is indecomposable if and only if it is simple. By Remark 4.10, the algebra $A = S_1 \oplus \ldots \oplus S_k$ is a direct sum of finitely many simple A-modules. In particular, A has finite length as an A-module and then every indecomposable (that is, simple) A-module is isomorphic to one of the finitely many A-modules $S_1, \ldots, S_k$, by Theorem 3.19. In particular, there are only finitely many finite-dimensional indecomposable A-modules, and A has finite representation type.

Note that in this case it may happen that the algebra A is not finite-dimensional, for example it could just be an infinite-dimensional division algebra.

(2) The polynomial algebra $K[X]$ has infinite representation type.

To see this, take some integer $m \geq 2$ and consider the (finite-dimensional) $K[X]$-module $W_m = K[X]/(X^m)$. This is indecomposable: In Remark 7.2 we have seen that it is indecomposable as a module for the factor algebra $K[X]/(X^m)$, and the argument we gave works here as well: every non-zero $K[X]$-submodule of W_m must contain the coset $X^{m-1} + (X^m)$. So W_m cannot be expressed as a direct sum of two non-zero submodules. The module W_m has dimension m, and hence different W_m are not isomorphic.

(3) For any $n \in \mathbb{N}$ the algebra $A := K[X]/(X^n)$ has finite representation type.

Recall that a finite-dimensional A-module is of the form V_α where V is a finite-dimensional K-vector space, and α is a linear transformation of V such that $\alpha^n = 0$ (here α describes the action of the coset of X). Since $\alpha^n = 0$, the only eigenvalue of α is 0, so the field K contains all eigenvalues of α. This means that there exists a Jordan canonical form for α over K. That is, V is the direct sum $V = V_1 \oplus \ldots \oplus V_r$, where each V_i is invariant under α, and each V_i

has a basis such that the matrix of α on V_i is a Jordan block matrix of the form

$$\begin{pmatrix} 0 & & & & \\ 1 & 0 & & & \\ & 1 & \ddots & & \\ & & \ddots & \ddots & \\ & & & 1 & 0 \end{pmatrix}.$$

Since V_i is invariant under α, it is an A-submodule. So if V_α is indecomposable then there is only one such direct summand. The matrix of α on V_α is then just one such Jordan block. Such a module V_α is indeed indecomposable (see Exercise 8.1). The Jordan block has size m with $m \leq n$ since $\alpha^n = 0$, and each m with $1 \leq m \leq n$ occurs. The summands for different m are not isomorphic, since they have different dimensions. This shows that there are precisely n indecomposable A-modules, up to isomorphism.

More generally, we can determine the representation type of any algebra $K[X]/(f)$ when $f \in K[X]$ is a non-constant polynomial. We consider a module of the form V_α where α is a linear transformation with $f(\alpha) = 0$. The following exercise shows that if the minimal polynomial of α is a power g^t of some irreducible polynomial $g \in K[X]$ and V_α is a cyclic A-module (see Definition 4.4), then V_α is indecomposable. A solution can be found in the appendix.

Exercise 8.1. Assume $A = K[X]/(f)$, where $f \in K[X]$ is a non-constant polynomial, and let V_α be a finite-dimensional A-module.

(a) Assume that V_α is a cyclic A-module, and assume the minimal polynomial of α is equal to g^t, where g is an irreducible polynomial in $K[X]$.

   (i) Explain why the map $T : V_\alpha \to V_\alpha$, $v \mapsto \alpha(v)$, is an A-module homomorphism, and why this also has minimal polynomial g^t.
   (ii) Let $\phi : V_\alpha \to V_\alpha$ be an A-module homomorphism. Show that ϕ is a polynomial in T, that is, ϕ can be written in the form $\phi = \sum_i a_i T^i$ with $a_i \in K$.
   (iii) Show that if $\phi^2 = \phi$ then $\phi = 0$ or $\phi = \mathrm{id}_V$. Deduce that V_α is an indecomposable A-module.

(b) Suppose that with respect to some basis, α has matrix $J_n(\lambda)$, the Jordan block of size n with eigenvalue λ. Explain why the previous implies that V_α is indecomposable.

Let $A = K[X]/(f)$ as above and assume that V_α is a cyclic A-module, generated by a vector b, say. We recall from linear algebra how to describe such a module. Let $h \in K[X]$ be the minimal polynomial of α, we can write it as $h = X^d + \sum_{i=0}^{d-1} c_i X^i$ where $c_i \in K$. Then the set of vectors $\{b, \alpha(b), \ldots, \alpha^{d-1}(b)\}$ is a K-basis of V_α.

The matrix of α with respect to this basis is given explicitly by

$$
C(h) := \begin{pmatrix}
0 & \cdots & & 0 & -c_0 \\
1 & 0 & & 0 & -c_1 \\
0 & 1 & 0 & 0 & -c_2 \\
\vdots & \ddots & \ddots & \ddots & \vdots & \vdots \\
\vdots & & \ddots & \ddots & 0 & -c_{d-2} \\
0 & \cdots\cdots & & 0 & 1 & -c_{d-1}
\end{pmatrix}.
$$

This is sometimes called 'the companion matrix' of h. More generally, one defines the companion matrix for an arbitrary (not necessarily irreducible) monic polynomial h in the same way. Then the minimal polynomial of $C(h)$ is equal to h, by linear algebra.

We can now describe all indecomposable modules for the algebras of the form $K[X]/(f)$.

Lemma 8.4. *Assume* $A = K[X]/(f)$, *where* $f \in K[X]$ *is a non-constant polynomial.*

(a) *The finite-dimensional indecomposable A-modules are, up to isomorphism, precisely given by* V_α, *where* α *has matrix* $C(g^t)$ *with respect to a suitable basis, and where* $g \in K[X]$ *is an irreducible polynomial such that* g^t *divides* f *for some* $t \in \mathbb{N}$.

(b) *A has finite representation type.*

Proof. (a) Assume V_α is an A-module such that α has matrix $C(g^t)$ with respect to some basis, and where g is an irreducible polynomial such that g^t divides f. Then from the shape of $C(g^t)$ we see that V_α is a cyclic A-module, generated by the first vector in this basis. As remarked above, the minimal polynomial of $C(g^t)$, and hence of α, is equal to g^t. By Exercise 8.1 we then know that V_α is indecomposable.

For the converse, take a finite-dimensional indecomposable A-module V_α, so V is a finite-dimensional K-vector space and α is a linear transformation with $f(\alpha) = 0$. We apply two decomposition theorems from linear algebra, a good reference for these is the book by Blyth and Robertson in this series.[1] First, the Primary Decomposition Theorem (see for example §3 in the book by Blyth and Robertson) shows that any non-zero A-module can be written as direct sum of submodules, such that the restriction of α to a summand has minimal polynomial of the form g^m with g an irreducible polynomial and g^m divides f. Second, the Cyclic Decomposition Theorem (see for example §6 in the book by Blyth and Robertson) shows that an A-module whose minimal polynomial is g^m as above, must be a direct

---

[1]T.S. Blyth, E.F. Robertson, *Further Linear Algebra*. Springer Undergraduate Mathematics Series. Springer-Verlag London, Ltd., 2002.

sum of cyclic A-modules. Then the minimal polynomial of a cyclic summand must divide g^m and hence is of the form g^t with $t \le m$.

We apply these results to V_α. Since V_α is assumed to be indecomposable we see that V_α must be cyclic and α has minimal polynomial g^t where g is an irreducible polynomial such that g^t divides f. By the remark preceding the lemma, we know that the matrix of α with respect to some basis is of the form $C(g^t)$.

(b) By part (a), every finite-dimensional indecomposable A-module is of the form V_α, where α has matrix $C(g^t)$ with respect to a suitable basis, and where g is irreducible and g^t divides f. Note that two such modules for the same factor g^t are isomorphic (see Example 2.23 for modules over the polynomial algebra, but the argument carries over immediately to $A = K[X]/(f)$).

There are only finitely many factors of f of the form g^t with g an irreducible polynomial in $K[X]$. Hence A has only finitely many finite-dimensional indecomposable A-modules, that is, A has finite representation type. □

One may ask what a small algebra of infinite representation type might look like. The following is an example:

Lemma 8.5. *For any field K, the 3-dimensional commutative K-algebra*

$$A := K[X, Y]/(X^2, Y^2, XY)$$

has infinite representation type.

Proof. Any A-module V is specified by two K-linear maps $\alpha_X, \alpha_Y : V \to V$, describing the action on V of the cosets of X and Y in A, respectively. These K-linear maps must satisfy the equations

$$\alpha_X^2 = 0, \ \alpha_Y^2 = 0 \ \text{and} \ \alpha_X \alpha_Y = 0 = \alpha_Y \alpha_X. \tag{8.1}$$

For any $n \in \mathbb{N}$ and $\lambda \in K$, we define a $2n$-dimensional A-module $V_n(\lambda)$ as follows. Take the vector space K^{2n}, and let the cosets of X and Y act by the K-linear maps given by the block matrices

$$\alpha_X := \left(\begin{array}{c|c} 0 & 0 \\ \hline E_n & 0 \end{array} \right) \ \text{and} \ \alpha_Y := \left(\begin{array}{c|c} 0 & 0 \\ \hline J_n(\lambda) & 0 \end{array} \right)$$

where E_n is the $n \times n$ identity matrix, and

$$J_n(\lambda) = \begin{pmatrix} \lambda & & & & \\ 1 & \lambda & & & \\ & 1 & \ddots & & \\ & & \ddots & \ddots & \\ & & & 1 & \lambda \end{pmatrix}$$

is a Jordan block of size n for the eigenvalue λ. One checks that α_X and α_Y satisfy the equations in (8.1), that is, this defines an A-module $V_n(\lambda)$.

We will now show that $V_n(\lambda)$ is an indecomposable A-module; note that because of the different dimensions, for a fixed λ, the $V_n(\lambda)$ are pairwise non-isomorphic. (Even more, for fixed n, the modules $V_n(\lambda)$ and $V_n(\mu)$ are not isomorphic for different λ, μ in K; see Exercise 8.5.)

By Lemma 7.3 it suffices to show that the only idempotents in the endomorphism algebra $\text{End}_A(V_n(\lambda))$ are the zero and the identity map. So let $\varphi \in \text{End}_A(V_n(\lambda))$ be an idempotent element. In particular, φ is a K-linear map of $V_n(\lambda)$, so we can write it as a block matrix $\tilde{\varphi} = \left(\begin{array}{c|c} A_1 & A_2 \\ \hline A_3 & A_4 \end{array} \right)$ where A_1, A_2, A_3, A_4 are $n \times n$-matrices over K. Then φ is an A-module homomorphism if and only if this matrix commutes with the matrices α_X and α_Y. By using that $\tilde{\varphi}\alpha_X = \alpha_X\tilde{\varphi}$ we deduce that $A_2 = 0$ and $A_1 = A_4$. Moreover, since $\tilde{\varphi}\alpha_Y = \alpha_Y\tilde{\varphi}$, we get that $A_1 J_n(\lambda) = J_n(\lambda)A_1$. So $\tilde{\varphi} = \left(\begin{array}{c|c} A_1 & 0 \\ \hline A_3 & A_1 \end{array} \right)$, where A_1 commutes with the Jordan block $J_n(\lambda)$.

Assume $\tilde{\varphi}^2 = \tilde{\varphi}$, then in particular $A_1^2 = A_1$. We exploit that A_1 commutes with $J_n(\lambda)$, namely we want to apply Exercise 8.1. Take $f = (X - \lambda)^n$, then A_1 is an endomorphism of the $K[X]/(f)$-module V_α, where α is given by $J_n(\lambda)$. This is a cyclic $K[X]/(f)$-module (generated by the first basis vector). We have $A_1^2 = A_1$, therefore by Exercise 8.1, A_1 is the zero or the identity matrix. In both cases, since $\tilde{\varphi}^2 = \tilde{\varphi}$ it follows that $A_3 = 0$ and hence $\tilde{\varphi}$ is zero or is the identity. This means that 0 and $\text{id}_{V_n(\lambda)}$ are the only idempotents in $\text{End}_A(V_n(\lambda))$. $\qquad\square$

In general, it may be a difficult problem to determine the representation type of a given algebra. There are some methods which reduce this problem to smaller algebras, one of these is the following.

Proposition 8.6. *Let A be a K-algebra and $I \subset A$ a two-sided ideal with $I \neq A$. If the factor algebra A/I has infinite representation type then A has infinite representation type.*

Proof. Note that if $I = 0$ then there is nothing to do.

We have seen in Lemma 2.37 that the A/I-modules are in bijection with those A-modules M such that $IM = 0$. Note that under this bijection, the underlying K-vector spaces remain the same, and the actions are related by $(a + I)m = am$ for all $a \in A$ and $m \in M$. From this it is clear that for any such module M, the A/I-submodules are the same as the A-submodules. In particular, M is an indecomposable A/I-module if and only if it is an indecomposable A-module, and M has finite dimension as an A/I-module if and only if it has finite dimension as an A-module. Moreover, two such modules are isomorphic as A/I-modules if and only if they are isomorphic as A-modules, roughly since they are not changed but just viewed differently. (Details for this are given in Exercise 8.8.) By assumption there are infinitely many pairwise non-isomorphic indecomposable A/I-modules of finite dimension. By the above remarks they also yield infinitely many pairwise non-isomorphic indecomposable A-modules of finite dimension, hence A has infinite representation type. $\qquad\square$

Example 8.7.

(1) Consider the commutative 4-dimensional K-algebra $A = K[X, Y]/(X^2, Y^2)$. Let I be the ideal of A generated by the coset of XY. Then A/I is isomorphic to the algebra $K[X, Y]/(X^2, Y^2, XY)$; this has infinite representation type by Lemma 8.5. Hence A has infinite representation type by Proposition 8.6.

(2) More generally, consider the commutative K-algebra $A = K[X, Y]/(X^r, Y^r)$ for $r \geq 2$, this has dimension r^2. Let I be the ideal generated by the cosets of X^2, Y^2 and XY. Then again A/I is isomorphic to the algebra $K[X, Y]/(X^2, Y^2, XY)$ and hence A has infinite representation type, as in part (1).

The representation type of a direct product of algebras can be determined from the representation type of its factors.

Proposition 8.8. *Let $A = A_1 \times \ldots \times A_r$ be the direct product of K-algebras $A_1, \ldots, A_r$. Then A has finite representation type if and only if all the algebras $A_1, \ldots, A_r$ have finite representation type.*

Proof. We have seen that every A-module M can be written as a direct sum $M = M_1 \oplus \ldots \oplus M_r$ where $M_i = \varepsilon_i M$, with $\varepsilon_i = (0, \ldots, 0, 1_{A_i}, 0, \ldots, 0)$, and M_i is an A-submodule of M (see Lemma 3.30).

Now, assume that M is an indecomposable A-module. So there exists a unique i such that $M = M_i$ and $M_j = 0$ for $j \neq i$. Let $I = A_1 \times \ldots \times A_{i-1} \times 0 \times A_{i+1} \times \ldots \times A_r$, this is an ideal of A and $A/I \cong A_i$. The ideal I acts as zero on M. Hence M is the inflation of an A_i-module (see Remark 2.38). By Lemma 2.37 the submodules of M as an A-module are the same as the submodules as an A_i-module. Hence the indecomposable A-module M is also an indecomposable A_i-module.

Conversely, every indecomposable A_i-module M clearly becomes an indecomposable A-module by inflation. Again, since A-submodules are the same as A_i-submodules, M is indecomposable as an A-module.

So we have shown that the indecomposable A-modules are in bijection with the union of the sets of indecomposable A_i-modules, for $1 \leq i \leq r$. Moreover, one sees that under this bijection, isomorphic modules correspond to isomorphic modules, and modules of finite dimension correspond to modules of finite dimension. Therefore (see Definition 8.1), A has finite representation type if and only if each A_i has finite representation type, as claimed. $\square$

8.2 Representation Type for Group Algebras

In this section we give a complete characterization of group algebras of finite groups which have finite representation type. This is a fundamental result in the representation theory of finite groups. It turns out that the answer is completely determined by the structure of the Sylow p-subgroups of the group. More precisely, a group algebra KG over a field K has finite representation type if and only if K

has characteristic 0 or K has prime characteristic $p > 0$ and a Sylow p-subgroup of G is cyclic. We recall what we need about Sylow p-subgroups of G. Assume the order of G is $|G| = p^a m$, where p does not divide m. Then a Sylow p-subgroup of G is a subgroup of G of order p^a. Sylow's Theorem states that such subgroups exist and any two such subgroups are conjugate in G and in particular are isomorphic.

For example, let G be the symmetric group on three letters, it can be generated by s, r where $s = (1\ 2)$ and $r = (1\ 2\ 3)$ and we have $s^2 = 1, r^3 = 1$ and $srs = r^2$. The order of G is 6. There is one Sylow 3-subgroup, which is the cyclic group generated by r. There are three Sylow 2-subgroups, each of them is generated by a 2-cycle.

If the field K has characteristic zero then the group algebra KG is semisimple, by Maschke's theorem (see Theorem 6.3). But semisimple algebras have finite representation type, see Example 8.3.

We will now deal with group algebras over fields with characteristic $p > 0$. We need to analyse in more detail a few group algebras of p-groups, that is, of groups whose order is a power of p. We start with cyclic p-groups.

Lemma 8.9. *Let G be the cyclic group $G = C_{p^a}$ of order p^a where p is prime and $a \geq 1$, and let K be a field of characteristic p. Then $KG \cong K[T]/(T^{p^a})$.*

Proof. Recall from Example 1.27 that for an arbitrary field K the group algebra KG is isomorphic to $K[X]/(X^{p^a} - 1)$. If K has characteristic p then $X^{p^a} - 1 = (X-1)^{p^a}$, this follows from the usual binomial expansion. If we substitute $T = X - 1$ then $KG \cong K[T]/(T^{p^a})$.   □

Next, we consider the product of two cyclic groups of order p.

Lemma 8.10. *Let K be a field.*

(a) The group algebra of the direct product $C_p \times C_p$ has the form

$$K(C_p \times C_p) \cong K[X_1, X_2]/(X_1^p - 1, X_2^p - 1).$$

(b) Suppose K has characteristic $p > 0$. Then we have

$$K(C_p \times C_p) \cong K[X, Y]/(X^p, Y^p).$$

Proof. (a) We choose $g_1, g_2 \in C_p \times C_p$ generating the two factors of the direct product. Moreover, we set $I := (X_1^p - 1, X_2^p - 1)$, the ideal generated by the polynomials $X_1^p - 1$ and $X_2^p - 1$. Then we consider the following map defined by

$$\Psi : K[X_1, X_2] \to K(C_p \times C_p),\ X_1^{a_1} X_2^{a_2} \mapsto (g_1^{a_1}, g_2^{a_s})$$

and linear extension to arbitrary polynomials. One checks that this map is an algebra homomorphism. Moreover, each $X_i^p - 1$ is contained in the kernel of Ψ, since $g_i^p = 1$ in the cyclic group C_p. Hence, $I \subseteq \ker(\Psi)$. On the other hand, Ψ is clearly surjective, so the isomorphism theorem for algebras (see Theorem 1.26) implies that

$$K[X_1, X_2]/\ker(\Psi) \cong K(C_p \times C_p).$$

Now we compare the dimensions (as K-vector spaces). The group algebra on the right has dimension p^2. The factor algebra $K[X_1, X_2]/I$ also has dimension p^2, the cosets of the monomials $X_1^{a_1} X_2^{a_2}$ with $0 \le a_i \le p - 1$ form a basis. But since $I \subseteq \ker(\Psi)$ these equal dimensions force $I = \ker(\Psi)$, which proves the desired isomorphism.

(b) Let K have characteristic $p > 0$. Then by the binomial formula we have $X_i^p - 1 = (X_i - 1)^p$ for $i = 1, 2$. This means that substituting $X_1 - 1$ for X and $X_2 - 1$ for Y yields a well-defined isomorphism

$$K[X, Y]/(X^p, Y^p) \to K[X_1, X_2]/(X_1^p - 1, X_2^p - 1).$$

By part (a) the algebra on the right-hand side is isomorphic to the group algebra $K(C_p \times C_p)$ and this completes the proof of the claim in (b). □

As a first step towards our main goal we can now find the representation type in the case of finite p-groups, the answer is easy to describe:

Theorem 8.11. *Let p be a prime number, K a field of characteristic p and let G be a finite p-group. Then the group algebra KG has finite representation type if and only if G is cyclic.*

To prove this, we will use the following property, which characterizes when a p-group is not cyclic.

Lemma 8.12. *If a finite p-group G is not cyclic then it has a factor group which is isomorphic to $C_p \times C_p$.*

Proof. In the case when G is abelian, we can deduce this from the general description of a finite abelian group. Indeed, such a group can be written as the direct product of cyclic groups, and if there are at least two factors, both necessarily p-groups, then there is a factor group as in the lemma. For the proof in the general case, we refer to the worked Exercise 8.6. □

Proof of Theorem 8.11. Suppose first that G is cyclic, that is, $G = C_{p^a}$ for some $a \in \mathbb{N}$. Then by Lemma 8.9, we have $KG \cong K[T]/(T^{p^a})$. We have seen in Example 8.3 that this algebra has finite representation type. So KG also has finite representation type, by Remark 8.2.

Conversely, suppose G is not cyclic. We must show that KG has infinite representation type. By Lemma 8.12, the group G has a normal subgroup N such that the factor group $\bar{G} := G/N$ is isomorphic to $C_p \times C_p$. We construct a surjective algebra homomorphism $\psi : KG \to K\bar{G}$, by taking $\psi(g) := gN$ and extending this to linear combinations. This is an algebra homomorphism, that is, it is compatible with products since $(g_1 N)(g_2 N) = g_1 g_2 N$ in the factor group G/N. Clearly ψ is surjective. Let $I = \ker(\psi)$, then $KG/I \cong K\bar{G}$ by the isomorphism theorem of algebras (see Theorem 1.26). We have seen in Lemma 8.10 that

$$K\bar{G} \cong K(C_p \times C_p) \cong K[X, Y]/(X^p, Y^p).$$

By Example 8.7, the latter algebra is of infinite representation type. Then the isomorphic algebras $K\bar{G}$ and KG/I also have infinite representation type, by Remark 8.2. Since the factor algebra KG/I has infinite representation type, by Proposition 8.6 the group algebra KG also has infinite representation type.   □

In order to determine precisely which group algebras have finite representation type, we need new tools to relate modules of a group to modules of a subgroup. They are known as 'restriction' and 'induction', and they are used extensively in the representation theory of finite groups.

The setup is as follows. Assume G is a finite group, and H is a subgroup of G. Take a field K. Then the group algebra KH is a subalgebra of KG (see Example 1.16). One would therefore like to relate KG-modules and KH-modules. (Restriction) If M is any KG-module then by restricting the action of KG to the subalgebra KH, the space M becomes a KH-module, called the restriction of M to KH.

(Induction) There is also the process of induction, this is described in detail in Chap. A, an appendix on induced modules. We briefly sketch the main construction.

Let W be a KH-module. Then we can form the tensor product $KG \otimes_K W$ of vector spaces; this becomes a KG-module, called the induced module, by setting $x \cdot (g \otimes w) = xg \otimes w$ for all $x, g \in G$ and $w \in W$ (and extending linearly). We then consider the K-subspace

$$\mathcal{H} = \text{span}\{gh \otimes w - g \otimes hw \mid g \in G, h \in H, w \in W\}$$

and this is a KG-submodule of $KG \otimes_K W$. The factor module

$$KG \otimes_H W := (KG \otimes_K W)/\mathcal{H}$$

is called the KG-module induced from the KH-module W. For convenience one writes

$$g \otimes_H w := g \otimes w + \mathcal{H} \in KG \otimes_H W.$$

The action of KG on this induced module is given by $x \cdot (g \otimes_H w) = xg \otimes_H w$ and by the choice of $\mathcal{H}$ we have

$$gh \otimes_H w = g \otimes_H hw \text{ for all } g \in G, h \in H \text{ and } w \in W.$$

This can be made explicit. We take a system T of representatives of the left cosets of H in G, that is, $G = \bigcup_{t \in T} tH$ is a disjoint union; and we can assume that the identity element 1 of the group G is contained in T. Suppose W is finite-dimensional, and let $\{w_1, \ldots, w_m\}$ be a basis of W as a K-vector space. Then, as a K-vector space, the induced module $KG \otimes_H W$ has a basis

$$\{t \otimes_H w_i \mid t \in T, i = 1, \ldots, m\};$$

a proof of this fact is given in Proposition A.5 in the appendix.

Example 8.13. Let $G = S_3$, the group of permutations of $\{1, 2, 3\}$. It can be generated by s, r where $s = (1\ 2)$ and $r = (1\ 2\ 3)$. Let H be the subgroup generated by r, so that H is cyclic of order 3. As a system of representatives, we can take $T = \{1, s\}$. Let W be a 1-dimensional KH-module, and fix a non-zero element $w \in W$. Then $rw = \alpha w$ with $\alpha \in K$ and $\alpha^3 = 1$. By the above, the induced module $M := KG \otimes_H W$ has dimension 2. It has basis $\{1 \otimes_H w, s \otimes_H w\}$. We can write down the matrices for the action of s and r on M with respect to this basis. Noting that $rs = sr^2$ we have

$$s \mapsto \begin{pmatrix} 0 & 1 \\ 1 & 0 \end{pmatrix}, \quad r \mapsto \begin{pmatrix} \alpha & 0 \\ 0 & \alpha^2 \end{pmatrix}.$$

We now collect a few properties of restricted and induced modules, and we assume for the rest of this chapter that all modules are finite-dimensional (to identify the representation type, this is all we need, see Definition 8.1).

For the rest of this section we do not distinguish between direct sums and direct products (or external direct sums) of modules and always write $\oplus$, to avoid notational overload (see Definition 2.17 for direct sums of modules which are not a priori submodules of a given module).

Lemma 8.14. *Let K be a field, let G be a finite group and H a subgroup of G.*

(a) *If M and N are finite-dimensional KG-modules then the restriction of the direct sum $M \oplus N$ to KH is the direct sum of the restrictions of M and of N to KH.*

(b) *If W is some finite-dimensional KH-module then it is isomorphic to a direct summand of the restriction of $KG \otimes_H W$ to KH.*

(c) *If M is any finite-dimensional KG-module, then the multiplication map*

$$\mu : KG \otimes_H M \to M, \quad g \otimes_H m \mapsto gm$$

is a well-defined surjective homomorphism of KG-modules.

Proof. (a) This is clear since both M and N are invariant under the action of KH.

(b) As we have seen above, the induced module $KG \otimes_H W$ has a basis

$$\{t \otimes_H w_i \mid t \in T, i = 1, \ldots, m\},$$

where T is a system of representatives of the left cosets of H in G with $1 \in T$, and where $\{w_1, \ldots, w_m\}$ is a K-basis of W.

We now collect suitable basis elements which span KH-submodules. Namely, for $t \in T$ we set $W_t := \text{span}\{t \otimes_H w_i \mid i = 1, \ldots, m\}$. Recall that the KG-action on the induced module is by multiplication on the first factor. When we restrict this action to KH we get for any $h \in H$ that $h \cdot (t \otimes_H w_i) = ht \otimes_H w_i$. The element $ht \in G$ appears in precisely one left coset, that is, there exist unique $s \in T$ and $\tilde{h} \in H$ such that $ht = s\tilde{h}$; note that for the identity element $t = 1$ also $s = 1$. This implies that

$$h \cdot (t \otimes_H w_i) = ht \otimes_H w_i = s\tilde{h} \otimes_H w_i = s \otimes_H \tilde{h}w_i$$

and then one checks that W_1 and $\sum_{t \in T \setminus \{1\}} W_t$ are KH-submodules of $KG \otimes_H W$.

Since these are spanned by elements of a basis of $KG \otimes_H W$ we obtain a direct sum decomposition of KH-modules

$$KG \otimes_H W = W_1 \oplus \left(\sum_{t \in T \setminus \{1\}} W_t \right).$$

Moreover, $W_1 = \mathrm{span}\{1 \otimes_H w_i \mid i = 1, \ldots, m\}$ is as a KH-module isomorphic to W, an isomorphism $W \to W_1$ is given by mapping $w_i \mapsto 1 \otimes_H w_i$ (and extending linearly).

Thus, W is isomorphic to a direct summand (namely W_1) of the restriction of $KG \otimes_H W$ to KH (that is, of $KG \otimes_H W$ considered as a KH-module), as claimed.

(c) Recall the definition of the induced module, namely $KG \otimes_H M = (KG \otimes_K M)/\mathcal{H}$. If $\{m_1, \ldots, m_r\}$ is a K-basis of M then a basis of the tensor product of vector spaces $KG \otimes_K M$ is given by $\{g \otimes m_i \mid g \in G, i = 1, \ldots, r\}$ (see Definition A.1). As is well-known from linear algebra, one can uniquely define a linear map on a basis and extend it linearly. In particular, setting $g \otimes m_i \mapsto gm_i$ defines a linear map $KG \otimes_K M \to M$. To see that this extends to a well-defined linear map $KG \otimes_H M \to M$ one has to verify that the subspace $\mathcal{H}$ is mapped to zero; indeed, for the generators of $\mathcal{H}$ we have $gh \otimes m - g \otimes hm \mapsto (gh)m - g(hm)$, which is zero because of the axioms of a module. This shows that we have a well-defined linear map $\mu : KG \otimes_H M \to M$ as given in the lemma. The map μ is clearly surjective, since for every $m \in M$ we have $\mu(1 \otimes_H m) = 1 \cdot m = m$. Finally, μ is a KG-module homomorphism since for all $x, g \in G$ and $m \in M$ we have

$$\mu(x \cdot (g \otimes_H m)) = \mu(xg \otimes_H m) = (xg)m = x(gm) = x\mu(g \otimes_H m).$$

This completes the proof of the lemma. $\square$

We can now state our first main result in this section, relating the representation types of the group algebra of a group G and the group algebra of a subgroup H.

Theorem 8.15. *Let K be a field and let H be a subgroup of a finite group G. Then the following holds:*

(a) *If KG has finite representation type then KH also has finite representation type.*

(b) *Suppose that the index $|G : H|$ is invertible in K. Then we have the following.*

    (i) *Every finite-dimensional KG-module M is isomorphic to a direct summand of the induced KG-module $KG \otimes_H M$.*

    (ii) *If KH has finite representation type then KG also has finite representation type.*

Remark 8.16. Note that the statement in (ii) above does not hold without the assumption that the index is invertible in the field K. As an example take a field K of prime characteristic $p > 0$ and consider the cyclic group $H = C_p$ as a subgroup of the direct product $G = C_p \times C_p$. Then KC_p has finite representation type, but $K(C_p \times C_p)$ is of infinite representation type, by Theorem 8.11.

Proof. (a) By assumption, KG has finite representation type, so let $M_1, \ldots, M_t$ be representatives for the isomorphism classes of finite-dimensional indecomposable KG-modules. When we restrict these to KH, each of them can be expressed as a direct sum of indecomposable KH-modules. From this we obtain a list of finitely many indecomposable KH-modules.

We want to show that any finite-dimensional indecomposable KH-module occurs in this list (up to isomorphism). So let W be a finite-dimensional indecomposable KH-module. By Lemma 8.14 we know that W is isomorphic to a direct summand of the restriction of $KG \otimes_H W$ to KH. We may write the KG-module $KG \otimes_H W$ as $M_{i_1} \oplus \ldots \oplus M_{i_s}$, where each M_{i_j} is isomorphic to one of $M_1, \ldots, M_t$. We know W is isomorphic to a summand of $M_{i_1} \oplus \ldots \oplus M_{i_s}$ restricted to KH, and this is the direct sum of the restrictions of the M_{i_j} to KH, by the first part of Lemma 8.14. Since W is an indecomposable KH-module, the Krull–Schmidt theorem (Theorem 7.18) implies that there is some j such that W is isomorphic to a direct summand of M_{i_j} restricted to KH. Hence W is one of the modules in our list, as required.

(b) We take as in the proof of Lemma 8.14 a system $T = \{g_1, \ldots, g_r\}$ of representatives for the left cosets of H in G, and $1 \in T$.

(i) Let M be a finite-dimensional KG-module. By part (c) of Lemma 8.14, the multiplication map $\mu : KG \otimes_H M \to M$ is a surjective KG-module homomorphism. The idea for the proof is to construct a suitable KG-module homomorphism $\kappa : M \to KG \otimes_H M$ such that $\mu \circ \kappa = \mathrm{id}_M$ and then to apply Lemma 2.30. We first consider the following map

$$\sigma : M \to KG \otimes_H M, \quad m \mapsto \sum_{i=1}^{r} g_i \otimes_H g_i^{-1} m.$$

(The reader might wonder how to get the idea to use this map. It is not too hard to see that we have an injective KH-module homomorphism $i : M \to KG \otimes_H M$, $m \mapsto 1 \otimes_H m$; the details are given in Proposition A.6 in the appendix. To make it a KG-module homomorphism, one mimics the averaging formula from the proof of Maschke's theorem, see Lemma 6.1, leading to the above map σ.)

We first show that σ is independent of the choice of the coset representatives $g_1, \ldots, g_r$. In fact, any other set of representatives has the form $g_1 h_1, \ldots, g_r h_r$ for some $h_1, \ldots, h_r \in H$. Then the above image of m under σ reads

$$\sum_{i=1}^{r} g_i h_i \otimes_H h_i^{-1} g_i^{-1} m = \sum_{i=1}^{r} g_i h_i h_i^{-1} \otimes_H g_i^{-1} m = \sum_{i=1}^{r} g_i \otimes_H g_i^{-1} m = \sigma(m),$$

where for the first equation we have used the defining relations in the induced module (coming from the subspace $\mathcal{H}$ in the definition of $KG \otimes_H M$).

Next, we show that σ is a KG-module homomorphism. In fact, for every $g \in G$ and $m \in M$ we have

$$\sigma(gm) = \sum_{i=1}^{r} g_i \otimes_H g_i^{-1} gm = \sum_{i=1}^{r} g_i \otimes_H (g^{-1} g_i)^{-1} m.$$

Setting $\tilde{g}_i := g^{-1} g_i$ we get another set of left coset representatives $\tilde{g}_1, \ldots, \tilde{g}_r$. This implies that the above sum can be rewritten as

$$\sigma(gm) = \sum_{i=1}^{r} g_i \otimes_H (g^{-1} g_i)^{-1} m = \sum_{i=1}^{r} g \tilde{g}_i \otimes_H \tilde{g}_i^{-1} m = g(\sum_{i=1}^{r} \tilde{g}_i \otimes_H \tilde{g}_i^{-1} m) = g\sigma(m),$$

where the last equation holds since we have seen above that σ is independent of the choice of coset representatives.

For the composition $\mu \circ \sigma : M \to M$ we obtain for all $m \in M$ that

$$(\mu \circ \sigma)(m) = \mu(\sum_{i=1}^{r} g_i \otimes_H g_i^{-1} m) = \sum_{i=1}^{r} g_i g_i^{-1} m = rm = |G : H|m.$$

So far we have not used our assumption that the index $|G : H|$ is invertible in the field K, but now it becomes crucial. We set $\kappa := \frac{1}{|G:H|}\sigma : M \to M$. Then from the above computation we deduce that $\mu \circ \kappa = \mathrm{id}_M$.

As the final step we can now apply Lemma 2.30 to get a direct sum decomposition $KG \otimes_H M = \mathrm{im}(\kappa) \oplus \ker(\mu)$. The map κ is injective (since $\mu \circ \kappa = \mathrm{id}_M$), so $\mathrm{im}(\kappa) \cong M$ and the claim follows.

(ii) By assumption, KH has finite representation type; let $W_1, \ldots, W_s$ be the finite-dimensional indecomposable KH-modules, up to isomorphism. It suffices to show that every finite-dimensional indecomposable KG-module M is isomorphic to a direct summand of one of the KG-modules $KG \otimes_H W_i$ with $1 \leq i \leq s$. In fact, since these finitely many finite-dimensional modules have only finitely many indecomposable summands, it then follows that there are only finitely many possibilities for M (up to isomorphism), that is, KG has finite representation type.

Consider the KG-module M as a KH-module. We can express it as a direct sum of indecomposable KH-modules, that is, there exist $a_1, \ldots, a_s \in \mathbb{N}_0$ such that

$$M \cong \underbrace{(W_1 \oplus \ldots \oplus W_1)}_{a_1} \oplus \ldots \oplus \underbrace{(W_s \oplus \ldots \oplus W_s)}_{a_s}$$

as a KH-module. Since tensor products commute with finite direct sums (see the worked Exercise 8.9 for a proof in the special case of induced modules used here), we obtain

$$KG \otimes_H M \cong (KG \otimes_H W_1)^{\oplus a_1} \oplus \ldots \oplus (KG \otimes_H W_s)^{\oplus a_s}.$$

By part (i) we know that M is isomorphic to a direct summand of this induced module $KG \otimes_H M$. Since M is indecomposable, the Krull–Schmidt theorem (see Theorem 7.18) implies that M is isomorphic to a direct summand of $KG \otimes_H W_i$ for some $i \in \{1, \ldots, s\}$. $\square$

We apply this now, taking for H a Sylow p-subgroup of G, and we come to the main result of this section.

Theorem 8.17. *Let K be a field and let G be a finite group. Then the following statements are equivalent.*

(i) The group algebra KG has finite representation type.
(ii) Either K has characteristic 0, or K has characteristic $p > 0$ and G has a cyclic Sylow p-subgroup.

Proof. We first prove that (i) implies (ii). So suppose that KG has finite representation type. If K has characteristic zero, there is nothing to prove. So suppose that K has characteristic $p > 0$, and let H be a Sylow p-subgroup of G. Since KG has finite representation type, KH also has finite representation type by Theorem 8.15. But H is a p-group, so Theorem 8.11 implies that H is a cyclic group, which proves part (ii).

Conversely, suppose (ii) holds. If K has characteristic zero then the group algebra KG is semisimple by Maschke's theorem (see Theorem 6.3). Hence KG is of finite representation type by Example 8.3.

So we are left with the case that K has characteristic $p > 0$. By assumption (ii), G has a cyclic Sylow p-subgroup H. The group algebra KH has finite representation type, by Theorem 8.11. Since H is a Sylow p-subgroup, we have $|G| = p^a m$ and $|H| = p^a$, and the index $m = |G : H|$ is not divisible by p and hence invertible in K. Then by Theorem 8.15 (part (b)) we conclude that KG has finite representation type. $\square$

Remark 8.18. Note that in positive characteristic p the condition in part (ii) of the theorem is also satisfied when p does not divide the group order (a Sylow p-subgroup is then the trivial group, which is clearly cyclic). In this case the group algebra KG is semisimple, by Maschke's theorem, which also implies that KG has finite representation type.

Example 8.19.

(1) Consider the symmetric group $G = S_3$ on three letters. Since $|G| = 6$, the group algebra KS_3 is semisimple whenever the characteristic of K is not 2 or 3. In characteristic 2, a Sylow 2-subgroup is given by $\langle(1\,2)\rangle$, the subgroup generated by $(1\,2)$, which is clearly cyclic. Similarly, in characteristic 3, a Sylow 3-subgroup is given by $\langle(1\,2\,3)\rangle$ which is also cyclic. Hence KS_3 has finite representation type for all fields K.

(2) Consider the alternating group $G = A_4$ of order $|A_4| = 12 = 2^2 \cdot 3$. If the field K has characteristic not 2 and 3 then KA_4 is semisimple and hence has finite representation type. Suppose K has characteristic 3; a Sylow 3-subgroup is given by $\langle(1\,2\,3)\rangle$, hence it is cyclic and KA_4 is of finite representation type. However, a Sylow 2-subgroup of A_4 is the Klein four group $\langle(1\,2)(3\,4), (1\,3)(2\,4)\rangle \cong C_2 \times C_2$, which is not cyclic. Hence, KA_4 has infinite representation type if K has characteristic 2.

The strategy of the proof of Theorem 8.15 can be extended to construct all indecomposable modules for a group algebra KG from the indecomposable modules of a group algebra KH when H is a subgroup of G such that the index $|G : H|$ is invertible in the field K. We illustrate this important method with an example.

Example 8.20. Assume G is the symmetric group S_3; we use the notation as in Example 8.13. Assume also that K has characteristic 3. A Sylow 3-subgroup of G has order 3, in fact there is only one in this case, which is H generated by $r = (1\,2\,3)$. Recall from the proof of Theorem 8.15 that every indecomposable KG-module appears as a direct summand of an induced module $KG \otimes_H W$, for some indecomposable KH-module W.

The group algebra KH is isomorphic to $K[T]/(T^3)$, see Lemma 8.9. Recall from the proof of Lemma 8.9 that the isomorphism takes the generator r of H to $T + 1$. The algebra $K[T]/(T^3)$ has three indecomposable modules, of dimensions 1, 2, 3, and the action of the coset of T is given by Jordan blocks $J_n(0)$ of sizes $n = 1, 2, 3$ for the eigenvalue 0, see Example 8.3. If we transfer these indecomposable modules to indecomposable modules for the isomorphic algebra KH, the action of r is given by the matrices $J_n(0) + E_n = J_n(1)$, for $n = 1, 2, 3$, where E_n is the identity matrix.

(1) Let $W = \text{span}\{w\}$ be the 1-dimensional indecomposable KH-module. The action of r on W is given by the matrix $J_1(1) = E_1$, that is, W is the trivial KH-module. From this we see that H acts trivially on the induced KG-module $M = KG \otimes_H W$. In fact, for this it suffices to show that r acts trivially on M: recall from Example 8.13 that $M = \text{span}\{1 \otimes_H w, s \otimes_H w\}$. Clearly $r(1 \otimes_H w) = r \otimes_H w = 1 \otimes_H rw = 1 \otimes_H w$. Note that $rs = sr^{-1}$ in S_3, so $r(s \otimes_H w) = (sr^{-1} \otimes_H w) = s \otimes_H r^{-1}w = s \otimes_H w$.

So we can view M as a module for the group algebra $K(G/H)$. Since G/H has order 2, it is isomorphic to the subgroup $\langle s \rangle$ of S_3 generated by s. So we can also view M as a module for the group algebra $K\langle s \rangle$. As such, it is the direct sum of two 1-dimensional submodules, with K-basis $(1 + s) \otimes_H w$ and $(1 - s) \otimes_H w$, respectively. Thus, from $M = KG \otimes_H W$ we obtain two 1-dimensional (hence simple) KG-modules

$$U_1 := \mathrm{span}\{(1 + s) \otimes_H w\} \quad \text{and} \quad V_1 := \mathrm{span}\{(1 - s) \otimes_H w\},$$

where U_1 is the trivial KG-module and on V_1 the element s acts by multiplication with -1. These two KG-modules are not isomorphic (due to the different actions of s).

(2) Let W_2 be the 2-dimensional indecomposable KH-module. The action of r on W_2 is given by the matrix $J_2(1) = \begin{pmatrix} 1 & 0 \\ 1 & 1 \end{pmatrix}$, so we take a basis for W_2 as $\{b_1, b_2\}$, where $rb_1 = b_1 + b_2$ and $rb_2 = b_2$.

The induced module $KG \otimes_H W_2$ has dimension 4, and a K-basis is given by

$$\{1 \otimes_H b_1, 1 \otimes_H b_2, s \otimes_H b_1, s \otimes_H b_2\},$$

see Proposition A.5. One checks that $KG \otimes_H W_2$ has the two KG-submodules

$$U_2 := \mathrm{span}\{(1+s) \otimes b_1, (1-s) \otimes b_2\} \quad \text{and} \quad V_2 := \mathrm{span}\{(1-s) \otimes b_1, (1+s) \otimes b_2\}.$$

From the above basis it is clear that $KG \otimes_H W_2 = U_2 + V_2$. Moreover, one checks that U_2 and V_2 each have a unique 1-dimensional submodule, namely $\mathrm{span}\{(1-s) \otimes_H b_2\} \cong V_1$ for U_2 and $\mathrm{span}\{(1+s) \otimes_H b_2\} \cong U_1$ for V_2 (where U_1 and V_1 are the simple KG-modules from (1)). Since these are different it follows that $U_2 \cap V_2 = 0$ and hence we have a direct sum decomposition $KG \otimes_H W_2 = U_2 \oplus V_2$.

Moreover, it also implies that U_2 and V_2 are indecomposable KG-modules since the only possible direct sum decomposition would be into two different 1-dimensional submodules.

Finally, U_2 and V_2 are not isomorphic; in fact, an isomorphism would yield an isomorphism between the unique 1-dimensional submodules, but these are isomorphic to V_1 and U_1, respectively.

(3) This leaves us to compute $KG \otimes_H W_3$ where W_3 is the 3-dimensional indecomposable KH-module. Then we can take $W_3 = KH$ (in fact, $KH \cong K[T]/(T^3)$) and $A = K[T]/(T^3)$ is an indecomposable A-module, see Remark 7.2, so KH is an indecomposable KH-module).

We consider the submodules generated by $(1 + s) \otimes_H 1$ and $(1 - s) \otimes_H 1$. More precisely, again using that $rs = sr^{-1}$ in S_3 one can check that

$$U_3 := KG((1+s) \otimes_H 1)$$

$$= \mathrm{span}\{u := (1 + s) \otimes_H 1, v := 1 \otimes_H r + s \otimes_H r^2, w := 1 \otimes_H r^2 + s \otimes_H r\}$$

and

$$V_3 := KG((1 - s) \otimes_H 1)$$

$$= \text{span}\{x := (1 - s) \otimes_H 1, y := 1 \otimes_H r - s \otimes_H r^2, z := 1 \otimes_H r^2 - s \otimes_H r\}.$$

Note that the KG-action on U_3 is given on basis elements by

$$su = u, \quad sv = w, \quad sw = v \text{ and } ru = v, \quad rv = w, \quad rw = u$$

and similarly the KG-action on V_3 is given by

$$sx = -x, \quad sy = -z, \quad sz = -y \text{ and } rx = y, \quad ry = z, \quad rz = x.$$

From this one checks that U_3 and V_3 each have a unique 1-dimensional KG-submodule, namely span$\{u + v + w\} \cong U_1$ for U_3 and span$\{x + y + z\} \cong V_1$ for V_3. From this one deduces that there is a direct sum decomposition $KG = U_3 \oplus V_3$.

Moreover, we claim that U_3 and V_3 are indecomposable KG-modules. For this it suffices to show that they are indecomposable when considered as KH-modules (with the restricted action); in fact, any direct sum decomposition as KG-modules would also be a direct sum decomposition as KH-modules.

Note that U_3 as a KH-module is isomorphic to KH (an isomorphism is given on basis elements by $u \mapsto 1$, $v \mapsto r$ and $w \mapsto r^2$) and this is an indecomposable KH-module. Hence U_3 is an indecomposable KG-module. Similarly, V_3 is an indecomposable KG-module.

Finally, U_3 and V_3 are not isomorphic; in fact, an isomorphism would yield an isomorphism between the unique 1-dimensional submodules, but these are isomorphic to U_1 and V_1, respectively.

According to the proof of Theorem 8.15 we have now shown that the group algebra KS_3 for a field K of characteristic 3 has precisely six indecomposable modules up to isomorphism, two 1-dimensional modules, two 2-dimensional modules and two 3-dimensional modules. Among these only the 1-dimensional modules are simple (we have found that each of the other indecomposable modules has a 1-dimensional submodule).

EXERCISES

8.2. Let $A = K[X]/(f)$, where $f \in K[X]$ is non-constant. Write $f = f_1^{a_1} f_2^{a_2} \ldots f_r^{a_r}$, where the f_i are pairwise coprime irreducible polynomials, and $a_i \geq 1$. Find the number of indecomposable A-modules, up to isomorphism.

8.3. Let A be a K-algebra. Assume M is a finite-dimensional A-module with basis $\mathcal{B} = \mathcal{B}_1 \cup \mathcal{B}_2$, where $\mathcal{B}_1$ and $\mathcal{B}_2$ are non-empty and disjoint. Assume that for each $a \in A$, the matrix of the action of a with respect to this basis has block form

$$\begin{pmatrix} M_1(a) & 0 \\ 0 & M_2(a) \end{pmatrix}.$$

Show that then M is the direct sum $M = M_1 \oplus M_2$, where M_i is the space with basis $\mathcal{B}_i$, and the action of a on M_i is given by the matrix $M_i(a)$.

8.4. For any field K, let $A = K[X, Y]/(X^2, Y^2, XY)$ be the 3-dimensional commutative algebra in Lemma 8.5. Find all ideals of A. Show that any proper factor algebra of A has dimension one or two, and is of finite representation type.

8.5. As in the previous exercise, consider the 3-dimensional commutative K-algebra

$$A = K[X, Y]/(X^2, Y^2, XY).$$

For $n \in \mathbb{N}$ and $\lambda \in K$ let $V_n(\lambda)$ be the $2n$-dimensional A-module defined in the proof of Lemma 8.5. Show that for $\lambda \neq \mu \in K$ the A-modules $V_n(\lambda)$ and $V_n(\mu)$ are not isomorphic. (Hint: You might try the case $n = 1$ first.)

8.6. Assume G is a group of order p^n where p is a prime number and $n \geq 1$. Consider the centre of G, that is,

$$Z(G) := \{g \in G \mid gx = xg \text{ for all } x \in G\}.$$

(a) Show that $Z(G)$ has order at least p. (Hint: note that $Z(G)$ consists of the elements whose conjugacy class has size 1, and that the number of elements with this property must be divisible by p.)

(b) Show that if $Z(G) \neq G$ (that is, G is not abelian) then $G/Z(G)$ cannot be cyclic.

(c) Suppose G is not cyclic. Show that then G has a normal subgroup N such that G/N is isomorphic to $C_p \times C_p$. When G is abelian, this follows from the structure of finite abelian groups. Prove the general case by induction on n.

8.7. Let A be a K-algebra. Suppose $f : M' \to M$ and $g : M \to M''$ are A-module homomorphisms between A-modules such that f is injective, g is surjective, and $\text{im}(f) = \ker(g)$. This is called a *short exact sequence* and it is written as

$$0 \to M' \xrightarrow{f} M \xrightarrow{g} M'' \to 0.$$

Show that for such a short exact sequence the following statements are equivalent.

(i) There exists an A-submodule $N \subseteq M$ such that $M = \ker(g) \oplus N$.

(ii) There exists an A-module homomorphism $\sigma : M'' \to M$ with $g \circ \sigma = \mathrm{id}_{M''}$.

(iii) There exists an A-module homomorphism $\tau : M \to M'$ such that $\tau \circ f = \mathrm{id}_{M'}$.

A short exact sequence satisfying the equivalent conditions (i)–(iii) is called a *split* short exact sequence.

8.8. Assume A is an algebra and I is a two-sided ideal of A with $I \neq A$. Suppose M is an A/I-module, recall that we can view it as an A-module by $am = (a + I)m$ for $a \in A$ and $m \in M$. Assume also that N is an A/I-module. Show that a map $f : M \to N$ is an A-module homomorphism if and only if it is an A/I-module homomorphism. Deduce that M and N are isomorphic as A-modules if and only if they are isomorphic as A/I-modules.

8.9. Let K be a field, and G a finite group with a subgroup H. Show that for every finite-dimensional KH-modules V and W, we have a KG-module homomorphism

$$KG \otimes_H (V \oplus W) \cong (KG \otimes_H V) \oplus (KG \otimes_H W).$$

8.10. Let $G = \{\pm 1, \pm i, \pm j, \pm k\}$ be the quaternion group, as defined in Remark 1.9 (see also Exercise 6.8).

(a) Determine a normal subgroup N of G of order 2 and show that G/N is isomorphic to $C_2 \times C_2$.

(b) For which fields K does the group algebra KG have finite representation type?

8.11. For which fields K does the group algebra KG have finite representation type where G is:

(a) the alternating group $G = A_5$ of even permutations on five letters,

(b) the dihedral group $G = D_n$ of order $2n$ where $n \geq 2$, that is, the symmetry group of the regular n-gon.

Chapter 9
Representations of Quivers

We have seen representations of a quiver in Chap. 2 and we have also seen how to relate representations of a quiver Q over a field K and modules for the path algebra KQ, and that quiver representations and KQ-modules are basically the same (see Sect. 2.5.2). For some tasks, quiver representations are more convenient than modules. In this chapter we develop the theory further and study representations of quivers in detail. In particular, we want to exploit properties which come from the graph structure of the quiver Q.

9.1 Definitions and Examples

Throughout we fix a field K.

We fix a quiver Q, that is, a finite directed graph (see Definition 1.11). We will often assume that Q does not have oriented cycles; recall from Exercise 1.2 that this is equivalent to the path algebra KQ being finite-dimensional.

We consider representations of Q over K and recall the definition from Definition 2.44. From now on we restrict to finite-dimensional representations, so the following is slightly less general than Definition 2.44.

Definition 9.1. Let Q be a quiver with vertex set Q_0 and arrow set Q_1.

(a) A *representation* $\mathcal{M}$ *of* Q *over* K is given by the following data:

- a finite-dimensional K-vector space $M(i)$ for each vertex $i \in Q_0$,
- a K-linear map $M(\alpha) : M(i) \to M(j)$ for each arrow $\alpha \in Q_1$, where $i \xrightarrow{\alpha} j$.

We write such a representation $\mathcal{M}$ as a tuple $\mathcal{M} = ((M(i))_{i \in Q_0}, (M(\alpha))_{\alpha \in Q_1})$, or just $\mathcal{M} = ((M(i))_i, (M(\alpha))_\alpha)$.

© Springer International Publishing AG, part of Springer Nature 2018
K. Erdmann, T. Holm, *Algebras and Representation Theory*, Springer
Undergraduate Mathematics Series, https://doi.org/10.1007/978-3-319-91998-0_9

(b) The *zero representation* $\mathcal{O}$ of Q is the representation where to each vertex
$i \in Q_0$ we assign the zero vector space. A representation $\mathcal{M}$ of Q is called
non-zero if $M(i) \neq 0$ for at least one vertex $i \in Q_0$.

Example 9.2. Let Q be the quiver

$$1 \xrightarrow{\alpha} 2 \xrightarrow{\beta} 3 \xrightarrow{\gamma} 4$$

and define a representation $\mathcal{M}$ of Q over K by

$$M(1) = K, \quad M(2) = K, \quad M(3) = 0, \quad M(4) = K^2, \quad M(\alpha) = \mathrm{id}_K, \quad M(\beta) = 0, \quad M(\gamma) = 0.$$

We write $\mathcal{M}$ using quiver notation as

$$K \xrightarrow{\mathrm{id}_K} K \longrightarrow 0 \longrightarrow K^2.$$

Note that the maps starting or ending at a space which is zero can only be zero maps
and there is no need to write this down.

   In Sect. 2.5.2 we have seen how to view a representation of a quiver Q over the
field K as a module for the path algebra KQ, and conversely how to view a module
for KQ as a representation of Q over K. We recall these constructions here, using
the quiver Q as in Example 9.2.

Example 9.3. Let Q be the quiver as in Example 9.2.

(a) Let $\mathcal{M}$ be as in Example 9.2. We translate the representation $\mathcal{M}$ of Q into
a KQ-module M. The underlying vector space is 4-dimensional, indeed,
according to Proposition 2.46 (a) we take $M = \prod_{i=1}^{4} M(i) = K^4$, where

$$e_1 M = \{(x, 0, 0, 0) \mid x \in K\}$$
$$e_2 M = \{(0, y, 0, 0) \mid y \in K\}$$
$$e_4 M = \{(0, 0, z, w) \mid z, w \in K\}.$$

Then $\alpha(x, y, z, w) = (0, x, 0, 0)$ and β, γ act as zero maps. Note that M is the
direct sum of two KQ-submodules, the summands are $e_1 M \oplus e_2 M$ and $e_4 M$.

(b) We give now an example which starts with a module and constructs from this
a quiver representation. Start with the KQ-module $P = (KQ)e_2$, that is, the
submodule of KQ generated by the trivial path e_2. It has basis $\{e_2, \beta, \gamma\beta\}$.
According to Proposition 2.46 (b), the representation $\mathcal{P}$ of Q corresponding
to the KQ-module P has the following shape. For each $i = 1, 2, 3, 4$ we
set $P(i) = e_i P = e_i(KQ)e_2$. A basis of this K-vector space is given by
paths in Q starting at vertex 2 and ending at vertex i. So we get $P(1) = 0$,

$P(2) = \text{span}\{e_2\}$, $P(3) = \text{span}\{\beta\}$ and $P(4) = \text{span}\{\gamma\beta\}$. Moreover, $P(\beta)$ maps $e_2 \mapsto \beta$ and $P(\gamma)$ maps $\beta \mapsto \gamma\beta$. To arrive at a picture as in Example 9.2, we identify each of the one-dimensional spaces with K, and then we can take the above linear maps as the identity map:

$$0 \longrightarrow K \xrightarrow{\text{id}_K} K \xrightarrow{\text{id}_K} K.$$

The close relation between modules for path algebras and representations of quivers means that our definitions and results on modules can be translated to representations of quivers. In particular, module homomorphisms become the following.

Definition 9.4. Let $Q = (Q_0, Q_1)$ be a quiver, and let $\mathcal{M}$ and $\mathcal{N}$ be representations of Q over K. A *homomorphism* $\varphi : \mathcal{M} \to \mathcal{N}$ of representations consists of a tuple $(\varphi_i)_{i \in Q_0}$ of K-linear maps $\varphi_i : M(i) \to N(i)$ for each vertex $i \in Q_0$ such that for each arrow $i \xrightarrow{\alpha} j$ in Q_1, the following diagram commutes:

$$
\begin{array}{ccc}
M(i) & \xrightarrow{M(\alpha)} & M(j) \\
\varphi_i \downarrow & & \downarrow \varphi_j \\
N(i) & \xrightarrow{N(\alpha)} & N(j)
\end{array}
$$

that is,

$$\varphi_j \circ M(\alpha) = N(\alpha) \circ \varphi_i.$$

Such a homomorphism $\varphi = (\varphi_i)_{i \in Q_0}$ of representations is called an *endomorphism* if $\mathcal{M} = \mathcal{N}$. It is called an *isomorphism* if all φ_i are vector space isomorphisms, and if so, the representations $\mathcal{M}$ and $\mathcal{N}$ are said to be *isomorphic*.

Let $\mathcal{M}$ be a representation of Q, then there is always the endomorphism φ where φ_i is the identity map, for each vertex i. We call this the identity of $\mathcal{M}$; if $\mathcal{M}$ is translated into a module of the algebra KQ then φ corresponds to the identity map.

As an illustration, we compute the endomorphisms of an explicit representation. This will also be used later. It might be surprising that this representation does not have any endomorphisms except scalar multiples of the identity, given that the space at vertex 4 is 2-dimensional.

Lemma 9.5. *Let Q be the quiver*

$$
\begin{array}{ccc}
& 2 & \\
& \downarrow{\scriptstyle \alpha_2} & \\
1 \xrightarrow{\ \alpha_1\ } & 4 & \xleftarrow{\ \alpha_3\ } 3
\end{array}
$$

Moreover, let $\mathcal{M}$ be the representation of Q with $M(i) = K$ for $1 \leq i \leq 3$ and $M(4) = K^2$, and define

$$M(\alpha_1) : K \to K^2, \quad x \mapsto (x, 0)$$

$$M(\alpha_2) : K \to K^2, \quad x \mapsto (0, x)$$

$$M(\alpha_3) : K \to K^2, \quad x \mapsto (x, x).$$

Then every homomorphism $\varphi : \mathcal{M} \to \mathcal{M}$ is a scalar multiple of the identity.

Proof. Take a homomorphism of representations $\varphi : \mathcal{M} \to \mathcal{M}$, that is, $\varphi = (\varphi_i)_{1 \leq i \leq 4}$ where each $\varphi_i : M(i) \to M(i)$ is a K-linear map. For $i = 1, 2, 3$ the map $\varphi_i : K \to K$ is multiplication by a scalar, say $\varphi_i(x) = c_i x$ for $c_i \in K$. Consider the commutative diagram corresponding to the arrow α_1,

$$
\begin{array}{ccc}
M(1) & \xrightarrow{M(\alpha_1)} & M(4) \\
\varphi_1 \downarrow & & \varphi_4 \downarrow \\
M(1) & \xrightarrow[M(\alpha_1)]{} & M(4)
\end{array}
$$

then we have

$$(c_1 x, 0) = (M(\alpha_1) \circ \varphi_1)(x) = (\varphi_4 \circ M(\alpha_1))(x) = \varphi_4(x, 0).$$

Using the other arrows we similarly find $\varphi_4(0, x) = (0, c_2 x)$, and $\varphi_4(x, x) = (c_3 x, c_3 x)$. Using linearity,

$$(c_3 x, c_3 x) = \varphi_4(x, x) = \varphi_4(x, 0) + \varphi_4(0, x) = (c_1 x, 0) + (0, c_2 x) = (c_1 x, c_2 x).$$

Hence for all $x \in K$ we have $c_1 x = c_3 x = c_2 x$ and therefore $c_1 = c_2 = c_3 =: c$. Now we deduce for all $x, y \in K$ that

$$\varphi_4(x, y) = \varphi_4(x, 0) + \varphi_4(0, y) = (cx, 0) + (0, cy) = (cx, cy) = c(x, y),$$

so $\varphi_4 = c \, \mathrm{id}_{K^2}$. Thus we have proved that every homomorphism $\varphi : \mathcal{M} \to \mathcal{M}$ is a scalar multiple of the identity. $\square$

The analogue of a submodule is a 'subrepresentation'.

Definition 9.6. Let $\mathcal{M} = ((M(i))_{i \in Q_0}, (M(\alpha))_{\alpha \in Q_1})$ be a representation of a quiver Q.

(a) A representation $\mathcal{U} = ((U(i))_{i \in Q_0}, (U(\alpha))_{\alpha \in Q_1})$ of Q is a *subrepresentation* of $\mathcal{M}$ if the following holds:

(i) For each vertex $i \in Q_0$, the vector space $U(i)$ is a subspace of $M(i)$.

(ii) For each arrow $i \xrightarrow{\alpha} j$ in Q, the linear map $U(\alpha) : U(i) \longrightarrow U(j)$ is the restriction of $M(\alpha)$ to the subspace $U(i)$.

(b) A non-zero representation $\mathcal{S}$ of Q is called *simple* if its only subrepresentations are $\mathcal{O}$ and $\mathcal{S}$.

Example 9.7. Let Q be a quiver.

(1) Every representation $\mathcal{M}$ of Q has the trivial subrepresentations $\mathcal{O}$ and $\mathcal{M}$.

(2) For each vertex $j \in Q_0$ we have a representation $\mathcal{S}_j$ of Q over K given by

$$S_j(i) = \begin{cases} K & \text{if } i = j \\ 0 & \text{if } i \neq j \end{cases}$$

and $S_j(\alpha) = 0$ for all $\alpha \in Q_1$. Then $\mathcal{S}_j$ is a simple representation.

(3) We consider the quiver $1 \xrightarrow{\alpha} 2$ and the representation $\mathcal{M}$ given by $M(1) = K = M(2)$ and $M(\alpha) = \mathrm{id}_K$. We write this representation in the form

$$K \xrightarrow{\mathrm{id}_K} K.$$

Consider the representation $\mathcal{U}$ given by $U(1) = 0$, $U(2) = K$, $U(\alpha) = 0$. This is a subrepresentation of $\mathcal{M}$. But there is no subrepresentation $\mathcal{T}$ given by $T(1) = K$, $T(2) = 0$, $T(\alpha) = 0$, because the map $T(\alpha)$ is not the restriction of $M(\alpha)$ to $T(1)$.

Exercise 9.1. Let Q be the quiver in Example 9.2, and let $\mathcal{M}$ be the representation

$$K \xrightarrow{\mathrm{id}_K} K \xrightarrow{\mathrm{id}_K} K \xrightarrow{\mathrm{id}_K} K.$$

Find all subrepresentations of $\mathcal{M}$. How many of them are simple?

Translating the description of simple modules for finite-dimensional path algebras (see Sect. 3.4.2, in particular Theorem 3.26) we get the following result.

Theorem 9.8. *Let Q be a quiver without oriented cycles. Then every simple representation of Q is isomorphic to one of the simple representations $\mathcal{S}_j$, with $j \in Q_0$, as defined in Example 9.7 above. Moreover, the representations $\mathcal{S}_j$, with $j \in Q_0$, are pairwise non-isomorphic.*

Instead of translating Theorem 3.26, one could also prove the above theorem directly in the language of representations of quivers, see Exercise 9.7.

For modules of an algebra we have seen in Chap. 7 that the building blocks are the indecomposable modules. So we should spell out what this corresponds to for representations of quivers.

Definition 9.9. Let Q be a quiver and let K be a field.

(1) Let $\mathcal{M} = ((M(i))_i, (M(\alpha))_\alpha)$ be a representation of Q over K, and assume $\mathcal{U}$ and $\mathcal{V}$ are subrepresentations of $\mathcal{M}$. Then $\mathcal{M}$ is the *direct sum* of $\mathcal{U}$ and $\mathcal{V}$ if for each $i \in Q_0$ we have $M(i) = U(i) \oplus V(i)$ as vector spaces. We write $\mathcal{M} = \mathcal{U} \oplus \mathcal{V}$.
(2) A non-zero representation $\mathcal{M}$ of Q is called *indecomposable* if it cannot be expressed as the direct sum $\mathcal{M} = \mathcal{U} \oplus \mathcal{V}$ with non-zero subrepresentations $\mathcal{U}$ and $\mathcal{V}$ of $\mathcal{M}$.

Note that since $\mathcal{U}$ and $\mathcal{V}$ are required to be subrepresentations of $\mathcal{M}$ in the above definition, for each arrow $i \xrightarrow{\alpha} j$ in Q, the linear map $M(\alpha)$ takes $U(i)$ into $U(j)$, and it takes $V(i)$ into $V(j)$ (see Definition 9.6).

Remark 9.10. There is a related notion of a direct product (or external direct sum) of representations of a quiver Q, see Exercise 9.13.

Exercise 9.2. Let Q be the quiver $1 \xrightarrow{\alpha} 2$.

(i) Consider the representation $\mathcal{M}$ of Q as in Example 9.7, so $M(1) = M(2) = K$, and $M(\alpha) = \mathrm{id}_K$. Show that $\mathcal{M}$ is indecomposable.
(ii) Let $\mathcal{N}$ be the representation of Q with $N(1) = N(2) = K$ and $N(\alpha) = 0$. Show that $\mathcal{N}$ decomposes as the direct sum of the subrepresentations $K \longrightarrow 0$ and $0 \longrightarrow K$.

In Chap. 7 we have studied indecomposable modules in detail. In particular, we have established methods to check whether a certain module is indecomposable. We translate some of these to the language of representations of quivers.

The first result is a restatement of Lemma 7.3.

Lemma 9.11. *Let Q be a quiver and $\mathcal{M}$ a non-zero representation of Q over K. Then the representation $\mathcal{M}$ is indecomposable if and only if the only homomorphisms of representations $\varphi : \mathcal{M} \to \mathcal{M}$ with $\varphi^2 = \varphi$ are the zero homomorphism and the identity.*

The condition in this lemma is satisfied in particular if all endomorphisms of $\mathcal{M}$ are scalar multiples of the identity. This gives the following, which is also a consequence of Corollary 7.16 translated into the language of quiver representations. It is the special case where the endomorphism algebra is isomorphic to the field K, hence is a local algebra.

Lemma 9.12. *Let Q be a quiver and $\mathcal{M}$ a non-zero representation of Q over K. Suppose that every homomorphism of representations $\varphi : \mathcal{M} \to \mathcal{M}$ is a scalar multiple of the identity. Then the representation $\mathcal{M}$ is indecomposable.*

9.2 Representations of Subquivers

When studying representations of quivers it is often useful to relate representations of a quiver to representations of a 'subquiver', a notion we now define.

Definition 9.13. Assume Q is a quiver with vertex set Q_0 and arrow set Q_1.

(a) A *subquiver* of Q is a quiver $Q' = (Q'_0, Q'_1)$ such that $Q'_0 \subseteq Q_0$ and $Q'_1 \subseteq Q_1$.

(b) A subquiver Q' of Q as above is called a *full subquiver* if for any two vertices $i, j \in Q'_0$ all arrows $i \xrightarrow{\alpha} j$ of Q are also arrows in Q'.

Note that since Q' must be a quiver, it is part of the definition that in a subquiver the starting and end points of any arrow are also in the subquiver (see Definition 1.11). Thus one cannot choose arbitrary subsets $Q'_1 \subseteq Q_1$ in the above definition.

Example 9.14. Let Q be the quiver

$$1 \underset{\beta}{\overset{\alpha}{\rightrightarrows}} 2 \xleftarrow{\gamma} 3$$

We determine the subquivers Q' of Q with vertex set $Q'_0 = \{1, 2\}$. For the arrow set we have the following possibilities: $Q'_1 = \emptyset$, $Q'_1 = \{\alpha\}$, $Q'_1 = \{\beta\}$ and $Q'_1 = \{\alpha, \beta\}$. Of these, only the last quiver is a full subquiver. However, by the preceding remark we cannot choose $Q'_1 = \{\alpha, \gamma\}$ since the vertex 3 is not in Q'_0.

Given a quiver Q with a subquiver Q', we want to relate representations of Q with representations of Q'. For our purposes two constructions will be particularly useful. We first present the 'restriction' of a representation of Q to a representation of Q'. Starting with a representation of Q', we then introduce the 'extension by zero' which produces a representation of Q.

Definition 9.15. Let $Q = (Q_0, Q_1)$ be a quiver and $Q' = (Q'_0, Q'_1)$ a subquiver of Q. If

$$\mathcal{M} = ((M(i))_{i \in Q_0}, (M(\alpha))_{\alpha \in Q_1})$$

is a representation of Q then we can restrict $\mathcal{M}$ to Q'. That is, we define a representation $\mathcal{M}'$ of Q' by

$$\mathcal{M}' := ((M(i))_{i \in Q'_0}, (M(\alpha))_{\alpha \in Q'_1}).$$

The representation $\mathcal{M}'$ is called the *restriction* of $\mathcal{M}$ to Q'.

For example, if Q and $\mathcal{M}$ are as in Example 9.2, and Q' is the subquiver

$$2 \xrightarrow{\;\beta\;} 3 \xrightarrow{\;\gamma\;} 4$$

then $\mathcal{M}'$ is the representation

$$K \longrightarrow 0 \longrightarrow K^2.$$

Conversely, suppose we have a representation of a subquiver Q' of a quiver Q. Then we can extend it to a representation of Q, by assigning arbitrary vector spaces and linear maps to the vertices and arrows which are not in the subquiver Q'. So there are many ways to extend a representation of Q' to one of Q (if $Q' \neq Q$). Perhaps the easiest construction is to extend 'by zero'.

Definition 9.16. Suppose Q is a quiver, which has a subquiver $Q' = (Q'_0, Q'_1)$. Suppose

$$\mathcal{M}' = ((M'(i))_{i \in Q'_0}, (M'(\alpha))_{\alpha \in Q'_1})$$

is a representation of Q'. Then we define a representation $\mathcal{M}$ of Q by

$$M(i) := \begin{cases} M'(i) & \text{if } i \in Q'_0 \\ 0 & \text{if } i \notin Q'_0 \end{cases} \quad \text{and} \quad M(\alpha) := \begin{cases} M'(\alpha) & \text{if } \alpha \in Q'_1 \\ 0 & \text{if } \alpha \notin Q'_1. \end{cases}$$

This defines a representation of Q, which we call the *extension by zero* of $\mathcal{M}'$.

Remark 9.17. Let Q be a quiver and let Q' be a subquiver of Q. Suppose that $\mathcal{M}'$ is a representation of Q' and let $\mathcal{M}$ be its extension by zero to Q. It follows directly from Definitions 9.15 and 9.16 that the restriction of $\mathcal{M}$ to Q' is equal to $\mathcal{M}'$. In other words, extension by zero followed by restriction acts as identity on representations of Q'.

But in general, first restricting a representation $\mathcal{M}$ of Q to Q' and then extending it by zero does not give back $\mathcal{M}$, because on vertices which are not in Q' non-zero vector spaces in $\mathcal{M}$ are replaced by 0.

As we have seen, indecomposable representations are the building blocks for arbitrary representations of quivers (just translate the results on modules for path algebras in Chap. 7 to the language of representations of quivers). The extension by zero is particularly useful because it preserves indecomposability.

Lemma 9.18. *Assume Q' is a subquiver of a quiver Q.*

(a) Suppose $\mathcal{M}'$ is a representation of Q' and $\mathcal{M}$ is its extension by zero to Q. If $\mathcal{M}'$ is indecomposable then so is $\mathcal{M}$.

(b) Suppose $\mathcal{M}'$ and $\mathcal{N}'$ are non-isomorphic indecomposable representations of Q'. Let $\mathcal{M}$ and $\mathcal{N}$ be their extensions by zero. Then $\mathcal{M}$ and $\mathcal{N}$ are non-isomorphic representations of Q.

Proof. (a) Suppose we had $\mathcal{M} = \mathcal{U} \oplus \mathcal{V}$ with subrepresentations $\mathcal{U}, \mathcal{V}$ of $\mathcal{M}$. According to Definition 9.9 we have to show that one of $\mathcal{U}$ and $\mathcal{V}$ is the zero representation.

Note that the restriction of $\mathcal{M}$ to Q' is $\mathcal{M}'$ (see Remark 9.17). We call $\mathcal{U}'$ the restriction of $\mathcal{U}$ to Q' and similarly define $\mathcal{V}'$. Then $\mathcal{M}' = \mathcal{U}' \oplus \mathcal{V}'$ because direct sums are compatible with restriction. But $\mathcal{M}'$ is indecomposable, therefore one of $\mathcal{U}'$ or $\mathcal{V}'$ is the zero representation, that is, $U(i) = 0$ for all $i \in Q_0'$ or $V(i) = 0$ for all $i \in Q_0'$.

On the other hand, since $\mathcal{U}$ and $\mathcal{V}$ are subrepresentations and $M(i) = 0$ for $i \notin Q_0'$ (by Definition 9.16), also $U(i) = 0$ and $V(i) = 0$ for all $i \notin Q_0'$.

In total, we get that $U(i) = 0$ for all $i \in Q_0$ or $V(i) = 0$ for all $i \in Q_0$, that is, one of $\mathcal{U}$ or $\mathcal{V}$ is the zero representation and hence $\mathcal{M}$ is indecomposable.

(b) Assume for a contradiction that there is an isomorphism $\varphi : \mathcal{M} \to \mathcal{N}$ of representations. Then, again using Remark 9.17, if we restrict to vertices and arrows in Q' we get a homomorphism $\varphi' : \mathcal{M}' \to \mathcal{N}'$. Moreover, since each $\varphi_i : M(i) \to N(i)$ is an isomorphism, φ_i' is also an isomorphism for each vertex i of Q'. This means that φ' is an isomorphism of representations, and $\mathcal{M}'$ is isomorphic to $\mathcal{N}'$, a contradiction. $\qquad\square$

There is a very useful reduction: When studying representations of a quiver, it is usually enough to study quivers which are connected. This is a consequence of the following:

Lemma 9.19. *Assume Q is a quiver which can be expressed as a disjoint union $Q = Q' \cup Q''$ of subquivers with no arrows between Q' and Q''. Then the indecomposable representations of Q are precisely the extensions by zero of the indecomposable representations of Q' and the indecomposable representations of Q''.*

Proof. Let $\mathcal{M}'$ be an indecomposable representation of Q'. We extend it by zero and then get an indecomposable representation of Q, by Lemma 9.18. Similarly any indecomposable representation of Q'' extends to one for Q.

Conversely, let $\mathcal{M}$ be any representation of Q. We show that $\mathcal{M}$ can be expressed as a direct sum. By restriction of $\mathcal{M}$ (see Definition 9.15) we get representations $\mathcal{M}'$ of Q' and $\mathcal{M}''$ of Q''. Now let $\mathcal{U}$ be the extension by zero of $\mathcal{M}'$, a representation of Q, and let $\mathcal{V}$ be the extension by zero of $\mathcal{M}''$, also a representation of Q. We claim that $\mathcal{M} = \mathcal{U} \oplus \mathcal{V}$.

Take a vertex i in Q. If $i \in Q'$ then $U(i) = M'(i) = M(i)$ and $V(i) = 0$ and therefore $M(i) = U(i) \oplus V(i)$. Similarly, if $i \in Q''$ we have $U(i) = 0$ and $V(i) = M''(i) = M(i)$ and $M(i) = U(i) \oplus V(i)$. Moreover, if α is an arrow of Q then it is either in Q' or in Q'', since by assumption there are no arrows between Q' and Q''. If it is in Q' then $U(\alpha) = M'(\alpha) = M(\alpha)$ and $V(\alpha) = 0$, so the map $M(\alpha)$

is compatible with the direct sum decomposition, and the same holds if α is in Q''. This shows that $\mathcal{M} = \mathcal{U} \oplus \mathcal{V}$.

Assume now that $\mathcal{M}$ is an indecomposable representation of Q. By the above, we have $\mathcal{M} = \mathcal{U} \oplus \mathcal{V}$, therefore one of $\mathcal{U}$ or $\mathcal{V}$ must be the zero representation. Say $\mathcal{U}$ is the zero representation, that is, $\mathcal{M}$ is the extension by zero of $\mathcal{M}''$, a representation of Q''. We claim that $\mathcal{M}''$ must be indecomposable: Suppose we had $\mathcal{M}'' = \mathcal{X}'' \oplus \mathcal{Y}''$ with subrepresentations $\mathcal{X}''$ and $\mathcal{Y}''$. Then we extend $\mathcal{X}''$ and $\mathcal{Y}''$ by zero and obtain a direct sum decomposition $\mathcal{M} = \mathcal{X} \oplus \mathcal{Y}$. Since $\mathcal{M}$ is indecomposable, one of $\mathcal{X}$ or $\mathcal{Y}$ is the zero representation. But since these are obtained as extensions by zero this implies that one of $\mathcal{X}''$ or $\mathcal{Y}''$ is the zero representation. Therefore, $\mathcal{M}''$ is indecomposable, as claimed. $\square$

9.3   Stretching Quivers and Representations

There are further methods to relate representations of different quivers. We will now present a general construction which will be very useful later. This construction works for quivers without loops; for simplicity we consider from now on only quivers without oriented cycles. Recall that the corresponding path algebras are then finite-dimensional, see Exercise 1.2.

Consider two quivers Q and $\widetilde{Q}$ where $\widetilde{Q}$ is obtained from Q by replacing one vertex i of Q by two vertices i_1, i_2 and one arrow, $i_1 \overset{\gamma}{\longrightarrow} i_2$, and by distributing the arrows adjacent to i between i_1 and i_2. The following definition makes this construction precise.

Definition 9.20. Let Q be a quiver without oriented cycles and i a fixed vertex. Let T be the set of all arrows adjacent to i, and suppose $T = T_1 \cup T_2$, a disjoint union. Define $\widetilde{Q}$ to be the quiver obtained from Q as follows.

(i)   Replace vertex i by $i_1 \overset{\gamma}{\longrightarrow} i_2$ (where i_1, i_2 are different vertices);
(ii) Join the arrows in T_1 to i_1;
(iii) Join the arrows in T_2 to i_2.

In (ii) and (iii) we keep the original orientation of the arrows. We call the new quiver $\widetilde{Q}$ a *stretch* of Q.

By assumption, Q does not have loops, so any arrow adjacent to i either starts at i or ends at i but not both, and it belongs either to T_1 or to T_2. Note that if T is large then there are many possible stretches of a quiver Q at a given vertex i, coming from different choices of the sets T_1 and T_2.

We illustrate the general construction from Definition 9.20 with several examples.

Example 9.21.

(1) Let Q be the quiver $1 \longrightarrow 2$. We stretch this quiver at vertex 2 and take $T_2 = \emptyset$, and we get the quiver

$$1 \longrightarrow 2_1 \xrightarrow{\gamma} 2_2.$$

If we take $T_1 = \emptyset$ then we obtain

$$1 \longrightarrow 2_2 \xleftarrow{\gamma} 2_1.$$

(2) Let Q be the quiver $1 \longrightarrow 2 \longrightarrow 3$. Stretching Q at vertex 2, and choosing $T_2 = \emptyset$, we obtain the following quiver

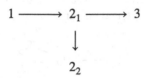

(3) Let Q be the quiver

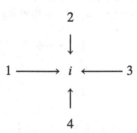

There are several stretches of Q at the vertex i, for example we can get the quiver

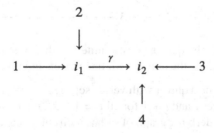

or the quiver

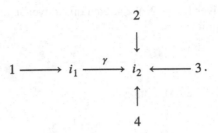

(4) Let Q be the Kronecker quiver $1 \overset{\alpha}{\underset{\beta}{\rightrightarrows}} 2$.

Stretching the Kronecker quiver at vertex 1 and choosing $T_1 = \emptyset$ and $T_2 = \{\alpha, \beta\}$ gives the stretched quiver

$$1_1 \overset{\gamma}{\longrightarrow} 1_2 \overset{\alpha}{\underset{\beta}{\rightrightarrows}} 2 .$$

Alternatively, if we stretch the Kronecker quiver at vertex 1 and choose $T_1 = \{\alpha\}$, $T_2 = \{\beta\}$ then we obtain the following triangle-shaped quiver

$$1_1 \overset{\alpha}{\longrightarrow} 2$$
$$\gamma \downarrow \quad \nearrow \beta$$
$$1_2$$

Exercise 9.3. Let $\widetilde{Q}$ be a quiver with vertex set $\{1, \ldots, n\}$ and $n - 1$ arrows such that for each i with $1 \leq i \leq n - 1$, there is precisely one arrow between vertices i and $i + 1$, with arbitrary orientation. That is, the underlying graph of $\widetilde{Q}$ has the shape

$$1 \longrightarrow 2 \longrightarrow 3 - \cdots - n - 1 \longrightarrow n$$

Show that one can get the quiver $\widetilde{Q}$ by a finite number of stretches starting with the one-vertex quiver, that is, the quiver with one vertex and no arrows.

Exercise 9.4. Let $\widetilde{Q}$ be a quiver with vertex set $\{1, \ldots, n + 1\}$ such that there is one arrow between vertices i and $i + 1$ for all $i = 1, \ldots, n$ and an arrow between $n + 1$ and 1. That is, the underlying graph of $\widetilde{Q}$ has a circular shape

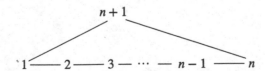

Suppose that the arrows of $\tilde{Q}$ are oriented so that $\tilde{Q}$ is not an oriented cycle. Show that one can obtain $\tilde{Q}$ by a finite number of stretches starting with the Kronecker quiver.

So far, stretching a quiver as in Definition 9.20 is a combinatorial construction which produces new quivers from given ones. We can similarly stretch representations. Roughly speaking, we replace the vector space $M(i)$ in a representation $\mathcal{M}$ of Q by two copies of $M(i)$ with the identity map between them, distributing the $M(\alpha)$ with α adjacent to i so that we get a representation of $\tilde{Q}$ and keeping the rest as it is.

Definition 9.22. Let Q be a quiver without oriented cycles and let $\tilde{Q}$ be the quiver obtained from Q by stretching at a fixed vertex i, with a new arrow $i_1 \xrightarrow{\gamma} i_2$ and where the arrows adjacent to i are the disjoint union $T = T_1 \cup T_2$, see Definition 9.20. Given a representation $\mathcal{M}$ of Q, define $\widetilde{\mathcal{M}}$ to be the representation of $\tilde{Q}$ by

$$\widetilde{M}(i_1) = M(i) = \widetilde{M}(i_2), \ \widetilde{M}(j) = M(j) \ \text{(for } j \neq i)$$

$$\widetilde{M}(\gamma) = \mathrm{id}_{M(i)}, \ \widetilde{M}(\alpha) = M(\alpha) \ \text{(for } \alpha \text{ any arrow of } Q).$$

Note that if α is in T_1 then $\widetilde{M}(\alpha)$ must start or end at vertex i_1, and similarly for α in T_2.

Example 9.23.

(1) As in Example 9.21 we consider the quiver Q of the form $1 \longrightarrow 2$, and the stretched quiver $\tilde{Q} : 1 \longrightarrow 2_1 \longrightarrow 2_2$. Moreover, let $\mathcal{M}$ be the representation $K \xrightarrow{\mathrm{id}_K} K$ of Q. Then the stretched representation $\widetilde{\mathcal{M}}$ of $\tilde{Q}$ has the form $K \xrightarrow{\mathrm{id}_K} K \xrightarrow{\mathrm{id}_K} K$. As another example, let $\mathcal{N}$ be the representation $K \longrightarrow 0$ of Q. Then the stretched representation $\widetilde{\mathcal{N}}$ of $\tilde{Q}$ has the form $K \longrightarrow 0 \longrightarrow 0$.

(2) Let Q be the Kronecker quiver and let $\mathcal{M}$ be the representation

$$K \underset{0}{\overset{\mathrm{id}_K}{\rightrightarrows}} K \ .$$

For the stretched quivers appearing in Example 9.21 we then obtain the following stretched representations:

$$K \xrightarrow{\mathrm{id}_K} K \underset{0}{\overset{\mathrm{id}_K}{\rightrightarrows}} K$$

and

$$K \xrightarrow{\text{id}_K} K$$

$$\text{id}_K \downarrow \quad \nearrow 0$$

$$K$$

Our main focus is on indecomposable representations. The following result is very useful because it shows that stretching representations preserves indecomposability.

Lemma 9.24. *Let Q be a quiver without oriented cycles and $\widetilde{Q}$ a quiver obtained by stretching Q. For any representation M of Q we denote by $\widetilde{M}$ the representation of $\widetilde{Q}$ obtained by stretching M (as in Definition 9.22). Then the following holds:*

(a) *If M is an indecomposable representation then $\widetilde{M}$ is also indecomposable.*
(b) *Suppose M and N are representations of Q which are not isomorphic. Then $\widetilde{M}$ and $\widetilde{N}$ are not isomorphic.*

Proof. Assume $\widetilde{Q}$ is obtained from Q by replacing vertex i by $i_1 \xrightarrow{\gamma} i_2$. Take two representations M and N of Q and a homomorphism $\widetilde{\varphi} : \widetilde{M} \to \widetilde{N}$ of the stretched representations. Since $\widetilde{\varphi}$ is a homomorphism of representations we have (see Definition 9.4):

$$\widetilde{\varphi}_{i_2} \circ \widetilde{M}(\gamma) = \widetilde{N}(\gamma) \circ \widetilde{\varphi}_{i_1}.$$

But $\widetilde{M}(\gamma)$ and $\widetilde{N}(\gamma)$ are identity maps by Definition 9.22, and hence $\widetilde{\varphi}_{i_1} = \widetilde{\varphi}_{i_2}$.

This means that we can define a homomorphism $\varphi : M \to N$ of representations by setting $\varphi_i := \widetilde{\varphi}_{i_1} = \widetilde{\varphi}_{i_2}$ and $\varphi_j := \widetilde{\varphi}_j$ for $j \neq i$. One checks that the relevant diagrams as in Definition 9.4 commute; this follows since the corresponding diagrams for $\widetilde{\varphi}$ commute, and since $\widetilde{\varphi}_{i_1} = \widetilde{\varphi}_{i_2}$.

With this preliminary observation we will now prove the two assertions.

(a) Consider the case $M = N$. To show that $\widetilde{M}$ is indecomposable it suffices by Lemma 9.11 to show that if $\widetilde{\varphi}^2 = \widetilde{\varphi}$ then $\widetilde{\varphi}$ is zero or the identity. By the above definition of the homomorphism φ we see that if $\widetilde{\varphi}^2 = \widetilde{\varphi}$ then also $\varphi^2 = \varphi$. By assumption, M is indecomposable and hence, again by Lemma 9.11, φ is zero or the identity. But then it follows directly from the definition of φ that $\widetilde{\varphi}$ is also zero or the identity.

(b) Assume M and N are not isomorphic. Suppose for a contradiction that $\widetilde{\varphi} : \widetilde{M} \to \widetilde{N}$ is an isomorphism of representations, that is, all linear maps $\widetilde{\varphi}_j : \widetilde{M}(j) \to \widetilde{N}(j)$ are isomorphisms. Then all linear maps $\varphi_j : M(j) \to N(j)$ are also isomorphisms and hence $\varphi : M \to N$ is an isomorphism, a contradiction.   □

9.4 Representation Type of Quivers

We translate modules over the path algebra to representations of quivers, and the Krull–Schmidt theorem translates as well. That is, every (finite-dimensional) representation of a quiver Q is a direct sum of indecomposable representations, unique up to isomorphism and labelling. Therefore it makes sense to define the representation type of a quiver.

Recall that we have fixed a field K and that we consider only finite-dimensional representations of quivers over K, see Definition 9.1. Moreover, we assume throughout that quivers have no oriented cycles; this allows us to apply the results of Sect. 9.3.

Definition 9.25. A quiver Q is said to be of *finite representation type* over K if there are only finitely many indecomposable representations of Q, up to isomorphism. Otherwise, we say that the quiver has *infinite representation type* over K.

By our Definition 9.1, a representation of Q always corresponds to a finite-dimensional KQ-module. In addition, we assume Q has no oriented cycles and hence KQ is finite-dimensional. Therefore the representation type of Q is the same as the representation type of the path algebra KQ, as in Definition 8.1.

In most situations, our arguments will not refer to a particular field K, so we often just speak of the representation type of a quiver, without mentioning the underlying field K explicitly.

For determining the representation type of quivers there are some reductions which follow from the work done in previous sections.

Given a quiver Q, since we have seen in Sect. 9.2 that we can relate indecomposable representations of its subquivers to indecomposable representations of Q, we might expect that there should be a connection between the representation type of subquivers with that of Q.

Lemma 9.26. *Assume Q' is a subquiver of a quiver Q. If Q' has infinite representation type then Q also has infinite representation type.*

Proof. This follows directly from Lemma 9.18. □

Furthermore, to identify the representation type, it is enough to consider connected quivers.

Lemma 9.27. *Assume a quiver Q is the disjoint union of finitely many subquivers $Q^{(1)}, \ldots, Q^{(r)}$. Then Q has finite representation type if and only if all subquivers $Q^{(1)}, \ldots, Q^{(r)}$ have finite representation type.*

Proof. This follows from Lemma 9.19, by induction on r. □

Example 9.28.

(1) The smallest connected quiver without oriented cycles consists of just one vertex. A representation of this quiver is given by assigning a (finite-dimensional)

vector space to this vertex. Any vector space has a basis, hence it is a direct sum of 1-dimensional subspaces; each subspace is a representation of the one-vertex quiver. So there is just one indecomposable representation of the one-vertex quiver, it is 1-dimensional. In particular, the one-vertex quiver has finite representation type.

(2) Let Q be the quiver $1 \xrightarrow{\alpha} 2$. We will determine explicitly its indecomposable representations. This will show that Q has finite representation type.

Let X be an arbitrary representation of Q, that is, we have two finite-dimensional vector spaces $X(1)$ and $X(2)$, and a linear map $T = X(\alpha) : X(1) \to X(2)$. We exploit the proof of the rank-nullity theorem from linear algebra.

Choose a basis $\{b_1, \ldots, b_n\}$ of the kernel $\ker(T)$, and extend it to a basis of $X(1)$, say by $\{c_1, \ldots, c_r\}$. Then the image $\operatorname{im}(T)$ has basis $\{T(c_1), \ldots, T(c_r)\}$, by the proof of the rank-nullity theorem. Extend this set to a basis of $X(2)$, say by $\{d_1, \ldots, d_s\}$. With this, we aim at expressing X as a direct sum of subrepresentations.

For each basis vector b_i of the kernel of T we get a subrepresentation $\mathcal{B}_i$ of X of the form

$$\operatorname{span}\{b_i\} \longrightarrow 0.$$

This is a subrepresentation, since the restriction of $T = X(\alpha)$ to $\operatorname{span}\{b_i\}$ maps b_i to zero. Each of these subrepresentations is isomorphic to the simple representation $\mathcal{S}_1$ as defined in Example 9.7.

For each basis vector c_i we get a subrepresentation $\mathcal{C}_i$ of X of the form

$$\operatorname{span}\{c_i\} \longrightarrow \operatorname{span}\{T(c_i)\}$$

where the map is given by T. Each of these representations is indecomposable, in fact it is isomorphic to the indecomposable representation $\mathcal{M}$ which we considered in Exercise 9.2.

For each basis vector d_i we get a subrepresentation $\mathcal{D}_i$ of X of the form

$$0 \longrightarrow \operatorname{span}\{d_i\}$$

which is isomorphic to the simple representation $\mathcal{S}_2$.

From the choice of our basis vectors we know that

$$X(1) = \operatorname{span}\{b_1, \ldots, b_n\} \oplus \operatorname{span}\{c_1, \ldots, c_r\}$$

and that

$$X(2) = \operatorname{span}\{T(c_1), \ldots, T(c_r)\} \oplus \operatorname{span}\{d_1, \ldots, d_s\}.$$

This implies that we have a decomposition

$$\mathcal{X} = \left(\bigoplus_{i=1}^{n} \mathcal{B}_i \right) \oplus \left(\bigoplus_{i=1}^{r} \mathcal{C}_i \right) \oplus \left(\bigoplus_{i=1}^{s} \mathcal{D}_i \right)$$

of $\mathcal{X}$ as a direct sum of subrepresentations.

Now assume that $\mathcal{X}$ is indecomposable, then there is only one summand in total, and hence $\mathcal{X}$ is isomorphic to one of $\mathcal{B}_i \cong \mathcal{S}_1, \mathcal{C}_i \cong \mathcal{M}$ or $\mathcal{D}_i \cong \mathcal{S}_2$.

By Theorem 9.8, the representations $\mathcal{S}_1$ and $\mathcal{S}_2$ are not isomorphic. By comparing dimensions, $\mathcal{M}$ is not isomorphic to $\mathcal{S}_1$ or $\mathcal{S}_2$. So we have proved that the quiver Q has precisely three indecomposable representations, up to isomorphism.

Example 9.29. Let Q be the Kronecker quiver,

$$1 \overset{\alpha}{\underset{\beta}{\rightrightarrows}} 2 \ .$$

We fix an element $\lambda \in K$, and we define a representation C_λ of Q as follows: $C_\lambda(1) = K, C_\lambda(2) = K, C_\lambda(\alpha) = \mathrm{id}_K$ and $C_\lambda(\beta) = \lambda \cdot \mathrm{id}_K$.

We claim that the following holds for these representations of the Kronecker quiver.

(a) C_λ is isomorphic to C_μ if and only if $\lambda = \mu$.
(b) For any $\lambda \in K$, the representation C_λ is indecomposable.

We start by proving (a). Let $\varphi : C_\lambda \to C_\mu$ be a homomorphism of representations, then we have a commutative diagram of K-linear maps

$$
\begin{array}{ccc}
C_\lambda(1) & \xrightarrow{C_\lambda(\alpha)} & C_\lambda(2) \\
\varphi_1 \downarrow & & \downarrow \varphi_2 \\
C_\mu(1) & \xrightarrow{C_\mu(\alpha)} & C_\mu(2)
\end{array}
$$

Since $C_\lambda(\alpha)$ and $C_\mu(\alpha)$ are identity maps, we have $\varphi_1 = \varphi_2$. We also have the commutative diagram

$$
\begin{array}{ccc}
C_\lambda(1) & \xrightarrow{C_\lambda(\beta)} & C_\lambda(2) \\
\varphi_1 \downarrow & & \downarrow \varphi_2 \\
C_\mu(1) & \xrightarrow{C_\mu(\beta)} & C_\mu(2)
\end{array}
$$

Since we already know that $\varphi_1 = \varphi_2$ we obtain that

$$\lambda \varphi_1 = \lambda \varphi_2 = \varphi_2 \circ C_\lambda(\beta) = C_\mu(\beta) \circ \varphi_1 = \mu \varphi_1.$$

If $\lambda \neq \mu$ then $\varphi_1 = 0$ and therefore we cannot have an isomorphism $C_\lambda \to C_\mu$. This proves claim (a).

We now prove (b). Let $\lambda = \mu$, then by the above we have computed an arbitrary homomorphism of representations $\varphi : C_\lambda \to C_\lambda$. Indeed, we have $\varphi_1 = \varphi_2$ and this is a scalar multiple of the identity (since $C_\lambda(1) = K = C_\lambda(2)$). Hence every homomorphism of C_λ is a scalar multiple of the identity homomorphism. This shows that C_λ is indecomposable, by Lemma 9.12.

The previous example shows already that if the field K is infinite, the Kronecker quiver Q has infinite representation type. However, Q has infinitely many indecomposable representations for arbitrary fields, as we will now show in the next example.

Example 9.30. For any $n \geq 1$, we define a representation C of the Kronecker quiver as follows. We take $C(1) = K^n$ and $C(2) = K^n$, and we define $C(\alpha) = \mathrm{id}_{K^n}$, and $C(\beta)$ is the linear map given by $J_n(1)$, the Jordan block of size n with eigenvalue 1. For simplicity, below we identify the maps with their matrices.

We will show that C is indecomposable. Since the above representations have different dimensions for different n, they are not isomorphic, and hence this will prove that the Kronecker quiver Q has infinite representation type for arbitrary fields.

Let $\varphi : C \to C$ be a homomorphism of representations, with K-linear maps φ_1 and φ_2 on K^n corresponding to the vertices of Q (see Definition 9.4).

Then we have a commutative diagram of K-linear maps

$$
\begin{array}{ccc}
C(1) & \xrightarrow{\ C(\alpha)\ } & C(2) \\
\varphi_1 \downarrow & & \varphi_2 \downarrow \\
C(1) & \xrightarrow{\ C(\alpha)\ } & C(2)
\end{array}
$$

Since $C(\alpha)$ is the identity map, it follows that $\varphi_1 = \varphi_2$. We also have a commutative diagram of K-linear maps

$$
\begin{array}{ccc}
C(1) & \xrightarrow{\ C(\beta)\ } & C(2) \\
\varphi_1 \downarrow & & \varphi_2 \downarrow \\
C(1) & \xrightarrow{\ C(\beta)\ } & C(2)
\end{array}
$$

and this gives

$$J_n(1) \circ \varphi_1 = \varphi_2 \circ J_n(1) = \varphi_1 \circ J_n(1).$$

That is, φ_1 is a linear transformation of $V = K^n$ which commutes with $J_n(1)$.

We assume that $\varphi^2 = \varphi$, that is, $\varphi_1^2 = \varphi_1$, and we want to show that φ is either zero or the identity map; then C is indecomposable by Lemma 9.11, and we are done.

We want to apply Exercise 8.1. Take $f = (X - 1)^n \in K[X]$, then $V_{C(\beta)}$ becomes a $K[X]/(f)$-module (see Theorem 2.10); in fact, one checks that $f(J_n(1)) = J_n(0)^n$ is the zero matrix. Note that this is a cyclic $K[X]/(f)$-module (generated by the first basis vector). Now, since φ_1 commutes with $J_n(1)$, the linear map φ_1 is even a $K[X]/(f)$-module homomorphism of $V_{C(\beta)}$. We have $\varphi_1^2 = \varphi_1$, therefore by Exercise 8.1, $\varphi_1 = 0$ or $\varphi_1 = \mathrm{id}_{K^n}$. This shows that φ is either zero or is the identity, and C is indecomposable, as observed above.

Suppose we know that a quiver Q has infinite representation type. Recall that in Sect. 9.3 we have defined how to stretch quivers and representations. We can exploit this and show that the stretch of Q also has infinite representation type.

Lemma 9.31. *Let Q be a quiver without oriented cycles and let $\tilde{Q}$ be a quiver which is a stretch of Q (as in Definition 9.20). If Q has infinite representation type then $\tilde{Q}$ also has infinite representation type.*

Proof. This follows immediately from Lemma 9.24. $\qquad\qquad\qquad\qquad\square$

Example 9.32. We have seen in Example 9.30 that the Kronecker quiver has infinite representation type over any field K. So by Lemma 9.31 every stretch of the Kronecker quiver also has infinite representation type. In particular, the quivers

$$1_1 \longrightarrow 1_2 \; \substack{\longrightarrow \\ \longrightarrow} \; 2$$

and

appearing in Example 9.21 have infinite representation type over any field K.

EXERCISES

9.5. Consider the representation defined in Example 9.2. Show that it is the direct sum of three indecomposable representations.

9.6. Let $Q = (Q_0, Q_1)$ be a quiver and let $\mathcal{M}$ be a representation of Q. Suppose that $\mathcal{M} = \mathcal{U} \oplus \mathcal{V}$ is a direct sum of subrepresentations. For each vertex $i \in Q_0$ let $\varphi_i : M(i) = U(i) \oplus V(i) \to U(i)$ be the linear map given by projecting onto the first summand, and let $\psi_i : U(i) \to M(i) = U(i) \oplus V(i)$ be the inclusion of $U(i)$ into $M(i)$. Show that $\varphi = (\varphi_i)_{i \in Q_0} : \mathcal{M} \to \mathcal{U}$ and $\psi = (\psi_i)_{i \in Q_0} : \mathcal{U} \to \mathcal{M}$ are homomorphisms of representations.

9.7. (This exercise gives an outline of an alternative proof of Theorem 9.8.) Let $Q = (Q_0, Q_1)$ be a quiver without oriented cycles. For each vertex $j \in Q_0$ let $\mathcal{S}_j$ be the simple representation of Q defined in Example 9.7.

    (i) Show that for $j \neq k \in Q_0$ the only homomorphism $\mathcal{S}_j \to \mathcal{S}_k$ of representations is the zero homomorphism. In particular, the different $\mathcal{S}_j$ are pairwise non-isomorphic.

Let $\mathcal{M}$ be a simple representation of Q.

    (ii) Show that there exists a vertex $k \in Q_0$ such that $M(k) \neq 0$ and $M(\alpha) = 0$ for all arrows $\alpha \in Q_1$ starting at k.

    (iii) Let $k \in Q_1$ be as in (ii). Deduce that $\mathcal{M}$ has a subrepresentation $\mathcal{U}$ with $U(k) = M(k)$ and $U(i) = 0$ for $i \neq k$.

    (iv) Show that $\mathcal{M}$ is isomorphic to the simple representation $\mathcal{S}_k$.

9.8. Let Q be a quiver. Let $\mathcal{M}$ be a representation of Q such that for a fixed vertex j of Q we have $M(i) = 0$ for all $i \neq j$. Show that $\mathcal{M}$ is isomorphic to a direct sum of $\dim_K M(j)$ many copies of the simple representation $\mathcal{S}_j$.

    Conversely, check that if a representation $\mathcal{M}$ of Q is isomorphic to a direct sum of copies of $\mathcal{S}_j$ then $M(i) = 0$ for all $i \neq j$.

9.9. Let Q be a quiver and j a sink of Q, that is, no arrow of Q starts at j. Let $\alpha_1, \ldots, \alpha_t$ be all the arrows ending at j. Let $\mathcal{M}$ be a representation of Q.

    (a) Show that $\mathcal{M}$ is a direct sum of subrepresentations, $\mathcal{M} = \mathcal{X} \oplus \mathcal{Y}$, where

        (i) $\mathcal{Y}$ satisfies $Y(k) = M(k)$ for $k \neq j$, and $Y(j) = \sum_{i=1}^{t} \mathrm{im}(M(\alpha_i))$ is the sum of the images of the maps $M(\alpha_i) : M(i) \to M(j)$,

        (ii) $\mathcal{X}$ is isomorphic to the direct sum of copies of the simple representation $\mathcal{S}_j$, and the number of copies is equal to $\dim_K M(j) - \dim_K Y(j)$.

    (b) If $\mathcal{M}$ has a direct summand isomorphic to $\mathcal{S}_j$ then $\sum_{i=1}^{t} \mathrm{im}(M(\alpha_i))$ is a proper subspace of $M(j)$.

9.10. Let Q' be a quiver and j a source of Q', that is, no arrow of Q' ends at j. Let $\beta_1, \ldots, \beta_t$ be the arrows starting at j. Let $\mathcal{N}$ be a representation of Q'.

(a) Consider the subspace $X(j) := \bigcap_{i=1}^{t} \ker(N(\beta_i))$ of $N(j)$. As a K-vector space we can decompose $N(j) = X(j) \oplus Y(j)$ for some subspace $Y(j)$. Show that $\mathcal{N}$ is a direct sum of subrepresentations, $\mathcal{N} = \mathcal{X} \oplus \mathcal{Y}$, where

(i) $\mathcal{Y}$ satisfies $Y(k) = N(k)$ for $k \neq j$, and $Y(j)$ is as above,
(ii) $\mathcal{X}$ is isomorphic to the direct sum of $\dim_K X(j)$ many copies of the simple representation $\mathcal{S}_j$.

(b) If $\mathcal{N}$ has a direct summand isomorphic to $\mathcal{S}_j$ then $\bigcap_{i=1}^{t} \ker(N(\beta_i))$ is a non-zero subspace of $N(j)$.

9.11. Let K be a field and let $Q = (Q_0, Q_1)$ be a quiver. For each vertex $i \in Q_0$ consider the KQ-module $P_i = KQe_i$ generated by the trivial path e_i.

(i) Interpret the KQ-module P_i as a representation $\mathcal{P}_i$ of Q. In particular, describe bases for the vector spaces $P_i(j)$ for $j \in Q_0$. (Hint: Do it first for the last quiver in (4) of Example 9.21 and use this as an illustration for the general case.)
(ii) Suppose that Q has no oriented cycles. Show that the representation $\mathcal{P}_i$ of Q is indecomposable.

9.12. Let $Q = (Q_0, Q_1)$ be a quiver without oriented cycles and suppose that $\mathcal{M} = ((M(i))_{i \in Q_0}, (M(\alpha))_{\alpha \in Q_1})$ is a representation of Q. Show that the following holds.

(a) The representation $\mathcal{M}$ is semisimple (that is, a direct sum of simple subrepresentations) if and only if $M(\alpha) = 0$ for each arrow $\alpha \in Q_1$.
(b) For each vertex $i \in Q_0$ we set

$$\mathrm{soc}_{\mathcal{M}}(i) = \bigcap_{s(\alpha)=i} \ker(M(\alpha))$$

(where $s(\alpha)$ denotes the starting vertex of the arrow α). Then

$$\mathrm{Soc}_{\mathcal{M}} = ((\mathrm{soc}_{\mathcal{M}}(i))_{i \in Q_0}, (S(\alpha) = 0)_{\alpha \in Q_1})$$

is a semisimple subrepresentation of $\mathcal{M}$, called the *socle* of $\mathcal{M}$.
(c) Every semisimple subrepresentation $\mathcal{U}$ of $\mathcal{M}$ is a subrepresentation of the socle $\mathrm{Soc}_{\mathcal{M}}$.

9.13. This exercise is an analogue of Exercise 2.15, in the language of representations of quivers.

Let $\mathcal{M}$ and $\mathcal{N}$ be representations of a quiver Q. The *direct product* (or external direct sum) $\mathcal{P} = \mathcal{M} \times \mathcal{N}$ is the representation of Q with $P(i) = M(i) \times N(i)$ for each vertex i of Q (this is the direct product of vector spaces, that is, the cartesian product with componentwise addition and scalar multiplication). For every arrow α in Q from i to j we set

$P(\alpha) : P(i) \to P(j), (m, n) \mapsto (M(\alpha)(m), N(\alpha)(n))$, and we sometimes also denote this map by $P(\alpha) = M(\alpha) \times N(\alpha)$.

(i) Verify that $\mathcal{P}$ is a representation of Q.

(ii) Check that the following defines a subrepresentation $\widetilde{\mathcal{M}}$ of $\mathcal{P}$. For every vertex i we set $\widetilde{M}(i) = \{(m, 0) \mid m \in M(i)\}$ and for every arrow α from i to j we set $\widetilde{M}(\alpha) : \widetilde{M}(i) \to \widetilde{M}(j), (m, 0) \mapsto (M(\alpha)(m), 0)$.

Similarly we get a subrepresentation $\widetilde{\mathcal{N}}$ from $\mathcal{N}$.

(iii) Show that $\mathcal{P} = \widetilde{\mathcal{M}} \oplus \widetilde{\mathcal{N}}$ is a direct sum of the subrepresentations.

(iv) Consider $\widetilde{\mathcal{M}}$ and $\widetilde{\mathcal{N}}$ as representations of Q. Show that $\widetilde{\mathcal{M}}$ is isomorphic to $\mathcal{M}$ and that $\widetilde{\mathcal{N}}$ is isomorphic to $\mathcal{N}$.

Chapter 10
Diagrams and Roots

Our aim is to determine when a quiver Q with no oriented cycles is of finite representation type. This is answered completely by Gabriel's celebrated theorem, which he proved in the 1970s, and the answer is given in terms of the underlying graph of Q. This graph is obtained by forgetting the orientation of the arrows in the quiver. In this chapter, we describe the relevant graphs and their properties we need; these graphs are known as Dynkin diagrams and Euclidean diagrams, and they occur in many parts of mathematics. We discuss further tools which we will use to prove Gabriel's theorem, such as the Coxeter transformations. The content of this chapter mainly involves basic combinatorics and linear algebra.

We fix a graph Γ; later this will be the underlying graph of a quiver. We sometimes write $\Gamma = (\Gamma_0, \Gamma_1)$, where Γ_0 is the set of vertices and Γ_1 is the set of (unoriented) edges of Γ. All graphs are assumed to be finite, that is, Γ_0 and Γ_1 are finite sets.

10.1 Dynkin Diagrams and Euclidean Diagrams

Gabriel's theorem (which will be proved in the next chapter) states that a connected quiver has finite representation type if and only if the underlying graph Γ is one of the *Dynkin diagrams* of types A_n for $n \geq 1$, D_n for $n \geq 4$, E_6, E_7, E_8, which we define in Fig. 10.1.

We have seen some small special cases of Gabriel's theorem earlier in the book. Namely, a quiver of type A_1 (that is, the one-vertex quiver) has only one indecomposable representation by Example 9.28; in particular, it is of finite representation type. Moreover, also in Example 9.28 we have shown that the quiver $1 \longrightarrow 2$ has finite representation type; note that this quiver has as underlying graph a Dynkin diagram of type A_2.

© Springer International Publishing AG, part of Springer Nature 2018
K. Erdmann, T. Holm, *Algebras and Representation Theory*, Springer
Undergraduate Mathematics Series, https://doi.org/10.1007/978-3-319-91998-0_10

Fig. 10.1 Dynkin diagrams
of types A, D, E. The index
gives the number of vertices
in each diagram

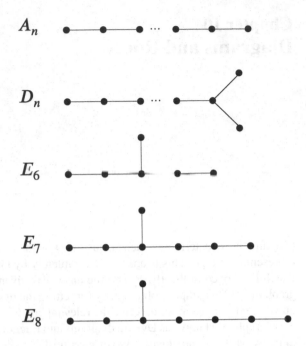

To deal with the case when Γ is not a Dynkin diagram, we will only need a small list of graphs. These are the *Euclidean diagrams*, sometimes also called *extended Dynkin diagrams*. They are shown in Fig. 10.2, and are denoted by $\widetilde{A}_n$ for $n \geq 1$, $\widetilde{D}_n$ for $n \geq 4$, and $\widetilde{E}_6$, $\widetilde{E}_7$, $\widetilde{E}_8$. For example, the Kronecker quiver is a quiver with underlying graph a Euclidean diagram of type $\widetilde{A}_1$; and we have seen already in Example 9.30 that the Kronecker quiver has infinite representation type.

We refer to graphs in Fig. 10.1 as graphs of type A, D, or E. We say that a quiver has Dynkin type if its underlying graph is one of the graphs in Fig. 10.1. Similarly, we say that a quiver has Euclidean type if its underlying graph belongs to the list in Fig. 10.2.

In analogy to the definition of a subquiver in Definition 9.13, a *subgraph* $\Gamma' = (\Gamma_0', \Gamma_1')$ of a graph Γ is a graph which consists of a subset $\Gamma_0' \subseteq \Gamma_0$ of the vertices of Γ and a subset $\Gamma_1' \subseteq \Gamma_1$ of the edges of Γ.

The following result shows that we might not need any other graphs than Dynkin and Euclidean diagrams.

Lemma 10.1. *Assume Γ is a connected graph. If Γ is not a Dynkin diagram then Γ has a subgraph which is a Euclidean diagram.*

Proof. Assume Γ does not have a Euclidean diagram as a subgraph, we will show that then Γ is a Dynkin diagram.

The Euclidean diagrams of type $\widetilde{A}_n$ are just the cycles; so Γ does not contain a cycle; in particular, it does not have a multiple edge. Since Γ is connected by assumption, it must then be a tree.

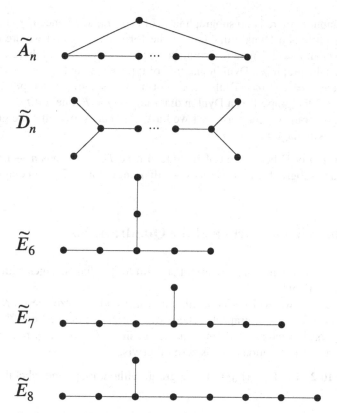

Fig. 10.2 The Euclidean diagrams of types $\widetilde{A}$, $\widetilde{D}$, $\widetilde{E}$. The index plus 1 gives the number of vertices in each diagram

The graph Γ does not have a subgraph of type $\widetilde{D}_4$ and hence every vertex in Γ is adjacent to at most three other vertices. Moreover, since there is no subgraph of type $\widetilde{D}_n$ for $n \geq 5$, at most one vertex in Γ is adjacent to three other vertices. In total, this means that the graph Γ is of the form

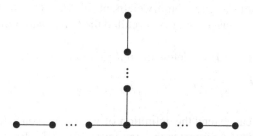

where we denote the numbers of vertices in the three 'arms' by $r, s, t \geq 0$ (the central vertex is not counted here, so the total number of vertices is $r + s + t + 1$). We may assume that $r \leq s \leq t$.

By assumption, there is no subgraph in Γ of type $\widetilde{E}_6$ and hence $r \leq 1$. If $r = 0$ then the graph Γ is a Dynkin diagram of the form A_{s+t+1}, and we are done. So assume now that $r = 1$. There is also no subgraph of type $\widetilde{E}_7$ and therefore $s \leq 2$. If $s = 1$ then the graph is a Dynkin diagram of type D_{t+3}, and again we are done.

So assume $s = 2$. Since Γ also does not have a subgraph of type $\widetilde{E}_8$ we get $t \leq 4$. If $t = 2$ the graph Γ is a Dynkin diagram of type E_6, next if $t = 3$ we have the Dynkin diagram E_7 and for $t = 4$ we have E_8. This shows that the graph Γ is indeed a Dynkin diagram. $\qquad\qquad\qquad\qquad\qquad\qquad\qquad\qquad\qquad\qquad\qquad\qquad\qquad\quad$ $\square$

Exercise 10.1. Let Γ be a graph of Euclidean type $\widetilde{D}_n$ (so Γ has $n + 1$ vertices). Show that any subgraph with n vertices is a disjoint union of Dynkin diagrams.

10.2   The Bilinear Form and the Quadratic Form

Let Γ be a graph, assume that it does not contain loops, that is, edges with the same starting and end point.

In this section we will define a bilinear form and analyze the corresponding quadratic form for such a graph Γ. These two forms are defined on $\mathbb{Z}^n$, by using the standard basis vectors ε_i which form a $\mathbb{Z}$-basis of $\mathbb{Z}^n$. We refer to ε_i as a 'unit vector', it has a 1 in position i and is zero otherwise.

Definition 10.2. Let $\Gamma = (\Gamma_0, \Gamma_1)$ be a graph without loops and label the vertices by $\Gamma_0 = \{1, 2, \ldots, n\}$.

(a) For any vertices $i, j \in \Gamma_0$ let d_{ij} be the number of edges in Γ between i and j. Note that $d_{ij} = d_{ji}$ (since edges are unoriented).

(b) We define a symmetric bilinear form $(-, -)_\Gamma : \mathbb{Z}^n \times \mathbb{Z}^n \to \mathbb{Z}$ on the unit vectors by

$$(\varepsilon_i, \varepsilon_j)_\Gamma = \begin{cases} -d_{ij} & \text{if } i \neq j \\ 2 & \text{if } i = j \end{cases}$$

and extend it bilinearly to arbitrary elements in $\mathbb{Z}^n \times \mathbb{Z}^n$. The $n \times n$-matrix G_Γ with (i, j)-entry equal to $(\varepsilon_i, \varepsilon_j)_\Gamma$ is called the *Gram matrix* of the bilinear form $(-, -)_\Gamma$.

(c) For each vertex j of Γ we define the *reflection map* s_j by

$$s_j : \mathbb{Z}^n \to \mathbb{Z}^n, \quad s_j(a) = a - (a, \varepsilon_j)_\Gamma \, \varepsilon_j.$$

Remark 10.3. We can extend the definition of s_j to a map on $\mathbb{R}^n$, and then we can write down a matrix with respect to the standard basis of $\mathbb{R}^n$. But for our application it is important that s_j preserves $\mathbb{Z}^n$, and we work mostly with $\mathbb{Z}^n$.

We record some properties of the above reflection maps, which also justify why they are called reflections. Let j be a vertex of the graph Γ.

(i) The map s_j is $\mathbb{Z}$-linear.
(ii) When applied to some vector $a \in \mathbb{Z}^n$, the map s_j only changes the j-th coordinate.
(iii) $s_j(\varepsilon_j) = -\varepsilon_j$.
(iv) $s_j^2(a) = a$ for each $a \in \mathbb{Z}^n$.
(v) If there is no edge between different vertices i and j then $s_j(\varepsilon_i) = \varepsilon_i$.

Exercise 10.2.

(i) Let Γ be the graph $1 \text{——} 2 \text{——} 3$ of Dynkin type A_3. Write down the Gram matrix for the bilinear form $(-, -)_\Gamma$. Compute the reflection s_2: show that $s_2(a_1, a_2, a_3) = (a_1, a_1 - a_2 + a_3, a_3)$.
(ii) Let Γ be the graph of Euclidean type $\tilde{A}_1$, that is, the graph with two vertices and two edges. The Gram matrix is

$$\begin{pmatrix} 2 & -2 \\ -2 & 2 \end{pmatrix}.$$

Compute a formula for the reflections s_1 and s_2. Check also that their matrices with respect to the standard basis of $\mathbb{R}^2$ are

$$s_1 = \begin{pmatrix} -1 & 2 \\ 0 & 1 \end{pmatrix}, \quad s_2 = \begin{pmatrix} 1 & 0 \\ 2 & -1 \end{pmatrix}.$$

Example 10.4. We compute explicitly the Gram matrices of the bilinear forms corresponding to Dynkin diagrams of type A, D and E, defined in Fig. 10.1. Note that the bilinear forms depend on the numbering of the vertices of the graph. It is convenient to fix some 'standard labelling'. For later use, we also fix an orientation of the arrows; but note that the bilinear form $(-, -)_\Gamma$ is independent of the orientation.

Type A_n

$$1 \longleftarrow 2 \longleftarrow \ \cdots\ \longleftarrow n-1 \longleftarrow n$$

Type D_n

$$1 \longleftarrow 3 \longleftarrow 4 \longleftarrow \ \cdots\ \longleftarrow n-1 \longleftarrow n$$
$$\downarrow$$
$$2$$

Type E_8

$$1 \longleftarrow 2 \longleftarrow 4 \longleftarrow 5 \longleftarrow 6 \longleftarrow 7 \longleftarrow 8$$

$$\downarrow$$

$$3$$

Then, for E_6 we take the subquiver with vertices $1, 2, \ldots, 6$ and similarly for E_7.

With this fixed standard labelling of the Dynkin diagrams, the Gram matrices of the bilinear forms are as follows.

Type A_n:

$$\begin{pmatrix} 2 & -1 & 0 & \ldots & \ldots & 0 \\ -1 & 2 & -1 & 0 & & 0 \\ 0 & -1 & 2 & -1 & \ddots & \vdots \\ \vdots & \ddots & \ddots & \ddots & \ddots & 0 \\ 0 & \ldots & 0 & -1 & 2 & -1 \\ 0 & \ldots & \ldots & 0 & -1 & 2 \end{pmatrix}$$

Type D_n:

$$\begin{pmatrix} 2 & 0 & -1 & 0 & 0 & \ldots & 0 \\ 0 & 2 & -1 & 0 & 0 & \ldots & 0 \\ -1 & -1 & 2 & -1 & 0 & & \\ 0 & 0 & -1 & 2 & -1 & \ddots & \vdots \\ \vdots & & & \ddots & \ddots & \ddots & 0 \\ 0 & \ldots & & 0 & -1 & 2 & -1 \\ 0 & \ldots & \ldots & & 0 & -1 & 2 \end{pmatrix}$$

Type E_n: We write down the Gram matrix for type E_8, which is

$$\begin{pmatrix} 2 & -1 & 0 & 0 & 0 & 0 & 0 & 0 \\ -1 & 2 & 0 & -1 & 0 & 0 & 0 & 0 \\ 0 & 0 & 2 & -1 & 0 & 0 & 0 & 0 \\ 0 & -1 & -1 & 2 & -1 & 0 & 0 & 0 \\ 0 & 0 & 0 & -1 & 2 & -1 & 0 & 0 \\ 0 & 0 & 0 & 0 & -1 & 2 & -1 & 0 \\ 0 & 0 & 0 & 0 & 0 & -1 & 2 & -1 \\ 0 & 0 & 0 & 0 & 0 & 0 & -1 & 2 \end{pmatrix}$$

The matrices for E_6 and E_7 are then obtained by removing the last two rows and columns for E_6 and the last row and column for E_7.

Remark 10.5. In the above example we have chosen a certain labelling for each of the Dynkin diagrams. In general, let Γ be a graph without loops, and let $\tilde{\Gamma}$ be the graph obtained from Γ by choosing a different labelling of the vertices. Choosing a different labelling means permuting the unit vectors $\varepsilon_1, \ldots, \varepsilon_n$, and hence the rows and columns of the Gram matrix G_Γ are permuted accordingly. In other words, there is a permutation matrix P (that is, a matrix with precisely one entry 1 in each row and column, and zero entries otherwise) describing the basis transformation coming from the permutation of the unit vectors, and such that $P G_\Gamma P^{-1}$ is the Gram matrix of the graph $\tilde{\Gamma}$. Note that any permutation matrix P is orthogonal, hence $P^{-1} = P^t$, the transposed matrix, and $P G_\Gamma P^{-1} = P G_\Gamma P^t$.

Given any bilinear form, there is an associated quadratic form. We want to write down explicitly the quadratic form associated to the above bilinear form $(-, -)_\Gamma$ for a graph Γ.

Definition 10.6. Let $\Gamma = (\Gamma_0, \Gamma_1)$ be a graph without loops, and let $\Gamma_0 = \{1, \ldots, n\}$.

(a) If G_Γ is the Gram matrix of the bilinear form $(-, -)_\Gamma$ as defined in Definition 10.2, the associated quadratic form is given as follows

$$q_\Gamma : \mathbb{Z}^n \to \mathbb{Z} , \quad q_\Gamma(x) = \frac{1}{2}(x, x)_\Gamma = \frac{1}{2} x \, G_\Gamma \, x^t = \sum_{i=1}^n x_i^2 - \sum_{i<j} d_{ij} x_i x_j .$$

That is, $x = (x_1, x_2, \ldots, x_n) \in \mathbb{Z}^n$ taken as a row vector, or an $n \times 1$-matrix, and then $x \, G_\Gamma \, x^t$ is a matrix product.

(b) The elements of the set $\Delta_\Gamma := \{x \in \mathbb{Z}^n \mid q_\Gamma(x) = 1\}$ are called the *roots* of q_Γ.

Exercise 10.3.

(a) Verify that the different expressions for $q_\Gamma(x)$ in Definition 10.6 are equal.
(b) Show that the unit vectors are roots, that is, $\varepsilon_i \in \Delta_\Gamma$ for all $i = 1, \ldots, n$.

Remark 10.7. Let $\tilde{\Gamma}$ be the graph obtained from Γ by choosing a different labelling of the vertices. For the Gram matrices we have $G_{\tilde{\Gamma}} = P G_\Gamma P^{-1}$ with a permutation matrix P, see Remark 10.5. The formulas for the corresponding quadratic forms q_Γ and $q_{\tilde{\Gamma}}$ are different; however the roots of $q_{\tilde{\Gamma}}$ are obtained from the roots of q_Γ by permuting coordinates. Namely, there is a bijection $\Delta_\Gamma \to \Delta_{\tilde{\Gamma}}$ given by $x \mapsto x P^{-1}$ (recall that we consider x as a row vector). In fact, for $x \in \mathbb{Z}^n$ we have

$$q_{\tilde{\Gamma}}(x P^{-1}) = \frac{1}{2}(x P^{-1}) G_{\tilde{\Gamma}}(x P^{-1})^t = \frac{1}{2}(x P^{-1}) P G_\Gamma P^{-1}(x P^{-1})^t$$

$$= \frac{1}{2} x (P^{-1} P) G_\Gamma (P^{-1}(P^{-1})^t) x^t = \frac{1}{2} x G_\Gamma x^t = q_\Gamma(x)$$

where we have used that $(P^{-1})^t = P$ since P is an orthogonal matrix. In particular, x is a root of q_Γ if and only if xP^{-1} is a root of $q_{\tilde\Gamma}$.

Example 10.8 (Roots in Dynkin Type A).   Let Γ be the Dynkin diagram of type A_n, with standard labelling as in Example 10.4. We will compute the set Δ_Γ and show

$$|\Delta_\Gamma| = n(n+1).$$

The quadratic form is (see Definition 10.6)

$$q_\Gamma(x) = \sum_{i=1}^{n} x_i^2 - \sum_{i=1}^{n-1} x_i x_{i+1}.$$

We complete squares, this gives

$$2q_\Gamma(x) = x_1^2 + \sum_{i=1}^{n-1}(x_i - x_{i+1})^2 + x_n^2.$$

We want to determine the set of roots Δ_Γ, that is, we have to find all $x \in \mathbb{Z}^n$ such that $2q_\Gamma(x) = 2$. If so, then $|x_i - x_{i+1}| \le 1$ for $1 \le i \le n-1$, and $|x_1|$ and $|x_n|$ also are ≤ 1 (recall that the x_i are integers). Precisely two of the numbers $|x_i - x_{i+1}|$, $|x_1|$, $|x_n|$ are equal to 1 and all others are zero.

Let $r \in \{1, \dots, n\}$ be minimal such that $x_r \ne 0$ (this exists since x cannot be the zero vector, otherwise $q_\Gamma(x) = 0$). So $x_r = \pm 1$ and $|x_{r-1} - x_r| = |x_r| = 1$. Then among $|x_i - x_{i+1}|$ with $r+1 \le i \le n-1$, and $|x_n|$ precisely one further 1 appears. So the only possibilities are $x = \varepsilon_r + \varepsilon_{r+1} + \dots + \varepsilon_s$ or $x = -\varepsilon_r - \varepsilon_{r+1} - \dots - \varepsilon_s$ for some $s \in \{r, \dots, n\}$.

Thus we have shown that the roots of a Dynkin diagram of type A_n are given by

$$\Delta_\Gamma = \{\pm(\varepsilon_r + \varepsilon_{r+1} + \dots + \varepsilon_s) \mid 1 \le r \le n \text{ and } r \le s \le n\}.$$

In particular, the total number of roots in Dynkin type A_n is

$$|\Delta_\Gamma| = 2(n + (n-1) + \dots + 2 + 1) = n(n+1).$$

Exercise 10.4. Write down the roots for Dynkin type A_3.

The set of roots has a lot of symmetry; among the many nice properties we show here that the set of roots is invariant under the reflections defined in Definition 10.2.

Lemma 10.9. *Let $\Gamma = (\Gamma_0, \Gamma_1)$ be a graph without loops and let j be a vertex of Γ. If $x \in \mathbb{Z}^n$ is a root of q_Γ then $s_j(x)$ is also a root of q_Γ.*

Proof. We will show that s_j preserves the bilinear form $(-, -)_\Gamma$: for $x \in \mathbb{Z}^n$ we have

$$(s_j(x), s_j(x))_\Gamma = (x - (x, \varepsilon_j)_\Gamma \varepsilon_j, x - (x, \varepsilon_j)_\Gamma \varepsilon_j)_\Gamma$$
$$= (x, x)_\Gamma - 2(x, \varepsilon_j)_\Gamma (\varepsilon_j, x)_\Gamma + (x, \varepsilon_j)_\Gamma^2 (\varepsilon_j, \varepsilon_j)_\Gamma$$
$$= (x, x)_\Gamma.$$

For the last equality we have used that the bilinear form is symmetric and that $(\varepsilon_j, \varepsilon_j)_\Gamma = 2$ by Definition 10.2. For the corresponding quadratic form we get

$$q_\Gamma(s_j(x)) = \frac{1}{2}(s_j(x), s_j(x))_\Gamma = \frac{1}{2}(x, x)_\Gamma = q_\Gamma(x).$$

Hence if x is a root, that is $q_\Gamma(x) = 1$, then $q_\Gamma(s_j(x)) = 1$ and $s_j(x)$ is a root. □

We want to show that there are only finitely many roots if Γ is a Dynkin diagram. To do so, we will prove that q_Γ is positive definite and we want to use tools from linear algebra. Therefore, we extend the bilinear form $(-, -)_\Gamma$ and the quadratic form q_Γ to $\mathbb{R}^n$. That is, for the standard basis we take the same formulae as in Definitions 10.2 and 10.6, and we apply them to arbitrary $x \in \mathbb{R}^n$.

Recall from linear algebra that a quadratic form $q : \mathbb{R}^n \to \mathbb{R}^n$ is called positive definite if $q(x) > 0$ for any non-zero $x \in \mathbb{R}^n$. Suppose the quadratic form comes from a symmetric bilinear form as in our case, where (see Definition 10.6)

$$q_\Gamma(x) = \frac{1}{2}(x, x)_\Gamma = \frac{1}{2} x \, G_\Gamma \, x^t.$$

Then the quadratic form is positive definite if and only if, for some labelling, the Gram matrix of the symmetric bilinear form is positive definite. Recall from linear algebra that a symmetric real matrix is positive definite if and only if all its leading principal minors are positive. The leading principal k-minor of an $n \times n$-matrix is the determinant of the submatrix obtained by deleting rows and columns $k + 1, k + 2, \ldots, n$. This is what we will use in the proof of the following result.

Proposition 10.10. *Assume Γ is a Dynkin diagram. Then the quadratic form q_Γ is positive definite.*

Proof. We have seen in Remarks 10.5 and 10.7 how the quadratic forms change when the labelling of the vertices is changed. With a different labelling, one only permutes the coordinates of an element in $\mathbb{R}^n$ but this does not affect the condition of whether $q_\Gamma(x) > 0$ for all non-zero $x \in \mathbb{R}^n$, that is, the condition of whether q_Γ is positive definite. So we can take the standard labelling as in Example 10.4, and it suffices to show that the Gram matrices given in Example 10.4 are positive definite. (1) We start with the Gram matrix in type A_n. Then the leading principal k-minor is the determinant of the Gram matrix of type A_k, and there is a recursion formula: Write $d(A_k)$ for the determinant of the matrix of type A_k. Then we have $d(A_1) = 2$

and $d(A_2) = 3$. Expanding the determinant by the last row of the matrix we find

$$d(A_k) = 2d(A_{k-1}) - d(A_{k-2}) \text{ for all } k \geq 3.$$

It follows by induction on n that $d(A_n) = n + 1$ for all $n \in \mathbb{N}$, and hence all leading principal minors are positive.

(2) Next, consider the Gram matrix of type D_n for $n \geq 4$. Again, the leading principal k-minor for $k \geq 4$ is the determinant of the Gram matrix of type D_k. When $k = 2$ we write D_2 for the submatrix obtained by removing rows and columns with labels ≥ 3, and similarly we define D_3. We write $d(D_k)$ for the determinant of the matrix D_k for $k \geq 2$. We see directly that $d(D_2) = 4$ and $d(D_3) = 4$. For $k \geq 4$ the same expansion of the determinant as for type A gives the recursion

$$d(D_k) = 2d(D_{k-1}) - d(D_{k-2}).$$

By induction, we find that for $k \geq 4$ we have $d(D_k) = 4$. In particular, all leading principal minors of the Gram matrix of type D_n are positive and hence the quadratic form is positive definite.

(3) Now consider the Gram matrix of type E_n for any $n = 6, 7, 8$. In Example 10.4 we have given the Gram matrix for E_8; the matrices for E_6 or E_7 are obtained by removing the last two rows and columns, or the last row and column. Direct calculations show that the values of the first five leading principal minors of the Gram matrix of type E_8 are 2, 3, 6, 5 and 4. Let $n \geq 6$. We expand using the first row and find, after one more step, that

$$d(E_n) = 2d(D_{n-1}) - d(A_{n-2}).$$

Using the above calculations in (1) and (2) this gives

$$d(E_n) = 2 \cdot 4 - (n - 1) = 9 - n \text{ for all } n \geq 6.$$

Hence for $n = 6, 7, 8$ all leading principal minors of the Gram matrix for types E_6, E_7 and E_8 are positive and hence the associated quadratic form q_Γ is positive definite.   □

Exercise 10.5. Let Γ be the graph of type $\widetilde{A}_1$, as in Exercise 10.2. Verify that the quadratic form is

$$q_\Gamma(x) = x_1^2 + x_2^2 - 2x_1x_2 = (x_1 - x_2)^2$$

hence it is not positive definite. However $q_\Gamma(x) \geq 0$ for all $x \in \mathbb{R}^2$ (that is, q_Γ is positive semidefinite).

Remark 10.11.

(1) Alternatively one could prove Proposition 10.10 by finding a suitable formula for $q_\Gamma(x)$ as a sum of squares. We have used this strategy for Dynkin type A_n in Example 10.8. The formula there,

$$2q_\Gamma(x) = x_1^2 + \sum_{i=1}^{n-1}(x_i - x_{i+1})^2 + x_n^2,$$

implies easily that $q_\Gamma(x) > 0$ for all non-zero $x \in \mathbb{R}^n$, that is, the quadratic form q_Γ is positive definite for Dynkin type A_n. Similarly, one can find suitable formulae for the other Dynkin types. See Exercise 10.6 for type D_n.

(2) Usually, the quadratic form of a graph is not positive definite. If Γ is as in Exercise 10.5 then obviously $q_\Gamma(x) = 0$ for $x = (a, a)$ and arbitrary a. We can see another example if we enlarge the E_8-diagram by more vertices and obtain E_n-diagrams for $n > 8$, then the computation in the above proof still gives $d(E_n) = 9 - n$; but this means that the quadratic form is not positive definite for $n > 8$.

(3) The previous remarks and Proposition 10.10 are a special case of a very nice result which characterises Dynkin and Euclidean diagrams by the associated quadratic forms. Namely, let Γ be a connected graph (without loops). Then the quadratic form q_Γ is positive definite if and only if Γ is a Dynkin diagram. Moreover, the quadratic form is positive semidefinite, but not positive definite, if and only if Γ is a Euclidean diagram. This is not very difficult but we do not need it for the proof of Gabriel's theorem.

Exercise 10.6. Let Γ be the Dynkin diagram of type D_n with standard labelling as in Example 10.4. Show that for the quadratic form q_Γ we have

$$4q_\Gamma(x) = (2x_1 - x_3)^2 + (2x_2 - x_3)^2 + 2\left(\sum_{i=3}^{n-1}(x_i - x_{i+1})^2\right) + 2x_n^2.$$

Deduce that the quadratic form q_Γ is positive definite.

We want to show that for a Dynkin diagram, the quadratic form has only finitely many roots. We have seen this already for Dynkin type A_n (with standard labelling) in Example 10.8, but now we give a unified proof for all Dynkin types. The crucial input is that the quadratic forms are positive definite, as shown in Proposition 10.10.

Proposition 10.12. *Let Γ be a Dynkin diagram. Then the set Δ_Γ of roots is finite.*

Proof. We have seen in Remark 10.7 that changing the labelling of the vertices only permutes the entries of the roots. So the statement on finiteness of Δ_Γ does not depend on the labelling of the diagram, so we can use the standard labelling, as in Example 10.4.

We view the Gram matrix G_Γ as a real matrix. It is symmetric, therefore, by linear algebra, all its eigenvalues are real and it is diagonalizable by an orthogonal matrix. Since q_Γ, and hence G_Γ, is positive definite, the eigenvalues must be positive. It follows that $G_\Gamma = PDP^t$, where P is orthogonal and D is diagonal with positive real diagonal entries. Using the definition of q_Γ we can thus write

$$q_\Gamma(x) = \frac{1}{2} x G_\Gamma x^t = \frac{1}{2} (xP)D(xP)^t \tag{10.1}$$

and we want to show that there are at most finitely many roots of q_Γ, that is, solutions with $x \in \mathbb{Z}^n$ such that $q_\Gamma(x) = 1$ (see Definition 10.6).

Suppose $q_\Gamma(x) = 1$ and write $xP = (\xi_1, \ldots, \xi_n)$. Then Equation (10.1) becomes

$$2 = 2 q_\Gamma(x) = \sum_{i=1}^{n} \xi_i^2 \lambda_i,$$

where the λ_i are the positive real diagonal entries of D.

Then the square of the length of $(\xi_1, \ldots, \xi_n)$ is bounded; for example, we must have $\xi_i^2 \leq 2\lambda_i^{-1}$ for each $i = 1, \ldots, n$. Now take $R = \max\{2\lambda_i^{-1} \mid 1 \leq i \leq n\}$, then we have

$$\sum_{i=1}^{n} \xi_i^2 \leq nR.$$

Since the matrix P is orthogonal, it preserves lengths, so we know

$$\sum_{i=1}^{n} x_i^2 = |x|^2 = |xP|^2 = \sum_{i=1}^{n} \xi_i^2 \leq nR.$$

Hence there are at most finitely many solutions for $(x_1, \ldots, x_n) \in \mathbb{Z}^n$ with $q_\Gamma(x) = 1$, that is, there are only finitely many roots for q_Γ. $\square$

In Example 10.8 we have determined the (finite) set of roots for a Dynkin diagram of type A_n. Exercise 10.14 asks to find the roots for a Dynkin diagram of type D_n. For most graphs, there are infinitely many roots.

Exercise 10.7. Consider the graph Γ of type $\tilde{A}_1$ as in Exercises 10.2 and 10.5 which is the underlying graph of the Kronecker quiver. Show that

$$\Delta_\Gamma = \{(a, a \pm 1) \mid a \in \mathbb{Z}\}.$$

For the Dynkin diagrams we refine the set Δ_Γ of roots, namely we divide the roots into 'positive' and 'negative' roots.

Lemma 10.13. *Let Γ be a Dynkin diagram. Suppose $x \in \mathbb{Z}^n$ is a root of q_Γ. Then $x \neq 0$, and either $x_t \geq 0$ for all t, or $x_t \leq 0$ for all t. In the first case we say that x is a* positive root, *and in the second case we call x a* negative root.

Proof. Assume $x = (x_1, \ldots, x_n) \in \mathbb{Z}^n$ and $x \in \Delta_\Gamma$ is a root, that is $q_\Gamma(x) = 1$. Note that we have $x \neq 0$ since $q_\Gamma(x) = 1$. We can sort positive and negative entries in x and write $x = x^+ + x^-$, where

$$x^+ = \sum_{s, x_s > 0} x_s \varepsilon_s, \quad \text{and} \quad x^- = \sum_{t, x_t < 0} x_t \varepsilon_t$$

(here the ε_i are the unit vectors). Moreover, by definition of q_Γ (see Definition 10.6) we have

$$q_\Gamma(x) = \frac{1}{2}(x, x)_\Gamma = \frac{1}{2}(x^+ + x^-, x^+ + x^-)_\Gamma = (x^+, x^-)_\Gamma + q_\Gamma(x^+) + q_\Gamma(x^-).$$

Using the definition of x^+ and x^- we expand $(x^+, x^-)_\Gamma$ and obtain

$$(x^+, x^-)_\Gamma = \sum_{s, x_s > 0} \sum_{t, x_t < 0} x_s x_t (\varepsilon_s, \varepsilon_t)_\Gamma = \sum_{s, x_s > 0} \sum_{t, x_t < 0} x_s x_t (-d_{st}) \geq 0$$

(for the last equality we used the definition of $(-, -)_\Gamma$, see Definition 10.2). Since Γ is one of the Dynkin diagrams, the quadratic form q_Γ is positive definite by Proposition 10.10. In particular, $q_\Gamma(x^+) \geq 0$ and $q_\Gamma(x^-) \geq 0$. But since $x = x^+ + x^-$ is non-zero, at least one of x^+ and x^- is non-zero and then $q_\Gamma(x^+) + q_\Gamma(x^-) > 0$, again by positive definiteness.

In summary, we get

$$1 = q_\Gamma(x) = (x^+, x^-)_\Gamma + q_\Gamma(x^+) + q_\Gamma(x^-) \geq q_\Gamma(x^+) + q_\Gamma(x^-) > 0.$$

Since the quadratic form has integral values, precisely one of $q_\Gamma(x^+)$ and $q_\Gamma(x^-)$ is 1 and the other is 0. Since q_Γ is positive definite, $x^+ = 0$ or $x^- = 0$, that is, $x = x^-$ or $x = x^+$, which proves the claim.   □

10.3   The Coxeter Transformation

In this section we will introduce a particular map, the Coxeter transformation, associated to a Dynkin diagram with standard labelling as in Example 10.4. This map will later be used to show that for Dynkin diagrams positive roots parametrize indecomposable representations.

Let Γ be one of the Dynkin diagrams, with standard labelling. We have seen in Lemma 10.9 that each reflection s_j, where j is a vertex of Γ, preserves the set Δ_Γ of roots. Then the set Δ_Γ of roots is also preserved by arbitrary products of reflections, that is, by any element in the group W, the subgroup of the automorphism group $\mathrm{Aut}(\mathbb{Z}^n)$ generated by the reflections s_j. The Coxeter transformation is an element of W and it has special properties.

Definition 10.14. Assume Γ is a Dynkin diagram with standard labelling as in Example 10.4. Let $s_j : \mathbb{Z}^n \to \mathbb{Z}^n$, $s_j(x) = x - (x, \varepsilon_j)_\Gamma \varepsilon_j$ be the reflections as in Definition 10.2. The *Coxeter transformation* C_Γ is the map

$$C_\Gamma = s_n \circ s_{n-1} \circ \ldots \circ s_2 \circ s_1 : \mathbb{Z}^n \to \mathbb{Z}^n.$$

The *Coxeter matrix* is the matrix of C_Γ with respect to the standard basis of $\mathbb{R}^n$.

Example 10.15 (Coxeter Transformation in Dynkin Type A). Let Γ be the Dynkin diagram of type A_n with standard labelling. We describe the Coxeter transformation and its action on the roots of q_Γ. To check some of the details, see Exercise 10.8 below. Let s_j be the reflection, as defined in Definition 10.2. Explicitly, we have for $x = (x_1, x_2, \ldots, x_n) \in \mathbb{R}^n$ that

$$s_j(x) = \begin{cases} (-x_1 + x_2, x_2, \ldots, x_n) & j = 1 \\ (x_1, \ldots, x_{j-1}, -x_j + x_{j-1} + x_{j+1}, x_{j+1}, \ldots, x_n) & 2 \le j \le n-1 \\ (x_1, \ldots, x_{n-1}, x_{n-1} - x_n) & j = n. \end{cases}$$

With this, we compute the Coxeter transformation

$$C_\Gamma(x) = s_n \circ s_{n-1} \circ \ldots \circ s_2 \circ s_1(x) = (-x_1 + x_2, -x_1 + x_3, \ldots, -x_1 + x_n, -x_1).$$

Consider the action of C_Γ on the set of roots. Recall from Example 10.8 that for the Dynkin diagram of type A_n the set of roots is given by

$$\Delta_\Gamma = \{\pm\alpha_{r,s} \mid 1 \le r \le s \le n\},$$

where $\alpha_{r,s} = \varepsilon_r + \varepsilon_{r+1} + \ldots + \varepsilon_s$. Consider the root $C_\Gamma(\alpha_{r,s})$. One checks the following formula:

$$C_\Gamma(\alpha_{r,s}) = \begin{cases} \alpha_{r-1,s-1} & r > 1 \\ -\alpha_{s,n} & r = 1. \end{cases}$$

Since also $C_\Gamma(-x) = -C_\Gamma(x)$, we see that C_Γ permutes the elements in Δ_Γ (in fact, this already follows from the fact that this holds for each reflection s_j). We also see that C_Γ can take positive roots to negative roots.

Exercise 10.8. Check the details when $n = 4$ in Example 10.15.

Exercise 10.9. Let Γ be the graph of type $\tilde{A}_1$ as in Exercise 10.2. Define $C_\Gamma := s_2 \circ s_1$. Show that with respect to the standard basis of $\mathbb{R}^2$ this has matrix

$$\begin{pmatrix} -1 & 2 \\ -2 & 3 \end{pmatrix}.$$

Show by induction on $k \geq 1$ that C_Γ^k has matrix

$$\begin{pmatrix} -(2k-1) & 2k \\ -2k & 2k+1 \end{pmatrix}.$$

For a vector $z = (z_1, \ldots, z_n) \in \mathbb{Z}^n$ we say that $z \geq 0$ if $z_i \geq 0$ for all $1 \leq i \leq n$, and otherwise we write $z \not\geq 0$.

Lemma 10.16. *Assume Γ is a Dynkin diagram with standard labelling, and let C_Γ be the Coxeter transformation. Then the following holds.*

(a) If $y \in \mathbb{Z}^n$ and $C_\Gamma(y) = y$ then $y = 0$.
(b) There is a positive number $h \in \mathbb{N}$ such that C_Γ^h is the identity map on $\mathbb{Z}^n$.
(c) For every $0 \neq x \in \mathbb{Z}^n$ there is some $r \geq 0$ such that $C_\Gamma^r(x) \not\geq 0$.

Part (a) of this lemma says that the Coxeter transformation has no fixed points on $\mathbb{Z}^n$ except zero. For Dynkin type A this can be deduced from the formulae in Example 10.15, see Exercise 10.12. In principle, one could prove this lemma case-by-case. But below, we give a unified proof which works for all Dynkin types. As we have noted, C_Γ preserves Δ_Γ and the elements of Δ_Γ are non-zero. Hence, by part (a), C_Γ does not fix any root.

Proof. We begin with some preliminary observations. Recall from Definition 10.2 that for each vertex i of Γ we have $s_i(y) = y - (y, \varepsilon_i)_\Gamma \varepsilon_i$. So $(y, \varepsilon_i)_\Gamma = 0$ if and only if $s_i(y) = y$. Recall also from Remark 10.3 that s_i^2 is the identity.
(a) We will show that $C_\Gamma(y) = y$ implies that $(y, \varepsilon_i)_\Gamma = 0$ for all i. Then for $y = \sum_{i=1}^n y_i \varepsilon_i$ we get

$$2q_\Gamma(y) = (y, y)_\Gamma = \sum_{i=1}^n y_i (y, \varepsilon_i)_\Gamma = 0$$

and since the quadratic form q_Γ is positive definite (by Proposition 10.10), it will follow that $y = 0$.

So suppose that $C_\Gamma(y) = y$. Since s_n^2 is the identity this implies $s_{n-1} \circ \ldots \circ s_1(y) = s_n(y)$. Since the reflection s_j only changes the j-th coordinate, the n-th coordinate of $s_{n-1} \circ \ldots \circ s_1(y)$ is y_n, and the n-th coordinate of $s_n(y)$ is $y_n - (y, \varepsilon_n)_\Gamma$. So we have $(y, \varepsilon_n)_\Gamma = 0$.

Now we proceed inductively. Since $(y, \varepsilon_n)_\Gamma = 0$ then also $s_n(y) = y$ by the introductory observation. So we have now that $s_{n-1} \circ \ldots \circ s_1(y) = y$. Then applying s_{n-1} and equating the $(n-1)$-th coordinate we get $y_{n-1} = y_{n-1} - (y, \varepsilon_{n-1})_\Gamma$ and hence $(y, \varepsilon_{n-1})_\Gamma = 0$.

Repeating the argument we eventually get $(y, \varepsilon_i)_\Gamma = 0$ for all i and then $y = 0$ as explained above.

(b) By Lemma 10.9 we know that each reflection s_j permutes the set Δ_Γ of roots. Hence $C_\Gamma = s_n \circ \ldots \circ s_1$ also permutes the roots. Since Γ is a Dynkin diagram, the set of roots is finite by Proposition 10.12. Hence there is some integer $h \geq 1$ such that C_Γ^h fixes each root. Then in particular $C_\Gamma^h(\varepsilon_i) = \varepsilon_i$ for all i, by Exercise 10.3. The ε_i form a basis for $\mathbb{Z}^n$, and C_Γ^h is $\mathbb{Z}$-linear (see Remark 10.3), therefore C_Γ^h is the identity map on $\mathbb{Z}^n$.

(c) If $x \not\geq 0$ then we can take $r = 0$. So assume now that $x \geq 0$. Let h be the minimal positive integer as in part (b) such that C_Γ^h is the identity. Then we set

$$y := \sum_{r=0}^{h-1} C_\Gamma^r(x) \in \mathbb{Z}^n.$$

For this particular vector we have

$$C_\Gamma(y) = C_\Gamma(x) + C_\Gamma^2(x) + \ldots + C_\Gamma^h(x) = C_\Gamma(x) + C_\Gamma^2(x) + \ldots + C_\Gamma^{h-1}(x) + x = y.$$

By part (a) we deduce $y = 0$. Now, $x \geq 0$ and x is non-zero. If we had $C_\Gamma^r(x) \geq 0$ for all r then it would follow that $0 \neq y \geq 0$, a contradiction. So there must be some $r \geq 1$ such that $C^r x \not\geq 0$, as required. $\square$

The properties in Lemma 10.16 depend crucially on the fact that Γ is a Dynkin diagram. For the Coxeter transformation in Exercise 10.9, each part of the lemma fails. Indeed, $C_\Gamma - \mathrm{id}$ is obviously singular, so part (a) does not hold. Furthermore, from the matrix of C_Γ^k in Exercise 10.9 one sees that no power of C_Γ can be the identity, so part (b) does not hold. In addition, for all $k \geq 0$ the matrix of C_Γ^k yields $C_\Gamma^k(\varepsilon_2) \geq 0$, so part (c) does not hold.

EXERCISES

10.10. Let C_Γ be the Coxeter transformation for Dynkin type A_n (with standard labelling), see Example 10.15; this permutes the set Δ_Γ of roots.

(a) Find the C_Γ-orbit of ε_n, show that it contains each ε_i, and that it has size $n + 1$.

(b) Show that each orbit contains a unique root of the form $\alpha_{t,n}$ for $1 \leq t \leq n$, compute its orbit, and verify that it has size $n + 1$.

(c) Deduce that C_Γ^{n+1} is the identity map of $\mathbb{R}^n$.

10.11. Let Γ be a Dynkin diagram with standard labelling. The *Coxeter number* of Γ is the smallest positive integer h such that C_Γ^h is the identity map of $\mathbb{R}^n$. Using the previous Exercise 10.10 show that the Coxeter number of the Dynkin diagram of type A_n is equal to $n + 1$.

10.12. Assume C_Γ is the Coxeter transformation for the Dynkin diagram Γ of type A_n with standard labelling. Show by using the formula in Example 10.15 that $C_\Gamma(y) = y$ for $y \in \mathbb{Z}^n$ implies that $y = 0$.

10.13. Consider the Coxeter transformation C_Γ for a Dynkin diagram of type A_n, with standard labelling. Show that its matrix with respect to the standard basis is given by

$$
C_n := \begin{pmatrix}
-1 & 1 & 0 & \cdots\cdots & 0 \\
-1 & 0 & 1 & 0 & \cdots & 0 \\
\vdots & \vdots & \vdots & \ddots & \ddots & \vdots \\
\vdots & \vdots & & 0 & 1 & 0 \\
-1 & 0 & & & 0 & 1 \\
-1 & 0 & \cdots\cdots & & & 0
\end{pmatrix}
$$

Let $f_n(x) = \det(C_n - xE_n)$, the characteristic polynomial of C_n.

(a) Check that $f_1(x) = -x - 1$ and $f_2(x) = x^2 + x + 1$.

(b) By using the expansion formula for the last row of the matrix $C_n - xE_n$, show that

$$
f_n(x) = (-1)^n - xf_{n-1}(x) \quad \text{(for } n \geq 3\text{)}.
$$

(c) By induction on n, deduce a formula for $f_n(x)$. Hence show that

$$
f_n(x) = (-1)^n \frac{x^{n+1} - 1}{x - 1}.
$$

(d) Deduce from this that C_n does not have an eigenvalue equal to 1. Hence deduce that $C_\Gamma(y) = y$ implies $y = 0$.

10.14. (Roots in Dynkin type D) Let Γ be the Dynkin diagram of Type D_n, where $n = 4$ and $n = 5$, with standard labelling as in Example 10.4. Use the formula for the quadratic form q_Γ given in Exercise 10.6 to determine all roots of q_Γ. (Hint: In total there are $2n(n - 1)$ roots.)

10.15. Compute the reflections and the Coxeter transformation for a Dynkin diagram Γ of type D_5 with standard labelling.

(a) Verify that

$$
C_\Gamma(x) = (x_3 - x_1, x_3 - x_2, x_3 + x_4 - x_1 - x_2, x_3 + x_5 - x_1 - x_2, x_3 - x_1 - x_2).
$$

(b) The map C_Γ permutes the roots. Find the orbits.

(c) Show that the Coxeter number of Γ (defined in Exercise 10.11) is equal to 8.

10.16. Let Γ be the Dynkin diagram of type A_n with standard labelling, and let W be the subgroup of $\text{Aut}(\mathbb{Z}^n)$ generated by $s_1, s_2, \ldots, s_n$. We want to show that W is isomorphic to the symmetric group S_{n+1}. The group S_{n+1} can be defined by generators and relations, where the generators are the transpositions $\tau_1, \tau_2, \ldots, \tau_n$, where τ_i interchanges i and $i+1$, and the relations are $\tau_i^2 = 1$ and the 'braid relations'

$$\tau_i \tau_{i+1} \tau_i = \tau_{i+1} \tau_i \tau_{i+1} \ (\text{for } 1 \le i < n), \quad \tau_i \tau_j = \tau_j \tau_i \ (\text{for } |i - j| > 1).$$

(a) Check that for the unit vectors we have

$$s_i(\varepsilon_i) = -\varepsilon_i$$
$$s_i(\varepsilon_{i+1}) = \varepsilon_{i+1} + \varepsilon_i$$
$$s_i(\varepsilon_{i-1}) = \varepsilon_{i-1} + \varepsilon_i$$

and that $s_i(\varepsilon_j) = \varepsilon_j$ for $j \notin \{i, i \pm 1\}$.

(b) Use (a) to show that the s_i satisfy the braid relations.

(c) By (b) and since $s_i^2 = 1$, there is a group homomorphism $\rho : S_{n+1} \to W$ such that $\rho(\tau_i) = s_i$. The kernel of ρ is a normal subgroup of S_{n+1}. Using the fact that the only normal subgroups of symmetric groups are the alternating groups, and the Klein 4-group when $n + 1 = 4$, show that the kernel of ρ must be the identity group. Hence ρ is an isomorphism.

Chapter 11
Gabriel's Theorem

Assume Q is a quiver without oriented cycles, then for any field K the path algebra KQ is finite-dimensional (see Exercise 1.2). We want to know when Q is of finite representation type; this is answered completely by Gabriel's theorem. Let $\bar{Q}$ be the underlying graph of Q, which is obtained by ignoring the orientation of the arrows. Gabriel's theorem states that KQ is of finite representation type if and only if $\bar{Q}$ is the disjoint union of Dynkin diagrams of type A, D and E. The relevant Dynkin diagrams are listed in Fig. 10.1. So the representation type of Q does not depend on the orientation of the arrows. Note that Gabriel's theorem holds, and is proved here, for an arbitrary field K.

Theorem 11.1 (Gabriel's Theorem). *Assume Q is a quiver without oriented cycles, and K is a field. Then Q has finite representation type if and only if the underlying graph $\bar{Q}$ is the disjoint union of Dynkin diagrams of types A_n for $n \geq 1$, or D_n for $n \geq 4$, or E_6, E_7, E_8.*

Moreover, if a quiver Q has finite representation type, then the indecomposable representations are parametrized by the set of positive roots (see Definition 10.6), associated to the underlying graph of Q. Dynkin diagrams and roots play a central role in Lie theory, and Gabriel's theorem connects representation theory with Lie theory.

11.1 Reflecting Quivers and Representations

Gabriel's theorem states implicitly that the representation type of a quiver depends only on the underlying graph but not on the orientation of the arrows. To prove this, we will use 'reflection maps', which relate representations of two quivers with the same underlying graph but where some arrows have different orientation. This

© Springer International Publishing AG, part of Springer Nature 2018
K. Erdmann, T. Holm, *Algebras and Representation Theory*, Springer
Undergraduate Mathematics Series, https://doi.org/10.1007/978-3-319-91998-0_11

construction will show that any two quivers with the same underlying graph Γ have the same representation type, if Γ is an arbitrary finite tree.

Throughout this chapter let K be an arbitrary field.

Definition 11.2. Let Q be a quiver. A vertex j of Q is called a *sink* if no arrows in Q start at j. A vertex k of Q is a *source* if no arrows in Q end at k.

For example, consider the quiver $1 \longrightarrow 2 \longleftarrow 3 \longleftarrow 4$. Then vertices 1 and 4 are sources, vertex 2 is a sink and vertex 3 is neither a sink nor a source.

Exercise 11.1. Let Q be a quiver without oriented cycles. Show that Q contains a sink and a source.

Definition 11.3. Let Q be a quiver and let j be a vertex in Q which is a sink or a source. We define a new quiver $\sigma_j Q$, this is the quiver obtained from Q by reversing all arrows adjacent to j, and keeping everything else unchanged. We call $\sigma_j Q$ the *reflection* of Q at the vertex j. Note that if a vertex j is a sink of Q then j is a source of $\sigma_j Q$, and if j is a source of Q then it is a sink of $\sigma_j Q$. We also have that $\sigma_j \sigma_j Q = Q$.

Example 11.4. Consider all quivers whose underlying graph is the Dynkin diagram of type A_4. Up to labelling of the vertices, there are four possible quivers,

$$Q_1 : 1 \longleftarrow 2 \longleftarrow 3 \longleftarrow 4$$

$$Q_2 : 1 \longrightarrow 2 \longleftarrow 3 \longleftarrow 4$$

$$Q_3 : 1 \longleftarrow 2 \longrightarrow 3 \longleftarrow 4$$

$$Q_4 : 1 \longleftarrow 2 \longleftarrow 3 \longrightarrow 4$$

Then $\sigma_1 Q_1 = Q_2$ and $\sigma_2 \sigma_1 Q_1 = Q_3$; and moreover $\sigma_3 \sigma_2 \sigma_1 Q_1 = Q_4$. Hence each Q_i can be obtained from Q_1 by applying reflections.

This observation is more general: We will now show that if we start with a quiver Q whose underlying graph is a tree, and repeatedly apply such reflections, we can get all quivers with the same underlying graph. That is, we can choose a vertex i_1 which is a sink or source of Q, then we get the quiver $\sigma_{i_1} Q$. Then one can find a sink or source i_2 of the quiver $\sigma_{i_1} Q$, and hence we get the quiver $\sigma_{i_2} \sigma_{i_1} Q$, and so on. If the underlying graph is a tree one can arrange this, and get an arbitrary quiver Q' with the same underlying graph. The idea is to organize this properly.

Proposition 11.5. *Let Q and Q' be quivers with the same underlying graph and assume that this graph is a tree. Then there exists a sequence $i_1, \ldots, i_r$ of vertices of Q such that the following holds:*

(i) *The vertex i_1 is a sink or source in Q.*

(ii) *For each j such that $1 < j < r$, the vertex i_j is a sink or source in the quiver $\sigma_{i_{j-1}} \ldots \sigma_{i_1} Q$ obtained from Q by successively reflecting at the vertices in the sequence.*

(iii) *We have $Q' = \sigma_{i_r} \ldots \sigma_{i_1} Q$.*

Proof. We prove this by induction on the number n of vertices of Q. Let Γ be the underlying graph. For $n = 1$ or $n = 2$ the statement is clear. So let $n \geq 3$ and assume, as an inductive hypothesis, that the statement holds for quivers with fewer than n vertices. Since Γ is a tree, it must have a vertex which is adjacent to only one other vertex (in fact, if each vertex has at least two neighbours one can find a cycle in the graph). We choose and fix such a vertex, and then we label the vertices so that this vertex is n and its unique neighbour is $n - 1$.

We remove the vertex n and the (unique) adjacent arrow, that is, the arrow between n and $n - 1$ from Q and Q'; this gives quivers $\widehat{Q}$ and $\widehat{Q}'$, each with $n - 1$ vertices and which have the same underlying graph $\widehat{\Gamma}$, which is also a tree.

By the inductive hypothesis, there exists a sequence $i_1, \dots, i_t$ of vertices of $\widehat{Q}$ such that for each j, the vertex i_j is a sink or source in $\sigma_{i_{j-1}} \dots \sigma_{i_1} \widehat{Q}$ and such that $\widehat{Q}' = \sigma_{i_t} \dots \sigma_{i_1} \widehat{Q}$. We want to extend this to Q but we have to be careful in cases when the vertex i_j is equal to $n - 1$.

At the first step, we have two cases: either i_1 is a sink or source in Q, or $i_1 = n-1$ but is not a sink or source in Q. In the second case, i_1 must be a sink or source in the quiver $\sigma_n Q$. We set $Q^{(1)} := \sigma_{i_1} Q$ in the first case, and in the second case we set $Q^{(1)} := \sigma_{i_1} \sigma_n Q$. Note that if we remove vertex n and the adjacent arrow from $Q^{(1)}$ we get $\sigma_{i_1} \widehat{Q}$.

If i_2 is a sink or source in $Q^{(1)}$ then set $Q^{(2)} := \sigma_2 Q^{(1)}$. Otherwise, $i_2 = n - 1$ and it is a sink or source of $\sigma_n Q^{(1)}$. In this case we set $Q^{(2)} := \sigma_{i_2} \sigma_n Q^{(1)}$. We note that if we remove vertex n and the adjacent arrow from $Q^{(2)}$ then we get $\sigma_{i_2} \sigma_{i_1} \widehat{Q}$.

We repeat this until we have after t steps a quiver $Q^{(t)}$, so that if we remove vertex n and the adjacent arrow from $Q^{(t)}$, then we get the quiver $\sigma_{i_t} \dots \sigma_{i_1} \widehat{Q} = \widehat{Q}'$. Then either $Q^{(t)} = Q'$, or if not then $\sigma_n Q^{(t)} = Q'$. In total we have obtained a sequence of reflections in sinks or sources which takes Q to Q'. The parameter r in the statement is then equal to t plus the number of times we have inserted the reflection σ_n. $\square$

Example 11.6. We illustrate the proof of Proposition 11.5. Let Q and Q' be the quivers

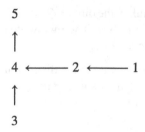

and

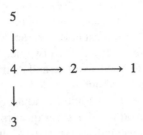

We write down $\widehat{Q}$, and $\widehat{Q}'$, obtained by removing vertex 5 and the adjacent arrow. Then we have $\sigma_4\sigma_1\widehat{Q} = \widehat{Q}'$. We extend the sequence to Q and we see that we must take twice a reflection at vertex 5, and get $\sigma_5\sigma_4\sigma_5\sigma_1(Q) = Q'$.

Starting with a quiver Q where vertex j is a sink or a source, we have obtained a new reflected quiver $\sigma_j Q$. We want to compare the representation type of these quivers, and want to construct from a representation $\mathcal{M}$ of Q a 'reflected representation' of the quiver $\sigma_j Q$.

11.1.1   The Reflection Σ_j^+ at a Sink

We assume that j is a sink of the quiver Q. For every representation $\mathcal{M}$ of Q we will construct from $\mathcal{M}$ a representation of $\sigma_j Q$, denoted by $\Sigma_j^+(\mathcal{M})$. The idea is to keep the vector space $M(r)$ as it is for any vertex $r \neq j$, and also to keep the linear map $M(\beta)$ as it is for any arrow β which does not end at j. We want to find a vector space $M^+(j)$, and for each arrow $\alpha_i : i \to j$ in Q, we want to define a linear map $M^+(\bar{\alpha}_i)$ from $M^+(j)$ to $M(i)$, to be constructed using only data from $\mathcal{M}$. We first fix some notation, and then we study small examples.

Definition 11.7. Let j be a sink in the quiver Q. We label the distinct arrows ending at j by $\alpha_1, \alpha_2, \ldots, \alpha_t$, say $\alpha_i : i \to j$. Then we write $\bar{\alpha}_i : j \to i$ for the arrows of $\sigma_j Q$ obtained by reversing the α_i.

Note that with this convention we have t distinct arrows ending at j, but if the quiver contains multiple arrows then some starting vertices of the arrows α_i may coincide; see also Remark 11.13.

Example 11.8.

(1) Let $t = 1$, and take the quivers Q and $\sigma_j Q$ as follows:

$$1 \xrightarrow{\alpha_1} j \quad \text{and} \quad 1 \xleftarrow{\bar{\alpha}_1} j.$$

We start with a representation $\mathcal{M}$ of Q,

$$M(1) \xrightarrow{M(\alpha_1)} M(j),$$

and we want to define a representation of $\sigma_j Q$, that is,

$$M(1) \xleftarrow{M^+(\bar{\alpha}_1)} M^+(j),$$

and this should only use information from $\mathcal{M}$. There is not much choice, we take $M^+(j) := \ker(M(\alpha_1))$, which is a subspace of $M(1)$, and we take $M^+(\bar{\alpha}_1)$ to be the inclusion map. This defines a representation $\Sigma_j^+(\mathcal{M})$ of $\sigma_j Q$.

(2) Let $t = 2$, and take the quivers Q and $\sigma_j Q$ as follows:

$$1 \xrightarrow{\alpha_1} j \xleftarrow{\alpha_2} 2 \quad \text{and} \quad 1 \xleftarrow{\bar{\alpha}_1} j \xrightarrow{\bar{\alpha}_2} 2.$$

We start with a representation $\mathcal{M}$ of Q,

$$M(1) \xrightarrow{M(\alpha_1)} M(j) \xleftarrow{M(\alpha_2)} M(2).$$

Here we can use the construction of the pull-back, which was introduced in Chap. 2 (see Exercise 2.16). This takes two linear maps to a fixed vector space and constructs from this a new space E, explicitly,

$$E = \{(m_1, m_2) \in M(1) \times M(2) \mid M(\alpha_1)(m_1) + M(\alpha_2)(m_2) = 0\}.$$

Denote the projection maps on E by $\pi_1(m_1, m_2) = m_1$, and $\pi_2(m_1, m_2) = m_2$. We define $M^+(j) := E$ and $M^+(\bar{\alpha}_1) := \pi_1$ and $M^+(\bar{\alpha}_2) := \pi_2$. That is, the representation $\Sigma_j^+(\mathcal{M})$ is then

$$M(1) \xleftarrow{\pi_1} E \xrightarrow{\pi_2} M(2).$$

(3) Let $t = 3$. We take the following quiver Q

$$2$$
$$\downarrow{\alpha_2}$$
$$1 \xrightarrow{\alpha_1} j \xleftarrow{\alpha_3} 3.$$

Then the quiver $\sigma_j Q$ is equal to

$$2$$

$$\uparrow \bar{\alpha}_2$$

$$1 \xleftarrow{\ \bar{\alpha}_1\ } j \xrightarrow{\ \bar{\alpha}_3\ } 3.$$

Suppose $\mathcal{M}$ is a representation of Q. We want to construct from $\mathcal{M}$ a new representation $\Sigma_j^+(\mathcal{M})$ of the quiver $\sigma_j Q$. We modify the idea of the previous example.

We set $M^+(i) = M(i)$ for $i = 1, 2, 3$, and we define

$$M^+(j) = \{(m_1, m_2, m_3) \in \prod_{i=1}^{3} M(i) \mid M(\alpha_1)(m_1) + M(\alpha_2)(m_2) + M(\alpha_3)(m_3) = 0\}.$$

As the required linear map $M^+(\bar{\alpha}_i) : M^+(j) \rightarrow M(i)$ we take the i-th projection, that is

$$M^+(\bar{\alpha}_i)((m_1, m_2, m_3)) = m_i \quad \text{for } i = 1, 2, 3.$$

Then $\Sigma_j^+(\mathcal{M})$ is a representation of the quiver $\sigma_j Q$.

We consider two explicit examples for this construction.

(i) Let $\mathcal{M}$ be the representation of Q with $M(i) = K$ for $1 \leq i \leq 3$ and $M(j) = K^2$, with maps $M(\alpha_i)$ given by

$$M(\alpha_1) : K \rightarrow K^2 , \quad x \mapsto (x, 0)$$

$$M(\alpha_2) : K \rightarrow K^2 , \quad x \mapsto (0, x)$$

$$M(\alpha_3) : K \rightarrow K^2 , \quad x \mapsto (x, x).$$

Then

$$M^+(j) = \{(m_1, m_2, m_3) \in K^3 \mid (m_1, 0) + (0, m_2) + (m_3, m_3) = 0\}$$

$$= \{(-x, -x, x) \mid x \in K\}$$

and $M^+(\bar{\alpha}_i)((-x, -x, x))$ is equal to $-x$, or to x, respectively.

(ii) Let $\mathcal{M}$ be the representation of Q with $M(1) = M(2) = M(3) = 0$ and $M(j) = K$, then $\Sigma_j^+(\mathcal{M})$ is the zero representation. Note that $\mathcal{M} = \mathcal{S}_j$, the simple representation for the vertex j (see Example 9.7).

The definition of $\Sigma_j^+(\mathcal{M})$ in general when j is a sink of a quiver Q is essentially the same as in the examples. Recall our notation, as in Definition 11.7. We also write $(M(\alpha_1), \ldots, M(\alpha_t))$ for the linear map from $\prod_{i=1}^{t} M(i)$ to $M(j)$ defined as

$$(m_1, \ldots, m_t) \mapsto M(\alpha_1)(m_1) + \ldots + M(\alpha_t)(m_t).$$

Definition 11.9. Let Q be a quiver and assume j is a sink in Q. For any representation $\mathcal{M}$ of Q we define a representation $\Sigma_j^+(\mathcal{M})$ of the quiver $\sigma_j Q$ as follows. We set

$$M^+(r) := \begin{cases} M(r) & (r \neq j), \\ \{(m_1, \ldots, m_t) \in \prod_{i=1}^t M(i) \mid (M(\alpha_1), \ldots, M(\alpha_t))(m_1, \ldots, m_t) = 0\} & (r = j). \end{cases}$$

If γ is an arrow of Q which does not end at the sink j then we set $M^+(\gamma) = M(\gamma)$. For $i = 1, \ldots, t$ we define $M^+(\bar{\alpha}_i) : M^+(j) \to M^+(i)$ to be the projection onto $M(i)$, that is, $M^+(\bar{\alpha}_i)(m_1, \ldots, m_t) = m_i$.

To compare the representation types of the quivers Q and $\sigma_j Q$, we need to keep track over direct sum decompositions. Fortunately, the construction of Σ_j^+ is compatible with taking direct sums:

Lemma 11.10. *Let Q be a quiver and let j be a sink in Q. Let $\mathcal{M}$ be a representation of Q such that $\mathcal{M} = \mathcal{X} \oplus \mathcal{Y}$ for subrepresentations $\mathcal{X}$ and $\mathcal{Y}$. Then we have*

$$\Sigma_j^+(\mathcal{M}) = \Sigma_j^+(\mathcal{X}) \oplus \Sigma_j^+(\mathcal{Y}).$$

We will prove this later, in Sect. 12.1.1, since the proof is slightly technical. With this lemma, we can focus on indecomposable representations of Q. We consider a small example.

Example 11.11. We consider the quiver $1 \xrightarrow{\alpha} 2$. In Example 9.28 we have seen that it has precisely three indecomposable representations (up to isomorphism), which are listed in the left column of the table below. We now reflect the quiver at the sink 2, and we compute the reflected representations $\Sigma_2^+(\mathcal{M})$, using Example 11.8. The representations $\Sigma_2^+(\mathcal{M})$ are listed in the right column of the following table.

$\mathcal{M}$	$\Sigma_2^+(\mathcal{M})$
$K \longrightarrow 0$	$K \xleftarrow{\text{id}_K} K$
$K \xrightarrow{\text{id}_K} K$	$K \longleftarrow 0$
$0 \longrightarrow K$	$0 \longleftarrow 0$

We see that Σ_2^+ permutes the indecomposable representations other than the simple representation S_2. Moreover, it takes S_2 to the zero representation, and S_2 does not appear as $\Sigma_2^+(\mathcal{M})$ for some $\mathcal{M}$.

We can generalize the last observation in the example:

Proposition 11.12. *Assume Q is a quiver and j is a sink in Q. Let M be a representation of Q.*

(a) $\Sigma_j^+(\mathcal{M})$ is the zero representation if and only if $M(r) = 0$ for all vertices $r \neq j$, equivalently $\mathcal{M}$ is isomorphic to a direct sum of copies of the simple representation S_j.

(b) $\Sigma_j^+(\mathcal{M})$ has no subrepresentation isomorphic to the simple representation S_j.

Proof. (a) Assume first that $M(r) = 0$ for all $r \neq j$ then it follows directly from Definition 11.9 that $\Sigma_j^+(\mathcal{M})$ is the zero representation. Conversely, if $\Sigma_j^+(\mathcal{M})$ is the zero representation then for $r \neq j$ we have $0 = M^+(r) = M(r)$. This condition means that $\mathcal{M}$ is isomorphic to a direct sum of copies of S_j, by Exercise 9.8.

(b) Suppose for a contradiction that $\Sigma_j^+(\mathcal{M})$ has a subrepresentation isomorphic to S_j. Then we have a non-zero element $m := (m_1, \ldots, m_t) \in M^+(j)$ with $M^+(\bar{\alpha}_i)(m) = 0$ for $i = 1, \ldots, t$. But by definition, the map $M^+(\bar{\alpha}_i)$ takes $(m_1, \ldots, m_t)$ to m_i. Therefore $m_i = 0$ for $i = 1, \ldots, t$ and $m = 0$, a contradiction. $\square$

Remark 11.13. Let j be a sink of Q, then we take in Definition 11.7 distinct arrows ending at j, but we are not excluding that some of these may start at the same vertex. For example, take the Kronecker quiver

$$1 \underset{\alpha_2}{\overset{\alpha_1}{\rightrightarrows}} j$$

then for a representation $\mathcal{M}$ of this quiver, to define $\Sigma_j^+(\mathcal{M})$ we must take

$$M^+(j) = \{(m, m') \in M(1) \times M(1) \mid M(\alpha_1)(m) + M(\alpha_2)(m') = 0\}.$$

We will not introduce extra notation for multiple arrows, since the only time we have multiple arrows is for examples using the Kronecker quiver.

11.1.2   The Reflection Σ_j^- at a Source

We assume that j is a source of the quiver Q'. For every representation $\mathcal{N}$ of Q' we will construct from $\mathcal{N}$ a representation of $\sigma_j Q'$, denoted by $\Sigma_j^-(\mathcal{N})$. The idea is to keep the vector space $N(r)$ as it is, for any vertex $r \neq j$, and also to keep the linear map $N(\gamma)$ as it is, for any arrow γ which does not start at j. We want to find a vector space $N^-(j)$, and for each arrow $\beta_i : j \to i$, we want to define a linear map $N^-(\bar{\beta}_i)$ from $N(i)$ to $N^-(j)$, to be constructed using only data from $\mathcal{N}$. We first fix some notation, and then we study small examples.

Definition 11.14. Let j be a source in the quiver Q'. We label the distinct arrows starting at j by $\beta_1, \beta_2, \ldots, \beta_t$, say $\beta_i : j \to i$. Then we write $\bar{\beta}_i : i \to j$ for the arrows of $\sigma_j Q'$ obtained by reversing the β_i.

Example 11.15.

(1) Let $t = 1$, and take the quivers Q' and $\sigma_j Q'$ as follows:

$$1 \xleftarrow{\beta_1} j \quad \text{and} \quad 1 \xrightarrow{\bar{\beta}_1} j.$$

We start with a representation $\mathcal{N}$ of Q',

$$N(1) \xleftarrow{N(\beta_1)} N(j),$$

and we want to define a representation of $\sigma_j Q'$, that is,

$$N(1) \xrightarrow{N^-(\bar{\beta}_1)} N^-(j),$$

and this should only use information from $\mathcal{N}$. There is not much choice, we take
$N^-(j) := N(1)/\mathrm{im}(N(\beta_1))$, which is a quotient space of $N(1)$, and we take
$N^-(\bar{\beta}_1)$ to be the canonical surjection. This defines the representation $\Sigma_j^-(\mathcal{N})$
of $\sigma_j Q'$.

(2) Let $t = 2$, and take the quivers Q' and $\sigma_j Q'$ as follows:

$$1 \xrightarrow{\beta_1} j \xrightarrow{\beta_2} 2 \quad \text{and} \quad 1 \xrightarrow{\bar{\beta}_1} j \xleftarrow{\bar{\beta}_2} 2.$$

We start with a representation $\mathcal{N}$ of Q',

$$N(1) \xleftarrow{N(\beta_1)} N(j) \xrightarrow{N(\beta_2)} N(2).$$

Here we can use the construction of the push-out, which was introduced in
Chap. 2 (see Exercise 2.17). This takes two linear maps starting at the same
vector space and constructs from this a new space, F, explicitly,

$$F = (N(1) \times N(2))/C, \quad \text{where} \quad C = \{(N(\beta_1)(x), N(\beta_2)(x)) \mid x \in N(j)\}.$$

We have then a canonical linear map $\mu_1 : N(1) \rightarrow F$ defined by
$\mu_1(n_1) = (n_1, 0) + C$, and similarly $\mu_2 : N(2) \rightarrow F$ is the map
$\mu_2(n_2) = (0, n_2) + C$. We define $N^-(j) := F$ and $N^-(\bar{\beta}_1) := \mu_1$, and
$N^-(\bar{\beta}_2) := \mu_2$. That is, the representation $\Sigma_j^-(\mathcal{N})$ is then

$$N(1) \xrightarrow{\mu_1} F \xleftarrow{\mu_2} N(2).$$

(3) Let $t = 3$. We take the following quiver Q'

$$2$$

$$\uparrow \beta_2$$

$$1 \xleftarrow{\quad} j \xrightarrow{\quad} 3$$
$$\quad\ \beta_1 \qquad\quad \beta_3$$

Then $\sigma_j Q'$ is the quiver

$$2$$

$$\downarrow \bar{\beta}_2$$

$$1 \xrightarrow{\quad} j \xleftarrow{\quad} 3$$
$$\quad\ \bar{\beta}_1 \qquad\quad \bar{\beta}_3$$

Let $\mathcal{N}$ be a representation of Q', we want to construct from $\mathcal{N}$ a representation $\Sigma_j^-(\mathcal{N})$ of $\sigma_j Q'$. We modify the idea of the previous example. We set $N^-(i) := N(i)$ for $i = 1, 2, 3$, and we define $N^-(j)$ to be the quotient space

$$N^-(j) := (N(1) \times N(2) \times N(3))/C_\mathcal{N},$$

where $C_\mathcal{N} := \{(N(\beta_1)(x), N(\beta_2)(x), N(\beta_3)(x)) \mid x \in N(j)\}$. As the required linear map $N^-(\bar{\beta}_1) : N(1) \to N^-(j)$ we take the canonical map

$$x \mapsto (x, 0, 0) + C_\mathcal{N},$$

and similarly we define $N^-(\bar{\beta}_i)$ for $i = 2, 3$. Then $\Sigma_j^-(\mathcal{N})$ is a representation of the quiver $\sigma_j Q'$.

We consider two explicit examples.
(i) Let $\mathcal{N}$ be the representation of Q' with $N(i) = K$ for $i = 1, 2, 3$ and with $N(j) = K^2$. Take

$$N(\beta_1)(x_1, x_2) := x_1$$
$$N(\beta_2)(x_1, x_2) := x_2$$
$$N(\beta_3)(x_1, x_2) := x_1 + x_2.$$

Then $N^-(j) = (K \times K \times K)/C_\mathcal{N}$ and

$$C_\mathcal{N} = \{(x_1, x_2, x_1 + x_2) \mid (x_1, x_2) \in K^2\}.$$

We see that $C_\mathcal{N}$ is 2-dimensional and $N^-(j)$ is therefore 1-dimensional, spanned for instance by $(1, 0, 0) + C_\mathcal{N}$.

(ii) Let $\mathcal{N}$ be the representation of Q' with $N(i) = 0$ for $i = 1, 2, 3$, and $N(j) = K$, then $\Sigma_j^-(\mathcal{N})$ is the zero representation. Note that $\mathcal{N} = \mathcal{S}_j$, the simple representation for the vertex j.

The definition of $\Sigma_j^-(\mathcal{N})$ in general, where j is a source of a quiver Q', is essentially the same as in the examples. Recall our notation, as it is fixed in Definition 11.14.

Definition 11.16. Let Q' be a quiver and assume j is a source of Q'. For any representation $\mathcal{N}$ of Q' we define a representation $\Sigma_j^-(\mathcal{N})$ of $\sigma_j Q'$ as follows. Given $\mathcal{N}$, then $\Sigma_j^-(\mathcal{N})$ is the representation such that

$$N^-(r) = \begin{cases} N(r) & \text{if } r \neq j \\ (N(1) \times \ldots \times N(t))/C_\mathcal{N} & \text{if } r = j, \end{cases}$$

where

$$C_\mathcal{N} := \{(N(\beta_1)(x), \ldots, N(\beta_t)(x)) \mid x \in N(j)\}.$$

Next, define $N^-(\gamma) = N(\gamma)$ if γ is an arrow which does not start at j, and for $1 \leq i \leq t$ define the linear map $N^-(\bar{\beta}_i) : N(i) \to N^-(j)$ by setting

$$N^-(\bar{\beta}_i)(n_i) := (0, \ldots, 0, n_i, 0, \ldots, 0) + C_\mathcal{N} \in N^-(j)$$

with $n_i \in N(i)$ in the i-th coordinate.

To compare the representation type of the quivers Q' and $\sigma_j Q'$, we need to keep track over direct sum decompositions. The construction of Σ_j^- is compatible with taking direct sums, but the situation is slightly more complicated than for Σ_j^+ in Lemma 11.10, since in general Σ_j^- does not take subrepresentations to subrepresentations, see Exercise 11.12 for an example.

The direct product of representations has been defined in Exercise 9.13, as an analogue of the direct product of modules.

Lemma 11.17. *Let Q' be a quiver and assume j is a source in Q'.*

(a) *Assume $\mathcal{N}$ is a representation of Q' and $\mathcal{N} = \mathcal{X} \oplus \mathcal{Y}$ with subrepresentations $\mathcal{X}$ and $\mathcal{Y}$. Then $\Sigma_j^-(\mathcal{N})$ is isomorphic to the direct product $\Sigma_j^-(\mathcal{X}) \times \Sigma_j^-(\mathcal{Y})$ of representations.*

(b) *Let $V = \Sigma_j^-(\mathcal{X}) \times \Sigma_j^-(\mathcal{Y})$ be the direct product representation in part (a). Then V has subrepresentations $\widetilde{\mathcal{X}}$ and $\widetilde{\mathcal{Y}}$ where $V = \widetilde{\mathcal{X}} \oplus \widetilde{\mathcal{Y}}$, and as representations of Q we have $\widetilde{\mathcal{X}} \cong \Sigma_j^-(\mathcal{X})$ and $\widetilde{\mathcal{Y}} \cong \Sigma_j^-(\mathcal{Y})$.*

(c) *Using the isomorphisms in (b) as identifications, we have*

$$\Sigma_j^-(\mathcal{N}) \cong \Sigma_j^-(\mathcal{X}) \oplus \Sigma_j^-(\mathcal{Y}).$$

This lemma will be proved in Sect. 12.1.1. Note that part (b) is a direct application of Exercise 9.13, and part (c) follows from parts (a) and (b). So it remains to prove part (a) of the lemma, and this is done in Sect. 12.1.1.

With this result, we focus on indecomposable representations, and we consider a small example.

Example 11.18. We consider the quiver $1 \xrightarrow{\alpha} 2$, and we reflect at the source 1. Recall that the quiver has three indecomposable representations, they are listed in the left column of the table below. We compute the reflected representations $\Sigma_1^-(\mathcal{N})$, using Example 11.15. The representations $\Sigma_1^-(\mathcal{N})$ are listed in the right column of the following table.

$$
\begin{array}{c|c}
\mathcal{N} & \Sigma_1^-(\mathcal{N}) \\
\hline
K \longrightarrow 0 & 0 \longleftarrow 0 \\
K \xrightarrow{\mathrm{id}_K} K & 0 \longleftarrow K \\
0 \longrightarrow K & K \xleftarrow{\mathrm{id}_K} K
\end{array}
$$

We see that Σ_1^- permutes the indecomposable representations other than the simple representation $\mathcal{S}_1$. Moreover, it takes $\mathcal{S}_1$ to the zero representation, and $\mathcal{S}_1$ does not appear as $\Sigma_1^-(\mathcal{N})$ for some $\mathcal{N}$.

We can generalize the last observation in this example.

Proposition 11.19. *Assume Q' is a quiver, and j is a source of Q'. Let $\mathcal{N}$ be a representation of Q'.*

(a) $\Sigma_j^-(\mathcal{N})$ *is the zero representation if and only if $N(i) = 0$ for all $i \neq j$, equivalently, $\mathcal{N}$ is isomorphic to a direct sum of copies of the simple representation $\mathcal{S}_j$.*

(b) $\Sigma_j^-(\mathcal{N})$ *has no direct summand isomorphic to the simple representation $\mathcal{S}_j$.*

Proof. (a) First, if $N(i) = 0$ for all $i \neq j$ then it follows directly from Definition 11.16 that $\Sigma_j^-(\mathcal{N})$ is the zero representation. Conversely, assume that $\Sigma_j^-(\mathcal{N})$ is the zero representation, that is $N^-(i) = 0$ for each vertex i. In particular, for $i \neq j$ we have $N(i) = N^-(i) = 0$. For the last part, see Exercise 9.8.

(b) Assume for a contradiction that $\Sigma_j^-(\mathcal{N}) = \mathcal{X} \oplus \mathcal{Y}$, where $\mathcal{X}$ is isomorphic to $\mathcal{S}_j$. Then $N(i) = N^-(i) = X(i) \oplus Y(i) = Y(i)$ for $i \neq j$ and $N^-(j) = X(j) \oplus Y(j)$ and $N^-(j) \neq Y(j)$ since $X(j)$ is non-zero. We get a contradiction if we show that $Y(j)$ is equal to $N^-(j)$.

By definition $Y(j) \subseteq N^-(j)$. Conversely, take an element in $N^-(j)$, it is of the form $(v_1, \dots, v_t) + C_{\mathcal{N}}$ with $v_i \in N(i)$. We can write it as

$$
((v_1, 0, \dots, 0) + C_{\mathcal{N}}) + ((0, v_2, 0, \dots, 0) + C_{\mathcal{N}}) + \dots + ((0, \dots, 0, v_t) + C_{\mathcal{N}}).
$$

Now $(0, \ldots, 0, v_i, 0, \ldots, 0) + C_\mathcal{N} = N^-(\bar{\beta}_i)(v_i) = Y(\bar{\beta}_i)(v_i)$ since $v_i \in N(i) = Y(i)$ and $\mathcal{Y}$ is a subrepresentation; moreover, this element lies in $Y(j)$ since $\mathcal{Y}$ is a subrepresentation. This holds for all $i = 1, \ldots, t$, and then the sum of these elements is also in $Y(j)$. We have proved $N^-(j) = Y(j)$, and have the required contradiction.   □

Remark 11.20. Let j be a source of the quiver Q'. In Definition 11.14 we take distinct arrows starting at j, but we have not excluded that some of these may end at the same vertex. For example, let Q' be the Kronecker quiver with two arrows β_1, β_2 from vertex j to vertex 1, then for a representation $\mathcal{N}$ of this quiver, to define $\Sigma_j^-(\mathcal{N})$ we must take $N^-(j) = (N(1) \times N(1))/C_\mathcal{N}$ and $C_\mathcal{N} = \{(N(\beta_1)(x), N(\beta_2)(x)) \mid x \in N(j)\}$. As for the case of a sink, we will not introduce extra notation since we only have a multiple arrow in examples with the Kronecker quiver but not otherwise.

11.1.3   Compositions $\Sigma_j^-\Sigma_j^+$ and $\Sigma_j^+\Sigma_j^-$

Assume j is a sink of a quiver Q. Then j is a source in the reflected quiver $\sigma_j Q$. If $\mathcal{M}$ is a representation of Q then $\Sigma_j^+(\mathcal{M})$, as in Definition 11.9, is a representation of $\sigma_j Q$. So it makes sense to apply Σ_j^- to this representation. We get

$$\Sigma_j^-\Sigma_j^+(\mathcal{M}) := \Sigma_j^-(\Sigma_j^+(\mathcal{M})),$$

which is a representation of Q (since $\sigma_j \sigma_j Q = Q$).

Similarly, if j is a source in a quiver Q' and $\mathcal{N}$ is a representation of Q' then we define $\Sigma_j^+\Sigma_j^-(\mathcal{N})$, which is a representation of Q'.

Example 11.21. We consider the quiver Q given by $1 \xrightarrow{\alpha} 2$. The vertex 2 is a sink in Q and hence a source in $\sigma_2 Q$. From the table in Example 11.11 we can find the composition $\Sigma_2^-\Sigma_2^+$,

$\mathcal{M}$	$\Sigma_2^+(\mathcal{M})$	$\Sigma_2^-\Sigma_2^+(\mathcal{M})$
$K \longrightarrow 0$	$K \xleftarrow{\mathrm{id}_K} K$	$K \longrightarrow 0$
$K \xrightarrow{\mathrm{id}_K} K$	$K \longleftarrow 0$	$K \xrightarrow{\mathrm{id}_K} K$
$0 \longrightarrow K$	$0 \longleftarrow 0$	$0 \longrightarrow 0$

Similarly, the vertex 1 is a source in Q and hence a sink in $\sigma_1 Q$. For the composition $\Sigma_1^+\Sigma_1^-$ we get the following, using the table in Example 11.18,

$\mathcal{N}$	$\Sigma_1^-(\mathcal{N})$	$\Sigma_1^+\Sigma_1^-(\mathcal{N})$
$K \longrightarrow 0$	$0 \longleftarrow 0$	$0 \longrightarrow 0$
$K \xrightarrow{\mathrm{id}_K} K$	$0 \longleftarrow K$	$K \xrightarrow{\mathrm{id}_K} K$
$0 \longrightarrow K$	$K \xleftarrow{\mathrm{id}_K} K$	$0 \longrightarrow K$

We observe in the first table that if $\mathcal{M}$ is not the simple representation $\mathcal{S}_2$ then $\Sigma_2^- \Sigma_2^+ (\mathcal{M})$ is isomorphic to $\mathcal{M}$. Similarly in the second table, if $\mathcal{N}$ is not the simple representation $\mathcal{S}_1$ then $\Sigma_1^+ \Sigma_1^- (\mathcal{N})$ is isomorphic to $\mathcal{N}$. We will see that this is not a coincidence.

Proposition 11.22. *Assume j is a sink of a quiver Q and let $\alpha_1, \ldots, \alpha_t$ be the arrows in Q ending at j. Suppose $\mathcal{M}$ is a representation of Q such that the linear map*

$$(M(\alpha_1), \ldots, M(\alpha_t)) : \prod_{i=1}^{t} M(i) \longrightarrow M(j)$$

is surjective. Then the representation $\Sigma_j^- \Sigma_j^+ (\mathcal{M})$ is isomorphic to $\mathcal{M}$.

This will be proved in the last chapter, see Sect. 12.1.2. The surjectivity condition is necessary. Otherwise $\mathcal{M}$ would have a direct summand isomorphic to $\mathcal{S}_j$, by Exercise 9.9, but $\Sigma_j^- \Sigma_j^+ (\mathcal{M})$ does not have such a summand, by Proposition 11.19.

Example 11.23. Let Q be the quiver of the form

$$1 \xrightarrow{\alpha_1} j \xleftarrow{\alpha_2} 2.$$

Suppose we take a representation $\mathcal{M}$ which satisfies the assumption of the above proposition. In Example 11.8 we have seen that $\Sigma_j^+ (\mathcal{M})$ is the representation

$$M(1) \xleftarrow{\pi_1} E \xrightarrow{\pi_2} M(2)$$

where

$$M^+(j) = E = \{(m_1, m_2) \in M(1) \times M(2) \mid M(\alpha_1)(m_1) + M(\alpha_2)(m_2) = 0\}$$

is the pull-back as in Exercise 2.16 and π_1, π_2 are the projection maps from E onto $M(1)$ and $M(2)$, respectively.

If $\mathcal{N} = \Sigma_j^- \Sigma_j^+ (\mathcal{M})$ then by Example 11.15 this representation has the form

$$M(1) \xrightarrow{\mu_1} F \xleftarrow{\mu_2} M(2)$$

where $N^-(j) = F = (M(1) \times M(2))/E$, the push-out as in Exercise 2.17, and μ_1 is given by $m_1 \mapsto (m_1, 0) + E$, and similarly for μ_2.

In Exercise 2.18 we have seen that F is isomorphic to $M(j)$, where an isomorphism $\varphi : F \to M(j)$ is given by $(m_1, m_2) + E \mapsto M(\alpha_1)(m_1) + M(\alpha_2)(m_2)$.

Then one checks that the tuple $(\text{id}_{M(1)}, \varphi, \text{id}_{M(2)})$ is a homomorphism of representations, and since all maps are isomorphisms, the representations $\Sigma_j^- \Sigma_j^+ (\mathcal{M})$ and $\mathcal{M}$ are isomorphic.

Returning to Example 11.21 we will now show that the observation on $\Sigma_1^+ \Sigma_1^- (\mathcal{N})$ is not a coincidence.

Proposition 11.24. *Assume j is a source of a quiver Q' and let $\beta_1, \ldots, \beta_t$ be the arrows in Q' starting at j. Suppose $\mathcal{N}$ is a representation of Q' such that $\bigcap_{i=1}^{t} \ker(N(\beta_i)) = 0$. Then the representations $\Sigma_j^+ \Sigma_j^- (\mathcal{N})$ and $\mathcal{N}$ are isomorphic.*

This is analogous to Proposition 11.22, and we will also prove it later in the final chapter, see Sect. 12.1.2. Again, the condition on the intersection of the kernels is necessary. If it does not hold then by Exercise 9.10, $\mathcal{N}$ has a direct summand isomorphic to $\mathcal{S}_j$. On the other hand, by Proposition 11.12, the representation $\Sigma_j^+ \Sigma_j^- (\mathcal{N})$ does not have such a direct summand.

The following exercise shows that if we are dealing with indecomposable (but not simple) representations, the assumptions of Propositions 11.22 and 11.24 are always satisfied.

Exercise 11.2. Assume Q is a quiver and vertex j of Q is a sink (or a source). Let $\alpha_1, \ldots, \alpha_t$ be the arrows in Q ending (or starting) in j. Suppose $\mathcal{M}$ is an indecomposable representation of Q which is not isomorphic to the simple representation $\mathcal{S}_j$.

   (a) Assume j is a sink. Show that then $M(j) = \sum_{i=1}^{t} \mathrm{im}(M(\alpha_i))$.
   (b) Assume j is a source. Show that then $\bigcap_{i=1}^{t} \ker(M(\alpha_i)) = 0$.

Hint: Apply Exercises 9.9 and 9.10.

For a worked solution of Exercise 11.2, see the Appendix.

We can now completely describe the action of the reflections Σ_j^+ and Σ_j^- on indecomposable representations.

Theorem 11.25. *Let Q be a quiver.*

(a) Assume j is a sink of Q and $\mathcal{M}$ is an indecomposable representation of Q not isomorphic to $\mathcal{S}_j$. Then $\Sigma_j^+ (\mathcal{M})$ is indecomposable and not isomorphic to $\mathcal{S}_j$.

(b) Assume j is a source in Q and $\mathcal{N}$ is an indecomposable representation of Q not isomorphic to $\mathcal{S}_j$. Then $\Sigma_j^- (\mathcal{N})$ is indecomposable and not isomorphic to $\mathcal{S}_j$.

(c) Assume j is a sink or a source of Q. The reflections $\Sigma_j^{\pm}$ give mutually inverse bijections between the indecomposable representations not isomorphic to $\mathcal{S}_j$ of Q and of $\sigma_j Q$.

(d) Assume j is a sink or a source of Q. Then Q has finite representation type if and only if so does $\sigma_j Q$.

Proof. (a) From the assumption and Exercise 11.2 (a) it follows that $M(j) = \sum_{i=1}^{t} \mathrm{im}(M(\alpha_i))$, where $\alpha_1, \ldots, \alpha_t$ are the arrows ending at the sink j. We apply Proposition 11.22 and obtain that $\Sigma_j^- \Sigma_j^+ (\mathcal{M}) \cong \mathcal{M}$. If we had $\Sigma_j^+ (\mathcal{M}) = \mathcal{U} \oplus \mathcal{V}$ for non-zero subrepresentations $\mathcal{U}$ and $\mathcal{V}$ then it would follow from Lemma 11.17 that

$$\mathcal{M} \cong \Sigma_j^- \Sigma_j^+ (\mathcal{M}) \cong \Sigma_j^- (\mathcal{U}) \oplus \Sigma_j^- (\mathcal{V}).$$

But $\mathcal{M}$ is indecomposable by assumption, so one of the summands, say $\Sigma_j^-(\mathcal{V})$, is the zero representation. By Proposition 11.19, if $\mathcal{V}$ is not the zero representation then it is isomorphic to a direct sum of copies of $\mathcal{S}_j$. On the other hand, $\mathcal{V}$ is a direct summand of $\Sigma_j^+(\mathcal{M})$ and therefore, by Proposition 11.12, it does not have a direct summand isomorphic to $\mathcal{S}_j$. It follows that $\mathcal{V}$ must be the zero representation, and we have a contradiction. This shows that $\Sigma_j^+(\mathcal{M})$ is indecomposable. The last part also follows from Proposition 11.12.

Part (b) is proved similarly; see Exercise 11.3 below.

(c) Say j is a sink of the quiver Q, and $Q' = \sigma_j Q$. If $\mathcal{M}$ is an indecomposable representation of Q which is not isomorphic to $\mathcal{S}_j$ then Exercise 11.2 and Proposition 11.22 give that $\mathcal{M} \cong \Sigma_j^- \Sigma_j^+(\mathcal{M})$. If $\mathcal{N}$ is an indecomposable representation of Q' not isomorphic to $\mathcal{S}_j$ then by Exercise 11.2 and Proposition 11.24 we have $\mathcal{N} \cong \Sigma_j^+ \Sigma_j^-(\mathcal{N})$. Part (c) is proved.

(d) This follows directly from (c).   □

Exercise 11.3. Write out a proof of part (b) of Theorem 11.25.

The following consequence of Theorem 11.25 shows that the representation type of a quiver does not depend on the orientation of the arrows, as long as the underlying graph is a tree.

Corollary 11.26. *Let Γ be a graph which is a tree. Then any two quivers with underlying graph Γ have the same representation type.*

Proof. By Proposition 11.5 we know that any two quivers with underlying graph Γ are related by a sequence of reflections in sinks or sources, so the corollary follows directly from Theorem 11.25.   □

11.2 Quivers of Infinite Representation Type

We will now prove that if the underlying graph of Q is not a union of Dynkin diagrams then Q has infinite representation type. This is one direction of Gabriel's theorem. As we have seen in Lemma 9.27, it is enough to consider connected quivers, and we should deal with smallest connected quivers whose underlying graph is not a Dynkin diagram (see Lemma 9.26).

Proposition 11.27. *Assume Q is a connected quiver with no oriented cycles. If the underlying graph of Q is not a Dynkin diagram, then Q has infinite representation type.*

The proof of Proposition 11.27 will take the entire section.

By Lemma 10.1 we know that a connected quiver Q whose underlying graph is not a Dynkin diagram must have a subquiver Q' whose underlying graph is a Euclidean diagram. By Lemma 9.26, it suffices to show that the subquiver Q' has infinite representation type. We will do this case-by-case going through the Euclidean diagrams listed in Fig. 10.2.

We start with Euclidean diagrams of type $\widetilde{A}_n$, which has almost been done already.

Proposition 11.28. *Assume Q' is a quiver without oriented cycles whose underlying graph is a Euclidean diagram of type $\widetilde{A}_n$. Then Q' is of infinite representation type.*

Proof. Let $n = 1$, then Q' is the Kronecker quiver, and we have seen in Example 9.30 that it has infinite representation type. Now assume $n > 1$. We will stretch the Kronecker quiver repeatedly as described in Definition 9.20; and Exercise 9.4 shows that Q' can be obtained from the Kronecker quiver by finitely many stretches. Now Lemma 9.31 implies that Q' has infinite representation type.
□

We will now deal with quivers whose underlying graphs are other Euclidean diagrams as listed in Fig. 10.2. We observe that each of them is a tree. Therefore, by Corollary 11.26, in each case we only need to show that it has infinite representation type just for one orientation, which we can choose as we like.

We will use a more general tool. This is inspired by the indecomposable representation of the quiver with underlying graph a Dynkin diagram D_4 where the space at the branch vertex is 2-dimensional, which we have seen a few times. In Lemma 9.5 we have proved that for this representation, any endomorphism is a scalar multiple of the identity. The following shows that this actually can be used to produce many representations of a new quiver obtained by just adding one vertex and one arrow.

Lemma 11.29. *Assume Q is a quiver and let K be a field. Suppose Q has a representation $\mathcal{M}$ such that*

(a) every endomorphism of $\mathcal{M}$ is a scalar multiple of the identity,
(b) there is a vertex k of Q such that $\dim_K \mathcal{M}(k) = 2$.

Let $\widehat{Q}$ be the quiver obtained from Q by adjoining a new vertex ω together with a new arrow $\omega \xrightarrow{\alpha} k$. Then for each $\lambda \in K$ there is an indecomposable representation $\mathcal{M}_\lambda$ of $\widehat{Q}$ which extends $\mathcal{M}$ (that is, $\mathcal{M}$ is the restriction of $\mathcal{M}_\lambda$ in the sense of Definition 9.15), and such that $\mathcal{M}_\lambda$ and $\mathcal{M}_\mu$ are not isomorphic for $\lambda \neq \mu$.

Definition 11.30. We call a representation $\mathcal{M}$ satisfying (a) and (b) of Lemma 11.29 a *special* representation.

Proof. We fix a basis for the vector space $M(k)$ and identify $M(k)$ with K^2. For any $\lambda \in K$ we extend $\mathcal{M}$ to a representation $\mathcal{M}_\lambda$ of $\widehat{Q}$ defined as follows:

$$M_\lambda(i) = \begin{cases} M(i) & \text{if } i \neq \omega \\ K & \text{if } i = \omega \end{cases}$$

and $M_\lambda(\beta) = M(\beta)$ for any arrow $\beta \neq \alpha$, and $M_\lambda(\alpha) : M_\lambda(\omega) \rightarrow M(k)$, that is, from K to K^2, is the map $x \mapsto (x, \lambda x)$. We want to show that M_λ is indecomposable, and that if $\lambda \neq \mu$ then $M_\lambda \not\cong M_\mu$.

Let $\varphi : M_\lambda \rightarrow M_\mu$ be a homomorphism of representations. Then the restriction of φ to vertices in Q is a homomorphism from M to M. By the assumption, this is a scalar multiple of the identity. In particular, at the vertex k we have $\varphi_k = c\,\mathrm{id}_{K^2}$ for some $c \in K$. Now, the space at vertex ω is the one-dimensional space K, so φ_ω is also a scalar multiple of the identity, say $\varphi_\omega = d\,\mathrm{id}_K$ with $d \in K$. Consider the commutative diagram

$$
\begin{array}{ccc}
M_\lambda(\omega) & \xrightarrow{\;M_\lambda(\alpha)\;} & M_\lambda(k) \\
\varphi_\omega \downarrow & & \downarrow \varphi_k \\
M_\mu(\omega) & \xrightarrow{\;M_\mu(\alpha)\;} & M_\mu(k)
\end{array}
$$

This gives that for $x \in M_\lambda(\omega) = K$ we have

$$
c(x, \lambda x) = (\varphi_k \circ M_\lambda(\alpha))(x) = (M_\mu(\alpha) \circ \varphi_\omega)(x) = d(x, \mu x).
$$

So for all $x \in M_\lambda(\omega) = K$ we have $cx = dx$ and $c\lambda x = d\mu x$.

If $\lambda = \mu$ then we get the only condition $cx = dx$ for all $x \in K$ and this clearly implies $c = d$. But since the restriction of φ to Q is already a scalar multiple of the identity we deduce that every homomorphism $\varphi : M_\lambda \rightarrow M_\lambda$ is a scalar multiple of the identity. This implies that M_λ is indecomposable, by Lemma 9.12.

If $\lambda \neq \mu$ then the above two conditions combine to

$$
c\lambda x = d\mu x = c\mu x \quad \text{and} \quad \lambda dx = \lambda cx = d\mu x,
$$

that is, $(\lambda - \mu)cx = 0$ and $(\lambda - \mu)dx = 0$. Since $\lambda - \mu \neq 0$ it follows that $cx = 0$ and $dx = 0$ for all $x \in K$, and hence $c = d = 0$. In other words, for $\lambda \neq \mu$ the only homomorphism $\varphi : M_\lambda \rightarrow M_\mu$ is the zero homomorphism. In particular, there is no isomorphism and $M_\lambda \not\cong M_\mu$. $\qquad\qquad\square$

We consider now quivers whose underlying graph is a Euclidean diagram of type $\widetilde{D}_n$. The main work will be for the case $n = 4$. We take the special representation M of the quiver Q with underlying graph of type D_4 as in Lemma 9.5,

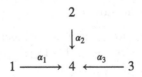

Recall that for the representation $\mathcal{M}$ we have $M(i) = K$ for $i \neq 4$ and $M(4) = K^2$. Now we take the extended quiver $\widehat{Q}$ as in Lemma 11.29. This is of the form

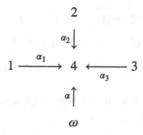

and the underlying graph is a Euclidean diagram of type $\widetilde{D}_4$. Using the representation $\mathcal{M}$ of Q from Lemma 9.5 we find by Lemma 11.29 pairwise non-isomorphic indecomposable representations $\mathcal{M}_\lambda$ of $\widehat{Q}$ for each $\lambda \in K$. In particular, the quiver $\widehat{Q}$ has infinite representation type over any infinite field K.

However, we want to prove Gabriel's theorem for arbitrary fields. We will therefore construct indecomposable representations of $\widehat{Q}$ of arbitrary dimension, which then shows that $\widehat{Q}$ always has infinite representation type, independent of the field. Roughly speaking, to construct these representations we take direct sums of the special representation, and glue them together at vertex 4 using a 'Jordan block' matrix. To set our notation, we denote by

$$
J_m = \begin{pmatrix} 1 & & & & \\ 1 & 1 & & & \\ & 1 & \ddots & & \\ & & \ddots & \ddots & \\ & & & 1 & 1 \end{pmatrix}
$$

the $m \times m$ Jordan block matrix with eigenvalue 1.

In the following, when writing K^m or similar, we always mean column vectors; for typographical convenience we sometimes write column vectors as transposed row vectors. We will write linear maps as matrices, applying them to the left.

Definition 11.31. Let K be a field and consider the quiver $\widehat{Q}$ of type $\widetilde{D}_4$ given above. Fix an integer $m \geq 1$ and take $V = K^m$, an m-dimensional vector space over K. Then take

$$
V(4) = \{(v_1, v_2)^t \mid v_i \in V\} = K^{2m}.
$$

Then we take each of the other spaces as subspaces of $V(4)$:

$$V(1) = \{(v, 0)^t \mid v \in V\}$$
$$V(2) = \{(0, v)^t \mid v \in V\}$$
$$V(3) = \{(v, v)^t \mid v \in V\}$$
$$V(\omega) = \{(v, J_m v)^t \mid v \in V\}.$$

The linear maps corresponding to the arrows of $\widehat{Q}$ are all taken as inclusions. This defines a representation $\mathcal{V}_m$ of the quiver $\widehat{Q}$.

Lemma 11.32. *The representation $\mathcal{V}_m$ of the quiver $\widehat{Q}$ is indecomposable.*

Proof. To prove this, we use the criterion from Lemma 9.11, that is, we show that the only endomorphisms $\varphi : \mathcal{V}_m \to \mathcal{V}_m$ of representations such that $\varphi^2 = \varphi$ are the zero and the identity homomorphism.

Since the spaces $V(i)$ are defined as subspaces of $V(4) = K^{2m}$ and all maps as inclusions, any endomorphism $\varphi : \mathcal{V}_m \to \mathcal{V}_m$ of representations is given by a linear map $V(4) \to V(4)$ which takes each $V(i)$ into $V(i)$, for $i = 1, 2, 3, \omega$. We take φ as a linear map $\varphi : K^{2m} \to K^{2m}$ and write it as a matrix in block form as

$$\begin{pmatrix} A & B \\ C & D \end{pmatrix}$$

with $m \times m$-matrices A, B, C, D. The linear map φ maps $V(1)$ to $V(1)$, therefore C is the zero matrix. Since it maps $V(2)$ to $V(2)$, the matrix B is also the zero matrix. Moreover, it maps $V(3)$ to $V(3)$, which implies $A = D$. It must also map $V(\omega)$ to $V(\omega)$, and this is the case if and only if A commutes with J_m. In fact, we have

$$\begin{pmatrix} A & 0 \\ 0 & A \end{pmatrix} \begin{pmatrix} v \\ J_m v \end{pmatrix} = \begin{pmatrix} Av \\ A J_m v \end{pmatrix}$$

and for this to be contained in $V(\omega)$ we must have $A J_m v = J_m A v$ for all $v \in K^m$; and if this holds for all v then $A J_m = J_m A$.

We can use the same argument as in the proof of Lemma 8.5 and also in Example 9.30, that is, we apply Exercise 8.1. The matrix A is an endomorphism of the module V_β for the algebra $K[X]/(f)$, where $f = (X - 1)^m$ and where β is given by J_m. This is a cyclic module, generated by the first basis element. By Exercise 8.1, if $A^2 = A$ then A is zero or the identity.

Therefore the only idempotent endomorphisms $\varphi : \mathcal{V}_m \to \mathcal{V}_m$ are zero and the identity. As mentioned at the beginning of the proof, Lemma 9.11 then gives that the representation $\mathcal{V}_m$ is indecomposable. $\qquad\square$

Note that if we had taken any non-zero $\lambda \in K$ as the eigenvalue of the Jordan block above we would still have obtained indecomposable representations. We chose $\lambda = 1$ since this lies in any field.

The above Lemma 11.32 shows that the quiver $\widehat{Q}$ has infinite representation type over any field K. Indeed, the indecomposable representations V_m are pairwise non-isomorphic since they have different dimensions. We will now use this to show that every quiver whose underlying graph is of type $\widetilde{D}_n$ for $n \geq 4$ has infinite representation type. Recall that we may choose the orientation as we like, by Corollary 11.26. For example, for type $\widetilde{D}_4$ it is enough to deal with the above quiver $\widehat{Q}$.

Proposition 11.33. *Every quiver whose underlying graph is a Euclidean diagram of type $\widetilde{D}_n$ for some $n \geq 4$ has infinite representation type.*

Proof. Assume first that $n = 4$. Then $\widehat{Q}$ has infinite representation type by Lemma 11.32. Indeed, if $m_1 \neq m_2$ then V_{m_1} and V_{m_2} cannot be isomorphic, as they have different dimensions. By Corollary 11.26, any quiver with underlying graph $\widetilde{D}_4$ has infinite representation type.

Now assume $n > 4$. Any quiver of type $\widetilde{D}_n$ can be obtained from the above quiver $\widehat{Q}$ by a finite sequence of stretches in the sense of Definition 9.20. When $n = 5$ this is Example 9.21, and for $n \geq 6$ one may replace the branching vertex of the $\widetilde{D}_4$ quiver by a quiver whose underlying graph is a line with the correct number of vertices. By Lemma 9.31, the stretched quiver has infinite representation type, and then by Corollary 11.26, every quiver of type $\widetilde{D}_n$ has infinite representation type.   □

So far we have shown that every quiver without oriented cycles whose underlying graph is a Euclidean diagram of type $\widetilde{A}_n$ $(n \geq 1)$ or $\widetilde{D}_n$ $(n \geq 4)$ has infinite representation type over any field K; see Propositions 11.28 and 11.33.

The only Euclidean diagrams in Fig. 10.2 we have not yet dealt with are quivers whose underlying graphs are of types $\widetilde{E}_6$, $\widetilde{E}_7$, and $\widetilde{E}_8$. The proof that these have infinite representation type over any field K will follow the same strategy as for type $\widetilde{D}_n$ above. However, the proofs are longer and more technical. Therefore, they are postponed to Sect. 12.2.

Taking these proofs for granted we have now completed the proof of Proposition 11.27 which shows that every quiver whose underlying graph is not a union of Dynkin diagrams has infinite representation type.

11.3 Dimension Vectors and Reflections

The next task is to show that any quiver whose underlying graph is a union of Dynkin diagrams has finite representation type. Recall that by Lemma 9.27 we need only to look at connected quivers. At the same time we want to parametrize the indecomposable representations. The appropriate invariants for this are the dimension vectors, which we will now define.

Again, we fix a field K and all representations are over K.

Definition 11.34. Let Q be a quiver and assume $\mathcal{M}$ is a representation of Q. Suppose Q has n vertices; we label the vertices by $1, 2, \ldots, n$. The *dimension vector* of the representation $\mathcal{M}$ is defined to be

$$\underline{\dim}\mathcal{M} := (\dim_K M(1), \ldots, \dim_K M(n)) \in \mathbb{Z}^n.$$

Note that by definition the dimension vector depends on the labelling of the vertices.

Example 11.35.

(1) Let Q be a quiver without oriented cycles. By Theorem 9.8, the simple representations of Q correspond to the vertices of Q. The simple representation S_j labelled by vertex j has dimension vector ε_j, the unit vector.

(2) In Example 9.28 we have classified the indecomposable representations of the quiver $1 \longrightarrow 2$, with underlying graph a Dynkin diagram A_2. We have seen that the three indecomposable representations have dimension vectors $\varepsilon_1 = (1, 0)$, $\varepsilon_2 = (0, 1)$ and $(1, 1)$.

Remark 11.36. Given two isomorphic representations $\mathcal{M} \cong \mathcal{N}$ of a quiver Q, then for each vertex i the spaces $M(i)$ and $N(i)$ are isomorphic, so they have the same dimension and hence $\underline{\dim}\mathcal{M} = \underline{\dim}\mathcal{N}$. That is, $\mathcal{M}$ and $\mathcal{N}$ have the same dimension vector. We will prove soon that for a Dynkin quiver the dimension vector actually determines the indecomposable representation.

However, this is not true for arbitrary quivers, and we have seen this already: in Example 9.29 we have constructed indecomposable representations C_λ for the Kronecker quiver. They all have dimension vector $(1, 1)$, but as we have shown, the representations C_λ for different values of $\lambda \in K$ are not isomorphic.

We consider quivers with a fixed underlying graph Γ. We will now analyse how dimension vectors of representations change if we apply the reflections defined in Sect. 11.1 to representations. We recall the definition of the bilinear form of Γ, and the definition of a reflection of $\mathbb{Z}^n$, from Definition 10.2. The bilinear form $(-, -)_\Gamma : \mathbb{Z}^n \times \mathbb{Z}^n \to \mathbb{Z}$ is defined on unit vectors by

$$(\varepsilon_i, \varepsilon_j)_\Gamma = \begin{cases} -d_{ij} & \text{if } i \neq j \\ 2 & \text{if } i = j \end{cases}$$

where d_{ij} is the number of edges between vertices i and j, and then extended bilinearly. Soon we will focus on the case when Γ is a Dynkin diagram, and then $d_{ij} = 0$ or 1, but we will also take Γ obtained from the Kronecker quiver, where $d_{12} = 2$.

For any vertex j of Γ the reflection map is defined as

$$s_j : \mathbb{Z}^n \to \mathbb{Z}^n, \quad s_j(a) = a - (a, \varepsilon_j)_\Gamma \varepsilon_j.$$

We will now see that the reflection maps precisely describe how the dimension vectors are changed when a representation is reflected.

Proposition 11.37.

(a) *Let Q be a quiver with underlying graph Γ. Assume j is a sink of Q and $\alpha_1, \ldots, \alpha_t$ are the arrows in Q ending at j. Let $\mathcal{M}$ be a representation of Q. If $\sum_{i=1}^{t} \operatorname{im}(M(\alpha_i)) = M(j)$ then $\underline{\dim}\Sigma_j^+(\mathcal{M}) = s_j(\underline{\dim}\mathcal{M})$. In particular, this holds if $\mathcal{M}$ is indecomposable and not isomorphic to the simple representation $\mathcal{S}_j$.*

(b) *Let Q' be a quiver with underlying graph Γ. Assume j is a source of Q' and $\beta_1, \ldots, \beta_t$ are the arrows in Q' starting at j. Let $\mathcal{N}$ be a representation of Q'. If $\bigcap_{i=1}^{t} \ker(N(\beta_i)) = 0$ then $\underline{\dim}\Sigma_j^-(\mathcal{N}) = s_j(\underline{\dim}\mathcal{N})$. In particular, this holds if $\mathcal{N}$ is indecomposable and not isomorphic to the simple representation $\mathcal{S}_j$.*

Proof. The second parts of the statements of (a) and (b) are part of Exercise 11.2; we include a worked solution in the appendix.

(a) We compare the entries in the vectors $\underline{\dim}\Sigma_j^+(\mathcal{M})$ and $s_j(\underline{\dim}\mathcal{M})$, respectively.

For vertices $i \neq j$ we have $M^+(i) = M(i)$ (see Definition 11.9). So the i-th entry in $\underline{\dim}\Sigma_j^+(\mathcal{M})$ is equal to $\dim_K M(i)$. On the other hand, the i-th entry in $s_j(\underline{\dim}\mathcal{M})$ also equals $\dim_K M(i)$ because s_j only changes the j-th coordinate (see Definition 10.2).

Now let $i = j$, then, by Definition 11.9, $M^+(j)$ is the kernel of the linear map

$$M(1) \times \ldots \times M(t) \to M(j), \quad (m_1, \ldots, m_t) \mapsto \sum_{i=1}^{t} M(\alpha_i)(m_i).$$

By rank-nullity, and using the assumption, we have

$$\dim_K M^+(j) = \dim_K(M(1) \times \ldots \times M(t)) - \dim_K M(j)$$

$$= \left(\sum_{i=1}^{t} \dim_K M(i) \right) - \dim_K M(j).$$

We set $a_i = \dim_K M(i)$ for abbreviation. Since $\alpha_1, \ldots, \alpha_t$ are the only arrows adjacent to the sink j, we have $d_{rj} = 0$ if a vertex r is not the starting point of one of the arrows $\alpha_1, \ldots, \alpha_t$. Recall that some of these arrows can start at the same vertex, so if r is the starting point of one of these arrows then we have d_{rj} arrows from r to j. So we can write

$$\dim_K M^+(j) = \left(\sum_{i=1}^{t} a_i \right) - a_j = \left(\sum_{r \in Q_0 \setminus \{j\}} d_{rj} a_r \right) - a_j.$$

This is the j-th coordinate of the vector $\underline{\dim} \Sigma_j^+(\mathcal{M})$. We compare this with the j-th coordinate of $s_j(\underline{\dim}\mathcal{M})$. By Definition 10.2 this is

$$a_j - (\underline{\dim}\mathcal{M}, \varepsilon_j)_\Gamma = a_j - \left(\sum_{r \in Q_0 \setminus \{j\}} -d_{rj}a_r + 2a_j \right),$$

which is the same.

(b) Now let j be a source. Similarly to part (a) it can be seen that for $i \neq j$ the i-th coordinates coincide in the two relevant vectors.

So it remains to consider the j-th coordinates. By Definition 11.16, the vector space $N^-(j)$ is the cokernel of the map

$$N(j) \to N(1) \times \ldots \times N(t), \quad x \mapsto (N(\beta_1)(x), \ldots, N(\beta_t)(x)).$$

By our assumption this linear map is injective, hence the image has dimension equal to $\dim_K N(j)$. We set $b_i = \dim_K N(i)$ for abbreviation and get

$$\dim_K N^-(j) = \left(\sum_{i=1}^t \dim_K N(i) \right) - \dim_K N(j) = \sum_{r \in Q_0 \setminus \{j\}} d_{rj}b_r - b_j,$$

which is the same formula as in part (a) and by what we have seen there, this is the j-th coordinate of $s_j(\underline{\dim}\mathcal{N})$.   □

We illustrate the above result with an example.

Example 11.38. As in Example 11.11 we consider the quiver Q of the form $1 \longrightarrow 2$ and reflect at the sink 2. The corresponding reflection map is equal to

$$s_2 : \mathbb{Z}^2 \to \mathbb{Z}^2, \quad a = (a_1, a_2) \mapsto a - (a, \varepsilon_2)_\Gamma \varepsilon_2 = a - (-a_1 + 2a_2)\varepsilon_2 = (a_1, a_1 - a_2).$$

In Example 11.11 we have listed the reflections of the three indecomposable representations of Q. In the following table we compare their dimension vectors with the vectors $s_2(\underline{\dim}\mathcal{M})$.

$\mathcal{M}$	$\underline{\dim}\mathcal{M}$	$s_2(\underline{\dim}\mathcal{M})$	$\Sigma_2^+(\mathcal{M})$
$K \longrightarrow 0$	$(1, 0)$	$(1, 1)$	$K \overset{\mathrm{id}_K}{\longleftarrow} K$
$K \overset{\mathrm{id}_K}{\longrightarrow} K$	$(1, 1)$	$(1, 0)$	$K \longleftarrow 0$
$0 \longrightarrow K$	$(0, 1)$	$(0, -1)$	$0 \longleftarrow 0$

This confirms what we have proved in Proposition 11.37. We also see that exluding the simple representation $\mathcal{S}_2$ in Proposition 11.37 is necessary, observing that $s_2(\underline{\dim}\mathcal{S}_2)$ is not a dimension vector of a representation.

11.4   Finite Representation Type for Dynkin Quivers

Assume Q is a quiver with underlying graph Γ. We assume Γ is a Dynkin diagram of one of the types A_n, D_n, E_6, E_7, E_8 (as in Fig. 10.1). Let q_Γ be the quadratic form associated to Γ, see Definition 10.2. We have studied the set of roots,

$$\Delta_\Gamma = \{x \in \mathbb{Z}^n \mid q_\Gamma(x) = 1\}$$

and we have proved that it is finite (see Proposition 10.12). We have also seen that a root x is either positive or negative, see Lemma 10.13. Recall that a non-zero $x \in \mathbb{Z}^n$ is positive if $x_i \geq 0$ for all i, and it is negative if $x_i \leq 0$ for all i.

In this section we will prove the following, which will complete the proof of Gabriel's Theorem:

Theorem 11.39. *Assume Q is a quiver whose underlying graph Γ is a union of Dynkin diagrams of type A_n ($n \geq 1$) or D_n ($n \geq 4$), or E_6, E_7, E_8. Then the following hold.*

(1) If $\mathcal{M}$ is an indecomposable representation of Q then $\underline{\dim}\mathcal{M}$ is in Δ_Γ.

(2) Every positive root is equal to $\underline{\dim}\mathcal{M}$ for a unique indecomposable representation $\mathcal{M}$ of Q.

In particular, Q has finite representation type.

Before starting with the proof, we consider some small examples.

Example 11.40.

(1) Let Q be the quiver $1 \longrightarrow 2$ with underlying graph the Dynkin diagram A_2. In Example 9.28 we have seen that this has three indecomposable representations, with dimension vectors $\varepsilon_1, \varepsilon_2$ and $(1, 1)$. We see that these are precisely the positive roots as described in Example 10.8.

(2) Let Q be the quiver $1 \longrightarrow 2 \longleftarrow 3$. The Exercises 11.8 and 11.9 prove using elementary linear algebra that the above theorem holds for Q. By applying reflections $\Sigma_1^\pm$ or $\Sigma_2^\pm$ (see Exercise 11.11), one deduces that the theorem holds for any quiver with underlying graph a Dynkin diagram of type A_3.

To prove Theorem 11.39 we first show that it suffices to prove it for connected quivers. This will follow from the lemma below, suitably adapted to finite unions of Dynkin diagrams.

Lemma 11.41. *Let $Q = Q' \cup Q''$ be a disjoint union of two quivers such that the underlying graph $\Gamma = \Gamma' \cup \Gamma''$ is a union of two Dynkin diagrams. Then Theorem 11.39 holds for Q if and only if it holds for Q' and for Q''.*

Proof. We label the vertices of Q as $\{1, \ldots, n'\} \cup \{n' + 1, \ldots, n' + n''\}$ where $\{1, \ldots, n'\}$ are the vertices of Q' and $\{n' + 1, \ldots, n' + n''\}$ are the vertices of Q''. So we can write every dimension vector in the form (x', x'') with $x' \in \mathbb{Z}^{n'}$ and $x'' \in \mathbb{Z}^{n''}$.

We have seen in Lemma 9.19 that the indecomposable representations of Q are precisely the extensions by zero of the indecomposable representations of Q' and of Q''. This means that the dimension vectors of indecomposable representations of Q are all of the form $(x', 0, \ldots, 0)$ or $(0, \ldots, 0, x'')$. The quadratic form for Γ is

$$q_\Gamma(x) = \sum_{i=1}^{n'+n''} x_i^2 - \sum_{i<j} d_{ij} x_i x_j$$

(see Definition 10.6). Since there are no edges between vertices of Γ' and Γ'', we see that for $x = (x', x'') \in \mathbb{Z}^{n'+n''}$ we have

$$q_\Gamma(x) = q_{\Gamma'}(x') + q_{\Gamma''}(x''). \tag{11.1}$$

In particular, $x = (x', x'')$ is a root of Γ if and only if $q_{\Gamma'}(x') + q_{\Gamma''}(x'') = 1$ (see Definition 10.6). By assumption, Γ' and Γ'' are Dynkin diagrams. Then by Proposition 10.10 the quadratic forms $q_{\Gamma'}$ and $q_{\Gamma''}$ are positive definite. Since the quadratic forms have integral values, the condition $q_{\Gamma'}(x') + q_{\Gamma''}(x'') = 1$ thus holds if and only if $q_{\Gamma'}(x') = 1$ and $q_{\Gamma''}(x'') = 0$, or vice versa. Thus, $x = (x', x'')$ is a root of Γ if and only if $x = (x', 0, \ldots, 0)$ with x' a root of Γ' or $x = (0, \ldots, 0, x'')$ with x'' a root of Γ''.

Together with the above description of indecomposable representations of Q this shows that statement (1) of Theorem 11.39 holds for Q if and only if it holds for Q' and Q''.

For statement (2) note that the positive roots of Γ are precisely the vectors of the form $x = (x', 0, \ldots, 0)$ with x' a positive root of Γ' or $x = (0, \ldots, 0, x'')$ with x'' a positive root of Γ''. From this and Lemma 9.19 one sees that (2) holds for Q if and only if it holds for Q' and Q''. $\square$

Remark 11.42. Observe that on the way the above lemma has proved that if $\Gamma = \Gamma' \cup \Gamma''$ is a disjoint union of Dynkin diagrams then the set of roots Δ_Γ can be interpreted as a union of $\Delta_{\Gamma'}$ and $\Delta_{\Gamma''}$, where the roots in $\Delta_{\Gamma'}$ and $\Delta_{\Gamma''}$ are extended by zeros.

Let Γ be a Dynkin diagram as in Fig. 10.1. We do not have to deal with an arbitrary quiver, instead it suffices to prove Theorem 11.39 for a fixed orientation of the arrows and a fixed labelling of the vertices. That is, we have the following reduction:

Lemma 11.43. *To prove (1) and (2) of Theorem 11.39, it is enough to take a quiver Q with standard labelling (as in Example 10.4).*

Proof. By Proposition 11.5, we know that any two quivers with underlying graph Γ a Dynkin diagram are related by a sequence of reflections in sinks or sources. So it suffices to show: if j is a sink or source of Q and the theorem holds for Q, then it also holds for $\sigma_j Q$. We will write down the details when j is a sink, the other case is analogous (we leave this for Exercise 11.4). Suppose j is a sink. We assume that (1) and (2) hold for Q.

First we show that (1) holds for $\sigma_j Q$. By Theorem 11.25 any indecomposable representation of $\sigma_j Q$ is either isomorphic to the simple representation $\mathcal{S}_j$, or is of the form $\Sigma_j^+(\mathcal{M})$ where $\mathcal{M}$ is an indecomposable representation of Q not isomorphic to $\mathcal{S}_j$. The dimension vector of $\mathcal{S}_j$ is ε_j, which is in Δ_Γ (see Exercise 10.3). Moreover, in the second case, the dimension vector of $\Sigma_j^+(\mathcal{M})$ is $s_j(\underline{\dim}\mathcal{M})$, by Proposition 11.37. We have $\underline{\dim}\mathcal{M}$ is in Δ_Γ by assumption, and by Lemma 10.9, s_j takes roots to roots. Moreover, $s_j(\underline{\dim}\mathcal{M})$ is positive since it is the dimension vector of a representation.

We show now that (2) holds for $\sigma_j Q$. Let $x \in \Delta_\Gamma$ be a positive root. If $x = \varepsilon_j$ then $x = \underline{\dim}\mathcal{S}_j$ and clearly this is the only possibility. So let $x \neq \varepsilon_j$, we must show that there is a unique indecomposable representation of $\sigma_j Q$ with dimension vector x. Since $x \neq \varepsilon_j$, we have $y := s_j(x) \neq \varepsilon_j$, and this is also a root. It is a positive root: Since x is positive and not equal to ε_j, there is some $k \neq j$ such that $x_k > 0$ (in fact, since $q_\Gamma(\lambda\varepsilon_j) = \lambda^2 q_\Gamma(\varepsilon_j)$, the only scalar multiples of ε_j which are in Δ_Γ are $\pm\varepsilon_j$; see also Proposition 12.16). The reflection map s_j changes only the j-th coordinate, therefore $y_k = x_k > 0$, and y is a positive root.

By assumption, there is a unique indecomposable representation $\mathcal{M}$ of Q, not isomorphic to $\mathcal{S}_j$, such that $\underline{\dim}\mathcal{M} = y$. Let $\mathcal{N} := \Sigma_j^+(\mathcal{M})$. This is an indecomposable representation, and $\underline{\dim}\mathcal{N} = s_j(y) = x$. To prove uniqueness, let $\mathcal{N}'$ be an indecomposable representation of $\sigma_j Q$ and $\underline{\dim}\mathcal{N}' = x$, then $\mathcal{N}' \not\cong \mathcal{S}_j$, hence by Theorem 11.25 there is a unique indecomposable representation $\mathcal{M}'$ of Q with $\Sigma_j^+(\mathcal{M}') = \mathcal{N}'$. Then we have $s_j(\underline{\dim}\mathcal{M}') = x$ and hence $\underline{\dim}\mathcal{M}' = s_j(x) = y$. Since (2) holds for Q we have $\mathcal{M} \cong \mathcal{M}'$ and then $\mathcal{N} \cong \mathcal{N}'$. This proves that (2) also holds for $\sigma_j Q$. $\qquad\square$

Exercise 11.4. Write down the details for the proof of Lemma 11.43 when the vertex j is a source of Q, analogous to the case when the vertex is a sink.

Assume from now that Q has standard labelling, as in Example 10.4. Then we have the following properties:

(i) Vertex 1 is a sink of Q, and for $2 \leq j \leq n$, vertex j is a sink of the quiver $\sigma_{j-1} \ldots \sigma_1 Q$;

(ii) $\sigma_n \sigma_{n-1} \ldots \sigma_1 Q = Q$.

Note that the corresponding sequence of reflections $s_n \circ s_{n-1} \circ \ldots \circ s_1$ is the Coxeter transformation C_Γ, where Γ is the underlying graph of Q (see Definition 10.14).

Exercise 11.5. Verify (i) and (ii) in detail when Q is of type A_n as in Example 10.4.

By Theorem 11.25, Σ_1^+ takes an indecomposable representation $\mathcal{M}$ of Q which is not isomorphic to $\mathcal{S}_1$ to an indecomposable representation of $\sigma_1 Q$ not isomorphic to $\mathcal{S}_1$. Similarly, Σ_j^+ takes an indecomposable representation $\mathcal{M}$ of $\sigma_{j-1} \ldots \sigma_1 Q$ which is not isomorphic to $\mathcal{S}_j$ to an indecomposable representation of $\sigma_j \sigma_{j-1} \ldots \sigma_1 Q$ not isomorphic to S_j. Moreover, if $x = \underline{\dim} \mathcal{M}$ then $s_j(x) = \underline{\dim} \Sigma_j^+(\mathcal{M})$, by Proposition 11.37.

We also want to consider compositions of these reflections. Let $\underline{\Sigma} = \Sigma_n^+ \ldots \Sigma_1^+$. If $\mathcal{M}$ is an indecomposable representation of Q and $\underline{\Sigma}(\mathcal{M})$ is non-zero then it is again an indecomposable representation of Q with dimension vector equal to $C_\Gamma(x)$ where C_Γ is the Coxeter transformation.

The following result proves part (1) of Gabriel's Theorem 11.39.

Proposition 11.44. *Let Q be a quiver whose underlying graph Γ is a union of Dynkin diagrams. Assume $\mathcal{M}$ is an indecomposable representation of Q. Then $\underline{\dim} \mathcal{M}$ belongs to Δ_Γ.*

Proof. By Lemma 11.41 we can assume that Γ is connected, and by Lemma 11.43 we can assume that Q has standard labelling.

Let $\mathcal{M}$ have dimension vector $x \in \mathbb{Z}^n$. We apply the Coxeter transformation $C = C_\Gamma$ to x. By Lemma 10.16, there is some $r \geq 0$ such that $C^r(x) \not\geq 0$. We must have $r \geq 1$ (since $x \geq 0$), and we take r minimal with this property. Then there is some j such that for

$$\tau := s_j \circ s_{j-1} \circ \ldots \circ s_1 \circ C^{r-1}$$

we have $\tau(x) \not\geq 0$; we also take j minimal. Then $s_{j-1} \circ \ldots \circ s_1 \circ C^{r-1}(x) \geq 0$ but applying s_j gives $\tau(x) \not\geq 0$. Let

$$\mathcal{M}' := \Sigma_{j-1}^+ \ldots \Sigma_1^+ (\underline{\Sigma}^{r-1}(\mathcal{M})).$$

By minimality of r and j and by Proposition 11.12, this representation is non-zero. (If at some intermediate step we obtained the zero representation then the corresponding dimension vector would be a multiple of the unit vector, and the associated reflection would produce a vector with non-positive entries, contradicting the minimality of r and j.) But the reflections take indecomposable representations to indecomposable representations, or to zero, see Theorem 11.25. That is, $\mathcal{M}'$ must be indecomposable. Then s_j takes the dimension vector of $\mathcal{M}'$ to $\tau(x) \not\geq 0$. This cannot be the dimension vector of a representation. Therefore Theorem 11.25 implies that $\mathcal{M}' \cong S_j$, and hence has dimension vector $\varepsilon_j \in \Delta_\Gamma$, which is a root. Then $\underline{\dim} \mathcal{M} = (s_1 s_2 \ldots s_n)^{r-1} s_1 \ldots s_{j-1}(\varepsilon_j)$ and hence it belongs to Δ_Γ, by Lemma 10.9. $\square$

Example 11.45. We give an example to illustrate the proof of Proposition 11.44. Consider the quiver Q given by

$$1 \longleftarrow 2 \longleftarrow 3$$

We have computed a formula for C_Γ of type A_n in Example 10.15, which in this case is

$$C_\Gamma(x_1, x_2, x_3) = (-x_1 + x_2, -x_1 + x_3, -x_1).$$

By definition $\underline{\Sigma} = \Sigma_3^+ \Sigma_2^+ \Sigma_1^+$. Consider an indecomposable representation $\mathcal{M}$ of Q with dimension vector $x = (0, x_2, x_3)$ and $x_2, x_3 \geq 1$. We have

$$C_\Gamma(x) = (x_2, x_3, 0) \geq 0, \quad C_\Gamma^2(x) = (-x_2 + x_3, -x_2, -x_2) \not\geq 0.$$

We know therefore that $\underline{\Sigma}(\mathcal{M}) =: \mathcal{N}$ is a non-zero indecomposable representation of Q with dimension vector $(x_2, x_3, 0)$ and $\underline{\Sigma}(\mathcal{N}) = 0$. Since $\mathcal{N} \not\cong S_1$ we can say that $\Sigma_1^+(\mathcal{N})$ must be indecomposable. This has dimension vector $s_1(x_2, x_3, 0) = (x_3 - x_2, x_3, 0)$.

We have $s_2(x_3 - x_2, x_3, 0) = (x_3 - x_2, -x_2, 0) \not\geq 0$. This is not a dimension vector, and it follows that $\Sigma_2^+(\Sigma_1^+(\mathcal{N})) = 0$ and therefore $\Sigma_1^+(\mathcal{N}) \cong S_2$. Hence $(x_3 - x_2, x_3, 0) = \varepsilon_2$ and $x_3 = 1$, $x_2 = 1$ and $\mathcal{M}$ has dimension vector $(0, 1, 1)$.

The following proposition proves part (2) of Gabriel's theorem, thus completing the proof of Theorem 11.39 (recall that part (1) has been proved in Proposition 11.44).

Proposition 11.46. *Let Q be a quiver whose underlying graph Γ is a union of Dynkin diagrams. For every positive root $x \in \Delta_\Gamma$ there is a unique indecomposable representation $\mathcal{M}$ of Q whose dimension vector is equal to x.*

Proof. By Lemma 11.41 we can assume that Γ is connected, and by Lemma 11.43 we can assume that Q has standard labelling.

As in the proof of Proposition 11.44 there are $r \geq 1$ and j with $1 \leq j \leq n$ such that $s_j \circ \ldots \circ s_1 \circ C^{r-1}(x) \not\geq 0$ and where r, j are minimal with these properties. Let $y := s_{j-1} \circ \ldots \circ s_1 \circ C^{r-1}(x)$, so that $y \geq 0$ but $s_j(y) \not\geq 0$.

Now, y and $s_j(y)$ are in Δ_Γ and it follows that $y = \varepsilon_j$ (in fact, y is the dimension vector of a representation, and this must be S_j since $s_j(y) \not\geq 0$). Take the representation S_j of the quiver $\sigma_{j-1} \ldots \sigma_1 Q$ and let

$$\mathcal{M} := (\underline{\Sigma}^{-1})^{r-1} \Sigma_1^- \ldots \Sigma_{j-1}^-(S_j).$$

This is an indecomposable representation of Q and has dimension vector x.

If $\mathcal{M}'$ is also an indecomposable representation of Q with dimension vector x then $\Sigma_{j-1}^+ \ldots \Sigma_1^+ \underline{\Sigma}^{r-1}(\mathcal{M}')$ has dimension vector ε_j, hence is isomorphic to S_j, and it follows that $\mathcal{M}'$ is isomorphic to $\mathcal{M}$. $\quad\square$

Gabriel's theorem allows one to determine all indecomposable representations of a quiver of Dynkin type, by knowing the roots of the associated quadratic form. In particular, this works for all possible orientations of the arrows of the quiver (see Lemma 11.43).

We illustrate this with an example.

Example 11.47. We have calculated the roots for the Dynkin diagram Γ of type A_n in Example 10.8. The positive roots are

$$\alpha_{r,s} = \varepsilon_r + \varepsilon_{r+1} + \ldots + \varepsilon_s \quad \text{for } 1 \leq r \leq s \leq n.$$

According to Theorem 11.39, these positive roots are in bijection with the indecomposable representations of any quiver Q with underlying graph Γ. In particular, any such quiver of Dynkin type A_n has precisely $\frac{n(n+1)}{2}$ indecomposable representations (up to isomorphism).

We can write down an indecomposable representation with dimension vector $\alpha_{r,s}$. This should exist independent of the orientation of Q. In fact, we can see this directly. Take the representation $\mathcal{M}$ where

$$M(i) = \begin{cases} K \text{ if } r \leq i \leq s \\ 0 \ \text{ else.} \end{cases}$$

Suppose α is an arrow $i \to j$ of Q so that $j = i \pm 1$. Then set

$$M(\alpha) = \begin{cases} \mathrm{id}_K \text{ if } r \leq i, j \leq s \\ 0 \ \ \text{ else.} \end{cases}$$

Then these representations are a complete set of indecomposable representations of a Dynkin quiver of type A_n.

Exercise 11.6. Let Q be the quiver of type A_n with standard labelling. Verify directly that the representation with dimension vector $\alpha_{r,s}$ (defined above) is indecomposable.

Let Q be a quiver which is the disjoint union of quivers Q_t for $1 \leq t \leq m$, and where each Q_t has an underlying graph which is a Dynkin diagram of type A, D or E. By Lemma 9.19 the indecomposable representations of Q are precisely the extensions by zero of the indecomposable representations of the Q_t. Now by Remark 11.42, the observation following Lemma 12.41, we deduce that the dimension vectors of these indecomposable representations are precisely the positive roots of the associated quadratic form.

We illustrate this with an example.

Example 11.48. Let Q be the (non-connected) quiver

$$1 \longleftarrow 2 \qquad 3 \longleftarrow 4 \longleftarrow 5$$

which is equal to $Q_1 \cup Q_2$ where Q_1 is $1 \longleftarrow 2$ and Q_2 is $3 \longleftarrow 4 \longleftarrow 5$. So the underlying graph $\Gamma = \Gamma_1 \cup \Gamma_2$ is a union of a Dynkin diagram of type A_2 and a Dynkin diagram of type A_3.

By Lemma 9.19 the indecomposable representations of Q are precisely the extensions by zero (see Definition 9.16) of the indecomposable representations of Q_1 and Q_2, and by Remark 11.42 the set of roots of Γ can be interpreted as the union of the set of roots of Γ_1 and of Γ_2.

Recall the Gram matrix of Γ, see Definition 10.2. This has block form,

$$G_\Gamma = \begin{pmatrix} G_1 & 0 \\ 0 & G_2 \end{pmatrix}$$

where

$$G_1 = \begin{pmatrix} 2 & -1 \\ -1 & 2 \end{pmatrix}, \quad G_2 = \begin{pmatrix} 2 & -1 & 0 \\ -1 & 2 & -1 \\ 0 & -1 & 2 \end{pmatrix}.$$

This shows that the quadratic form q_Γ is equal to

$$q_\Gamma(x_1, x_2, x_3, x_4, x_5) = q_{\Gamma_1}(x_1, x_2) + q_{\Gamma_2}(x_3, x_4, x_5).$$

For $x \in \mathbb{Z}^5$, we get $q_\Gamma(x) = 1$ if and only if one of $q_{\Gamma_1}(x_1, x_2)$ and $q_{\Gamma_2}(x_3, x_4, x_5)$ is equal to 1 and the other is zero, since the quadratic forms are positive definite (see Proposition 10.10). Hence a root of Γ is either a root of Γ_1 or a root of Γ_2.

Exercise 11.7. Explain briefly why this holds in general. That is, if Q is any quiver where all connected components are of Dynkin type A, D or E then the indecomposable representations of Q are in bijection with the positive roots of the quadratic form q_Γ of Q.

EXERCISES

11.8. Let Q be the quiver of Dynkin type A_3 with orientation

$$1 \overset{\alpha_1}{\longrightarrow} 2 \overset{\alpha_2}{\longleftarrow} 3$$

Assume $\mathcal{M}$ is an indecomposable representation of Q which is not simple.

(a) Show that $M(\alpha_1)$ and $M(\alpha_2)$ must be injective, and that

$$M(2) = \mathrm{im}(M(\alpha_1)) + \mathrm{im}(M(\alpha_2)).$$

(b) Determine all $\mathcal{M}$ such that $M(1) = 0$, by using the classification of indecomposable representations for Dynkin type A_2. Similarly determine all $\mathcal{M}$ such that $M(3) = 0$.

(c) Assume $M(1)$ and $M(3)$ are non-zero, deduce that then $M(2)$ is non-zero.

The following exercise uses the linear algebra proof of the dimension formula, $\dim_K(X + Y) = \dim_K X + \dim_K Y - \dim_K(X \cap Y)$ where X, Y are subspaces of some finite-dimensional vector space.

11.9. Let Q be as in Exercise 11.8. Take a representation $\mathcal{M}$ of Q which satisfies the conditions in part (a) of Exercise 11.8. Let $D := \mathrm{im}(M(\alpha_1)) \cap \mathrm{im}(M(\alpha_2))$, a subspace of $M(2)$.

   (a) Explain why $M(2)$ has a basis $\mathcal{B} = \{x_1, \ldots, x_d; v_1, \ldots, v_m; w_1, \ldots, w_n\}$ such that

   (i) $\{x_1, \ldots, x_d\}$ is a basis of D;
   (ii) $\{x_1, \ldots, x_d; v_1, \ldots, v_m\}$ is a basis of $\mathrm{im}(M(\alpha_1))$;
   (iii) $\{x_1, \ldots, x_d; w_1, \ldots, w_n\}$ is a basis of $\mathrm{im}(M(\alpha_2))$.

   (b) Explain why $M(1)$ has a basis of the form $\{a_1, \ldots, a_d, a_1', \ldots, a_m'\}$, where $M(\alpha_1)(a_i) = x_i$, and $M(\alpha_1)(a_j') = v_j$. Similarly, explain why $M(3)$ has a basis $\{b_1, \ldots, b_d, b_1', \ldots, b_n'\}$ such that $M(\alpha_2)(b_i) = x_i$, and $M(\alpha_2)(b_j') = w_j$.
   (c) Show that each x_i gives rise to an indecomposable direct summand of $\mathcal{M}$ of the form $K \longrightarrow K \longleftarrow K$. Moreover, show that each v_j gives rise to an indecomposable direct summand of $\mathcal{M}$ of the form $K \longrightarrow K \longleftarrow 0$. Similarly each w_j gives rise to an indecomposable direct summand of $\mathcal{M}$ of the form $0 \longrightarrow K \longleftarrow K$.

11.10. Let Q be the quiver of Dynkin type A_3 with the orientation as in Exercise 11.8. Explain how Exercises 11.8 and 11.9 classify the indecomposable representations of this quiver of type A_3. Confirm that the dimension vectors of the indecomposable representations are precisely the positive roots for the Dynkin diagram A_3.

11.11. Consider quivers whose underlying graph is the Dynkin diagram of type A_3. We have classified the indecomposable representations for the quiver

$$1 \xrightarrow{\alpha_1} 2 \xleftarrow{\alpha_2} 3$$

in the previous exercises. Explain how the general results on reflection maps $\Sigma_j^{\pm}$ imply Gabriel's theorem for the other two possible orientations.

11.12. The following shows that Σ_j^- does not take subrepresentations to subrepresentations in general. Let Q be the quiver of type A_3 with labelling

$$1 \xleftarrow{\beta_1} j \xrightarrow{\beta_2} 2$$

Define a representation $\mathcal{N}$ of Q as follows

$$\mathrm{span}\{f_1\} \longleftarrow \mathrm{span}\{e_1, e_2\} \longrightarrow \mathrm{span}\{f_2\}$$

and define the maps by $N(\beta_1)(e_1) = f_1$ and $N(\beta_1)(e_2) = 0$, and moreover $N(\beta_2)(e_1) = 0$ and $N(\beta_2)(e_2) = f_2$.

(a) Show that $C_{\mathcal{N}} = N(1) \times N(2)$ and hence that $N^-(j) = 0$.
(b) Let $\mathcal{U}$ be the subrepresentation of $\mathcal{N}$ given by

$$\text{span}\{f_1\} \longleftarrow \text{span}\{e_1 + e_2\} \longrightarrow \text{span}\{f_2\}.$$

Compute $C_{\mathcal{U}}$, and show that $U^-(j)$ is non-zero.

11.13. Let Q be a quiver and j a sink in Q. We denote by $\alpha_1, \ldots, \alpha_t$ the arrows of Q ending in j. Let $\mathcal{M}$ be an indecomposable representation of Q. Show that the following statements are equivalent.

   (i) We have $\sum_{i=1}^t \text{im}(M(\alpha_i)) = M(j)$, alternatively the map $(M(\alpha_1), \ldots, M(\alpha_t))$ is surjective.
   (ii) $\Sigma_j^- \Sigma_j^+(\mathcal{M}) \cong \mathcal{M}$.
   (iii) $\underline{\dim} \Sigma_j^+(\mathcal{M}) = s_j(\underline{\dim} \mathcal{M})$.
   (iv) $\mathcal{M}$ is not isomorphic to the simple representation $\mathcal{S}_j$.
   (v) $\Sigma_j^+(\mathcal{M})$ is indecomposable.

11.14. Formulate and verify an analogue of the previous exercise for the case when j is a source in a quiver Q'.

11.15. Consider a quiver with underlying graph Γ a Dynkin diagram of type D_n; assume that Q has standard labelling as in Example 10.4. For $n = 4$ we have seen in Exercise 10.14 that $(1, 1, 2, 1)$ is a root of Γ; a corresponding indecomposable representation has been constructed in Lemma 9.5 (but note that the labelling in Lemma 9.5 is not the standard labelling).

   Use the stretching method from Chap. 9, and show that a quiver of type D_n for $n \geq 4$ with standard labelling has indecomposable representations with dimension vectors

$$(1, 1, \underbrace{2, 2, \ldots, 2}_{a}, \underbrace{1, 1, \ldots, 1}_{b}, \underbrace{0, \ldots, 0}_{c})$$

for any positive integers a, b and any non-negative integer c such that $a + b + c = n - 2$.

   In the following exercises, we take the Kronecker quiver Q of the form

$$1 \underset{\alpha_2}{\overset{\alpha_1}{\rightrightarrows}} 2$$

to illustrate the various constructions and proofs in this chapter.

11.16. Assume M is an indecomposable representation of the Kronecker quiver Q which is not simple. Show that then

(i) $\ker(M(\alpha_1)) \cap \ker(M(\alpha_2)) = 0$,
(ii) $\operatorname{im}(M(\alpha_1)) + \operatorname{im}(M(\alpha_2)) = M(2)$.

(Hint: Apply Exercises 9.9 and 9.10.)

11.17. Assume M is a representation of the Kronecker quiver Q which satisfies

$$M(1) = \ker(M(\alpha_1)) \oplus \ker(M(\alpha_2)) \text{ and } M(2) = \operatorname{im}(M(\alpha_1)) = \operatorname{im}(M(\alpha_2)).$$

(a) Show that the restriction of $M(\alpha_1)$ to $\ker(M(\alpha_2))$ is an isomorphism from $\ker(M(\alpha_2))$ to $M(2)$ and also the restriction of $M(\alpha_2)$ is an isomorphism from $\ker(M(\alpha_1))$ to $M(1)$.
(b) Show that M is decomposable as follows. Take a basis $\{b_1, \ldots, b_r\}$ of $M(2)$ and take $v_1, \ldots, v_r$ in $\ker(M(\alpha_2))$ such that $M(\alpha_1)(v_i) = b_i$. Choose also $w_1, \ldots, w_r$ in $\ker(M(\alpha_1))$ such that $M(\alpha_2)(w_i) = b_i$.

(i) Show that $\{v_1, \ldots, v_r, w_1, \ldots, w_r\}$ is a basis for $M(1)$.
(ii) Deduce that M is a direct sum $\bigoplus_{i=1}^r \mathcal{N}_i$ of r representations, where $\mathcal{N}_i(1) = \operatorname{span}\{v_i, w_i\}$ and $\mathcal{N}_i(2) = \operatorname{span}\{b_i\}$.
(iii) Show that $\mathcal{N}_1$ is indecomposable, and each $\mathcal{N}_i$ is isomorphic to $\mathcal{N}_1$.

11.18. Assume M is a representation of the Kronecker quiver such that $\dim_K M(1) = 4$ and $\dim_K M(2) = 2$. Show by applying the previous exercises that M is decomposable.

With the methods of this chapter we can also classify indecomposable representations for the Kronecker quiver Q. Note that we must use the reflection Σ^- in the next exercise since vertex 1 is a source in Q (and vertex 2 is a source in $\sigma_1 Q$).

11.19. Let Q be the Kronecker quiver as above, and Γ its underlying graph, the Euclidean diagram of type $\tilde{A}_1$. The reflections s_1, s_2 and the Coxeter transformation C_Γ were computed in Exercises 10.2 and 10.9. Assume M is an indecomposable representation of Q which is not simple.

(a) Explain briefly why $\Sigma_2^- \Sigma_1^-(M)$ is either zero, or is an indecomposable representation of Q, and if it is non-zero, show that it has dimension vector $C_\Gamma \binom{a}{b}$ where $(a, b) = \underline{\dim} M$.
(b) Show also that $\Sigma_2^- \Sigma_1^-(M) = 0$ for M not simple precisely when M has dimension vector $(2, 1)$. Construct such an M, either directly, or by exploiting that $\Sigma_1^-(M) \cong S_2$.

11.20. The aim is to classify the indecomposable representations M of the Kronecker quiver Q with dimension vector (a, b), where $a > b \geq 0$.

(a) If $a > b \geq 0$, let $\binom{a_1}{b_1} := C_\Gamma \binom{a}{b}$. Check that $b_1 < a_1 < a$ and that $a_1 - b_1 = a - b$.

(b) Define a sequence $\binom{a_k}{b_k}$ where $\binom{a_k}{b_k} = C_\Gamma^k\binom{a}{b}$. Then $a > a_1 > a_2 > \ldots$ and $a_k > b_k$. Hence there is a largest $r \geq 0$ such that $a_r > b_r \geq 0$.

(c) Let $\Sigma = \Sigma_2^-\Sigma_1^-$. Explain why the representation $\mathcal{N} = \Sigma^r(\mathcal{M})$ is indecomposable with dimension vector $\binom{a_r}{b_r}$ and $\Sigma(\mathcal{N})$ is the zero representation.

(d) Deduce that either $\mathcal{N} = S_1$, or $\Sigma_1^-(\mathcal{N}) = S_2$. Hence deduce that $\mathcal{M}$ has dimension vector $(a, a-1)$ with $a \geq 1$. Note that this is a root of Γ. Explain briefly why this gives a classification of indecomposable representations of Q with dimension vector of the form $(a, a-1)$.

11.21. Suppose $\mathcal{M}$ is an indecomposable representation of the Kronecker quiver Q with dimension vector (a, b), where $0 \leq a < b$. Show that $(a, b) = (a, a+1)$ and hence is also a root. (Hint: this is analogous to Exercise 11.20.)

11.22. Let A be the 4-dimensional algebra

$$A := K[X, Y]/(X^2, Y^2).$$

It has basis the cosets of $1, X, Y, XY$; we identify $1, X, Y, XY$ with their cosets in A. Let $\mathcal{M}$ be a representation of the Kronecker quiver Q as above, we construct an A-module $V_\mathcal{M}$ with underlying space $V_\mathcal{M} = M(1) \times M(2)$. We define the action of X and of Y via matrices, written in block form

$$x = \begin{pmatrix} 0 & 0 \\ M(\alpha_1) & 0 \end{pmatrix}, \quad y = \begin{pmatrix} 0 & 0 \\ M(\alpha_2) & 0 \end{pmatrix}.$$

(a) Check that $x^2 = 0$, $y^2 = 0$ and that $xy = yx$, that is, $V_\mathcal{M}$ is an A-module.

(b) Assume $\mathcal{M}$ is an indecomposable representation, and at least one of the $M(\alpha_i)$ is surjective. Show that then $V_\mathcal{M}$ is an indecomposable A-module.

(c) Let $\mathcal{M}$ and $\mathcal{N}$ be representations of Q and let $\varphi : V_\mathcal{M} \to V_\mathcal{N}$ be an isomorphism. Assume at least one of the $M(\alpha_i)$ is surjective. Show that then $\mathcal{M}$ is isomorphic to $\mathcal{N}$.

11.23. Let A be as in Exercise 11.22. Assume that V is an indecomposable A-module on which XY acts as zero. Let $U := \{v \in V \mid Xv = 0 \text{ and } Yv = 0\}$. We assume $0 \neq U \neq V$.

(a) Show that U is an A-submodule of V.

(b) Write $V = W \oplus U$ for some subspace W of V. Check that $X(W) \subseteq U$ and $Y(W) \subseteq U$.

(c) Fix a basis of V which is a union of bases of W and of U, write the action of X and Y as matrices with respect to this basis. Check that these matrices have block form

$$x \mapsto \begin{pmatrix} 0 & 0 \\ A_1 & 0 \end{pmatrix}, \quad y \mapsto \begin{pmatrix} 0 & 0 \\ A_2 & 0 \end{pmatrix}.$$

Define $\mathcal{M} = (M(1), M(2), M(\alpha_1), M(\alpha_2))$ to be the representation of the Kronecker quiver where $M(1) = W$ and $M(2) = U$, and $M(\alpha_i) = A_i$.
(d) Show that $\mathcal{M}$ is an indecomposable representation.

11.24. Let Q be the quiver of type $\tilde{D}_4$,

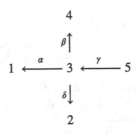

Let $A = KQ/I$ where I is the ideal spanned by $\{\beta\gamma, \alpha\gamma, \delta\gamma\}$. Identify the trivial path e_i with the coset $e_i + I$, and identify each arrow of Q with its coset in A. Then A is an algebra of dimension 9 where the product of any two arrows is zero in A.

(a) Explain briefly why A-modules are the same as representations $\mathcal{M}$ of Q which satisfy the paths in I, so that, for example, $M(\beta) \circ M(\gamma) = 0$.
(b) Let $\mathcal{M}$ be such a representation corresponding to an A-module.

   (i) Show that the kernel of $M(\gamma)$ gives rise to direct summands isomorphic to the simple representation S_5.
   (ii) Assume that $M(\gamma)$ is injective. Show that $\mathcal{M}$ is the direct sum of two representations, where one has dimension vector $(0, 0, d, 0, d)$, with $d = \dim_K M(5)$, and the other is zero at vertex 5. Moreover, the direct summand with dimension vector $(0, 0, d, 0, d)$ is the direct sum of d copies of an indecomposable representation with dimension vector $(0, 0, 1, 0, 1)$.

(c) Deduce that A has finite representation type.
(d) Why does this not contradict Gabriel's theorem?

Chapter 12
Proofs and Background

In this chapter we will give the proofs, and fill in the details, which were postponed in Chap. 11. First we will prove the results on reflection maps from Sect. 11.1, namely the compatibility of Σ_j^+ and Σ_j^- with direct sums (see Lemmas 11.10 and 11.17) and the fact that Σ_j^+ and Σ_j^- compose to the identity map (under certain assumptions), see Propositions 11.22 and 11.24. This is done in Sect. 12.1.

Secondly, in Sect. 11.2 we have shown that every connected quiver whose underlying graph is not a Dynkin diagram has infinite representation type. The crucial step is to prove this for a quiver whose underlying graph is a Euclidean diagram. We have done this in detail in Sect. 11.2 for types $\widetilde{A}_n$ and $\widetilde{D}_n$. In Sect. 12.2 below we will provide the technically more involved proofs that quivers of types $\widetilde{E}_6$, $\widetilde{E}_7$ and $\widetilde{E}_8$ have infinite representation type.

Finally, we have two sections containing some background and outlook. We give a brief account of root systems as they occur in Lie theory, and we show that the set of roots, as defined in Chap. 10, is in fact a root system in this sense. Then we provide an informal account of Morita equivalence.

12.1 Proofs on Reflections of Representations

In this section we give details for the results on the reflection maps $\Sigma_j^{\pm}$. Recall that we define these when j is a sink or a source in a quiver Q. We use the notation as before: If j is a sink of Q then we label the arrows ending at j by $\alpha_1, \ldots, \alpha_t$ (see Definition 11.7), and we let $\alpha_i : i \to j$ for $1 \le i \le t$. As explained in Remark 11.13, there may be multiple arrows, and if so then we identify the relevant vertices at which they start (rather than introducing more notation). If j is a source of Q then we label the arrows starting at vertex j by $\beta_1, \ldots, \beta_t$ (see Definition 11.14), where $\beta_i : j \to i$, and we make the same convention as above in the case of multiple arrows, see also Remark 11.20.

© Springer International Publishing AG, part of Springer Nature 2018

K. Erdmann, T. Holm, *Algebras and Representation Theory*, Springer Undergraduate Mathematics Series, https://doi.org/10.1007/978-3-319-91998-0_12

12.1.1 Invariance Under Direct Sums

The following is Lemma 11.10, which we restate for convenience.

Lemma (Lemma 11.10). *Suppose Q is a quiver with a sink j, and $\mathcal{M}$ is a representation of Q. Assume $\mathcal{M} = \mathcal{X} \oplus \mathcal{Y}$ is a direct sum of subrepresentations, then $\Sigma_j^+(\mathcal{M}) = \Sigma_j^+(\mathcal{X}) \oplus \Sigma_j^+(\mathcal{Y})$.*

Proof. We use the notation as in Definition 11.7. Recall that for any representation $\mathcal{V}$ of Q, the representation $\Sigma_j^{\,!}(\mathcal{V})$ is defined as follows (see Definition 11.9): we have $V^+(r) = V(r)$ for all vertices $r \neq j$, and $V^+(j)$ is the kernel of the linear map

$$(V(\alpha_1), \ldots, V(\alpha_t)) : \prod_{i=1}^{t} V(i) \to V(j) , \quad (v_1, \ldots, v_t) \mapsto \sum_{i=1}^{t} V(\alpha_i)(v_i).$$

Moreover, if γ is an arrow not adjacent to j then we set $V^+(\gamma) = V(\gamma)$; finally for $i = 1, \ldots, t$ the linear map $V^+(\bar{\alpha}_i) : V^+(j) \to V^+(i) = V(i)$ is defined by projecting onto the i-th coordinate. We apply this for $V = \mathcal{M}$ and for $V = \mathcal{X}$ or $\mathcal{Y}$.

(1) We show that $\Sigma_j^+(\mathcal{X})$ and $\Sigma_j^+(\mathcal{Y})$ are subrepresentations of $\Sigma_j^+(\mathcal{M})$:

We prove it for $\Sigma_j^+(\mathcal{X})$; the proof for $\Sigma_j^+(\mathcal{Y})$ is analogous. For the definition of subrepresentation, see Definition 9.6.

If $r \neq j$ then $X^+(r) = X(r)$ is a subspace of $M(r) = M^+(r)$, since $\mathcal{X}$ is a subrepresentation of $\mathcal{M}$. If γ is an arrow of Q not adjacent to j, say $\gamma : r \to s$, then we claim that $X^+(\gamma)$ is the restriction of $M^+(\gamma)$ to $X^+(r)$ and it maps into $X^+(s)$: By definition $X^+(\gamma) = X(\gamma)$ and, since $\mathcal{X}$ is a subrepresentation, it is the restriction of $M(\gamma)$ to $X(r)$ and maps into $X(s)$. In addition, $M^+(\gamma) = M(\gamma)$ and $X^+(r) = X(r)$, and $X^+(s) = X(s)$, hence the claim holds.

Let $r = j$, and take $w \in X^+(j)$, then $w = (x_1, \ldots, x_t)$ with $x_i \in X(i)$, such that $\sum_{i=1}^{t} X(\alpha_i)(x_i) = 0$. Since $\mathcal{X}$ is a subrepresentation of $\mathcal{M}$ we have $x_i \in X(i) \subseteq M(i)$, and

$$\sum_{i=1}^{t} M(\alpha_i)(x_i) = \sum_{i=1}^{t} X(\alpha_i)(x_i) = 0,$$

that is, $w \in M^+(j)$. Moreover, $X^+(\bar{\alpha}_i)(w) = x_i = M^+(\bar{\alpha}_i)(w)$, that is, $X^+(\bar{\alpha}_i)$ is the restriction of $M^+(\bar{\alpha}_i)$ to $X^+(j)$. We have now proved (1).

(2) We show now that $\Sigma_j^+(\mathcal{M})$ is the direct sum of $\Sigma_j^+(\mathcal{X})$ and $\Sigma_j^+(\mathcal{Y})$, that is, verify Definition 9.9. We must show that for each vertex r, we have $M^+(r) = X^+(r) \oplus Y^+(r)$ as vector spaces. This is clear for $r \neq j$, then since $\mathcal{M} = \mathcal{X} \oplus \mathcal{Y}$, we have

$$M^+(r) = M(r) = X(r) \oplus Y(r) = X^+(r) \oplus Y^+(r).$$

Let $r = j$. If $w \in X^+(j) \cap Y^+(j)$ then $w = (x_1, \ldots, x_t) = (y_1, \ldots, y_t)$ with $x_i \in X(i)$ and $y_i \in Y(i)$ for each i. So $x_i = y_i \in X(i) \cap Y(i) = 0$ and $w = 0$. Hence $X^+(j) \cap Y^+(j) = 0$.

We are left to show that $M^+(j) = X^+(j) + Y^+(j)$. Take an element $(m_1, \ldots, m_t) \in M^+(j)$, then $m_i \in M(i)$ and $\sum_{i=1}^t M(\alpha_i)(m_i) = 0$. Since $M(i) = X(i) \oplus Y(i)$ we have $m_i = x_i + y_i$ with $x_i \in X(i)$ and $y_i \in Y(i)$. Therefore

$$0 = \sum_{i=1}^t M(\alpha_i)(m_i) = \sum_{i=1}^t M(\alpha_i)(x_i + y_i) = \sum_{i=1}^t (M(\alpha_i)(x_i) + M(\alpha_i)(y_i))$$

$$= \sum_{i=1}^t M(\alpha_i)(x_i) + \sum_{i=1}^t M(\alpha_i)(y_i) = \sum_{i=1}^t X(\alpha_i)(x_i) + \sum_{i=1}^t Y(\alpha_i)(y_i).$$

Now, because $\mathcal{X}$ and $\mathcal{Y}$ are subrepresentations, we know that $\sum_{i=1}^t X(\alpha_i)(x_i)$ lies in $X(j)$ and that $\sum_{i=1}^t Y(\alpha_i)(y_i)$ lies in $Y(j)$. We assume that $M(j) = X(j) \oplus Y(j)$, so the intersection of $X(j)$ and $Y(j)$ is zero. It follows that both $\sum_{i=1}^t X(\alpha_i)(x_i) = 0$ and $\sum_{i=1}^t Y(\alpha_i)(y_i) = 0$. This means that $(x_1, \ldots, x_t) \in X^+(j)$ and $(y_1, \ldots, y_t) \in Y^+(j)$ and hence

$$(m_1, \ldots, m_t) = (x_1, \ldots x_t) + (y_1, \ldots, y_t) \in X^+(j) + Y^+(j).$$

The other inclusion follows from part (1) since $X^+(j)$ and $Y^+(j)$ are subspaces of $M^+(j)$. $\square$

We now give the proof of the analogous result for the reflection map Σ_j^-. This is Lemma 11.17, which has several parts. We explained right after Lemma 11.17 that it only remains to prove part (a) of that lemma, which we also restate here for convenience.

Lemma (Lemma 11.17(a)). *Suppose Q' is a quiver with a source j, and $\mathcal{N}$ is a representation of Q'. Assume $\mathcal{N} = \mathcal{X} \oplus \mathcal{Y}$ is a direct sum of subrepresentations, then $\Sigma_j^-(\mathcal{N})$ is isomorphic to the direct product $\Sigma_j^-(\mathcal{X}) \times \Sigma_j^-(\mathcal{Y})$ of representations.*

Proof. We use the notation as in Definition 11.14. We recall from Definition 11.16 how the reflection $\Sigma_j^-(\mathcal{V})$ is defined for any representation $\mathcal{V}$ of Q'. For vertices $r \neq j$ we set $V^-(r) = V(r)$, and $V^-(j)$ is the factor space $V^-(j) = (V(1) \times \ldots \times V(t))/C_\mathcal{V}$, where $C_\mathcal{V}$ is the image of the linear map

$$V(j) \to V(1) \times \ldots \times V(t) , \quad v \mapsto (V(\beta_1)(v), \ldots, V(\beta_t)(v)).$$

In this proof, we will take the direct product of the representations $\Sigma_j^-(\mathcal{X})$ and $\Sigma_j^-(\mathcal{Y})$, see Exercise 9.13.

We will first construct an isomorphism of vector spaces $N^-(j) \cong X^-(j) \times Y^-(j)$, and then use this to prove the lemma. Throughout, we write n_r for an element in $N(r)$, and we use that $n_r = x_r + y_r$ for unique elements $x_r \in X(r)$ and $y_r \in Y(r)$.

(1) We define a linear map $\psi : N^-(j) \to X^-(j) \times Y^-(j)$. We set

$$\psi((n_1, \ldots, n_t) + C_{\mathcal{N}}) := ((x_1, \ldots, x_t) + C_{\mathcal{X}}, (y_1, \ldots, y_t) + C_{\mathcal{Y}}).$$

This is well-defined: If $(n_1, \ldots, n_t) \in C_{\mathcal{N}}$ then there is an $n \in N(j)$ such that

$$(n_1, \ldots, n_t) = (N(\beta_1)(n), \ldots, N(\beta_t)(n)).$$

Since $\mathcal{N} = \mathcal{X} \oplus \mathcal{Y}$ we can write $n = x + y$ and $n_i = x_i + y_i$ with $x \in X(j)$ and $x_i \in X(i)$, and $y \in Y(j)$ and $y_i \in Y(i)$. Moreover, for each i we have

$$x_i + y_i = n_i = N(\beta_i)(n) = N(\beta_i)(x) + N(\beta_i)(y) = X(\beta_i)(x) + Y(\beta_i)(y).$$

We get $(x_1, \ldots, x_t) = (X(\beta_1)(x), \ldots, X(\beta_t)(x)) \in C_{\mathcal{X}}$ and similarly $(y_1, \ldots, y_t) \in C_{\mathcal{Y}}$.

The map ψ is an isomorphism: Suppose $\psi((n_1, \ldots, n_t) + C_{\mathcal{N}}) = 0$ then $(x_1, \ldots, x_t) \in C_{\mathcal{X}}$, that is, it is equal to $(X(\beta_1)(x), \ldots, X(\beta_t)(x))$ for some $x \in X(j)$, and similarly we have $(y_1, \ldots, y_t) = (Y(\beta_1)(y), \ldots, Y(\beta_t)(y))$ for some $y \in Y(j)$. It follows that

$$n_i = x_i + y_i = X(\beta_i)(x) + Y(\beta_i)(y) = N(\beta_i)(x) + N(\beta_i)(y) = N(\beta_i)(x + y)$$

and $(n_1, \ldots, n_t) = (N(\beta_1)(n), \ldots, N(\beta_t)(n))$ for $n = x + y$, hence it lies in $C_{\mathcal{N}}$. This shows that ψ is injective. The map ψ is clearly surjective, and we have now shown that it is an isomorphism.

(2) We define a homomorphism of representations $\theta := (\theta_r) : \Sigma_j^-(\mathcal{N}) \to \Sigma_j^-(\mathcal{X}) \times \Sigma_j^-(\mathcal{Y})$ as follows. Let $n \in N^-(r)$, and if $r \neq j$ write $n = x + y$ for $x \in X(r)$, $y \in Y(r)$, then

$$\theta_r(n) := \begin{cases} (x, y) & r \neq j \\ \psi(n) & r = j. \end{cases}$$

For each r, the map θ_r is a vector space isomorphism; for $r \neq j$ this holds because

$$N^-(r) = N(r) = X(r) \oplus Y(r) = X^-(r) \oplus Y^-(r)$$

and for $r = j$ it holds by (1).

We are left to check that the relevant diagrams commute (see Definition 9.4).

(i) Let $\gamma : r \to s$ be an arrow of Q which is not adjacent to j, then we must show that $\theta_s \circ N^-(\gamma) = (X^-(\gamma), Y^-(\gamma)) \circ \theta_r$. Take $n \in N^-(r) = N(r)$, then

$$(\theta_s \circ N^-(\gamma))(n) = (\theta_s \circ N(\gamma))(n) = \theta_s(N(\gamma))(x + y)$$

$$= \theta_s(X(\gamma)(x) + Y(\gamma)(y)) = (X(\gamma)(x), Y(\gamma)(y))$$

$$= (X(\gamma), Y(\gamma))(x, y) = ((X^-(\gamma), Y^-(\gamma)) \circ \theta_r)(n).$$

(ii) This leaves us to deal with an arrow $\bar{\beta}_i : i \to j$, hence we must show that we have $\theta_j \circ N^-(\bar{\beta}_i) = (X^-(\bar{\beta}_i), Y^-(\bar{\beta}_i)) \circ \theta_i$. Let $n_i \in N^-(i) = N(i) = X(i) \oplus Y(i)$, so that $n_i = x_i + y_i$ with $x_i \in X(i)$ and $y_i \in Y(i)$. Then

$$
\begin{aligned}
(\theta_j \circ N^-(\bar{\beta}_i))(n_i) &= \theta_j((0, \ldots, 0, n_i, 0, \ldots, 0) + C_{\mathcal{N}}) \\
&= \psi((0, \ldots, 0, n_i, 0, \ldots, 0) + C_{\mathcal{N}}) \\
&= ((0, \ldots, 0, x_i, 0, \ldots, 0) + C_{\mathcal{X}}, (0, \ldots, 0, y_i, 0, \ldots, 0) + C_{\mathcal{Y}}) \\
&= (X^-(\bar{\beta}_i)(x_i), Y^-(\bar{\beta}_i)(y_i)) \\
&= (X^-(\bar{\beta}_i), Y^-(\bar{\beta}_i))(x_i, y_i) \\
&= ((X^-(\bar{\beta}_i), Y^-(\bar{\beta}_i)) \circ \theta_i)(n_i),
\end{aligned}
$$

as required.   □

12.1.2   Compositions of Reflections

In this section we will give the proofs for the results on compositions of the reflection maps $\Sigma_j^{\pm}$. More precisely, we have stated in Propositions 11.22 and 11.24 that under certain assumptions on the representations $\mathcal{M}$ and $\mathcal{N}$ we have that $\Sigma_j^- \Sigma_j^+(\mathcal{M}) \cong \mathcal{M}$ and $\Sigma_j^+ \Sigma_j^-(\mathcal{N}) \cong \mathcal{N}$, respectively. For our purposes the most important case is when $\mathcal{M}$ and $\mathcal{N}$ are indecomposable (and not isomorphic to the simple representation $\mathcal{S}_j$), and then the assumptions are always satisfied (see Exercise 11.2). The following two propositions are crucial for the proof that Σ_j^+ and Σ_j^- give mutually inverse bijections as described in Theorem 11.25.

We start by proving Proposition 11.22, which we restate here.

Proposition (Proposition 11.22). *Assume j is a sink of a quiver Q and let $\alpha_1, \ldots, \alpha_t$ be the arrows in Q ending at j. Suppose $\mathcal{M}$ is a representation of Q such that the linear map*

$$
(M(\alpha_1), \ldots, M(\alpha_t)) : \prod_{i=1}^{t} M(i) \to M(j), \; (m_1, \ldots, m_t) \mapsto M(\alpha_1)(m_1) + \ldots + M(\alpha_t)(m_t)
$$

is surjective. Then the representation $\Sigma_j^- \Sigma_j^+(\mathcal{M})$ is isomorphic to $\mathcal{M}$.

Proof. We set $\mathcal{N} = \Sigma_j^+(\mathcal{M}) = \mathcal{M}^+$ and $\mathcal{N}^- = \Sigma_j^-(\mathcal{N})$, and we want to show that $\mathcal{N}^-$ is isomorphic to $\mathcal{M}$.

(1) We claim that $C_{\mathcal{N}}$ is equal to $M^+(j)$. By definition, we have

$$
C_{\mathcal{N}} = \{(N(\bar{\alpha}_1)(y), \ldots, N(\bar{\alpha}_t)(y)) \mid y \in N(j)\}.
$$

Now, $N(j) = M^+(j)$, and an element $y \in M^+(j)$ is in particular of the form

$$y = (m_1, \ldots, m_t) \in \prod_{i=1}^{t} M(i).$$

The map $N(\bar{\alpha}_i)$ is the projection onto the i-th coordinate, therefore

$$(N(\bar{\alpha}_1)(y), \ldots, N(\bar{\alpha}_t)(y)) = (m_1, \ldots, m_t) = y.$$

This shows that $C_{\mathcal{N}} = M^+(j)$.

(2) We define a homomorphism of representations $\varphi = (\varphi_r) : \mathcal{N}^- \to \mathcal{M}$ with linear maps $\varphi_r : N^-(r) \to M(r)$, as follows.

If $r \neq j$ then $N^-(r) = N(r) = M^+(r) = M(r)$ and we take φ_r to be the identity map, which is an isomorphism. Now let $r = j$, then

$$N^-(j) = \left(\prod_{i=1}^{t} N(i) \right) / C_{\mathcal{N}} = \left(\prod_{i=1}^{t} M(i) \right) / M^+(j)$$

by part (1), and since $N(i) = M^+(i) = M(i)$. We define $\varphi_j : N^-(j) \to M(j)$ by

$$\varphi_j((m_1, \ldots, m_t) + C_{\mathcal{N}}) = \sum_{i=1}^{t} M(\alpha_i)(m_i) \in M(j).$$

Then φ_j is well-defined: If $(m_1, \ldots, m_t) \in C_{\mathcal{N}} = M^+(j)$ then by definition we have $\sum_{i=1}^{t} M(\alpha_i)(m_i) = 0$. Moreover, φ_j is injective: indeed, if $\sum_{i=1}^{t} M(\alpha_i)(m_i) = 0$ then $(m_1, \ldots, m_t) \in M^+(j) = C_{\mathcal{N}}$. Furthermore, φ_j is surjective, by assumption, and we have shown that φ_j is an isomorphism.

Finally, we check that φ is a homomorphism of representations. If $\gamma : r \to s$ is an arrow not adjacent to j then $N^-(\gamma) = M(\gamma)$, and both φ_r and φ_s are the identity maps, so the relevant square commutes. This leaves us to consider the maps corresponding to the arrows $\alpha_i : i \to j$. They are the maps in the diagram

$$
\begin{array}{ccc}
M(i) = N^-(i) & \xrightarrow{\ N^-(\alpha_i)\ } & N^-(j) \\[4pt]
{\scriptstyle \varphi_i} \downarrow & & \downarrow {\scriptstyle \varphi_j} \\[4pt]
M(i) & \xrightarrow{\ M(\alpha_i)\ } & M(j)
\end{array}
$$

Since φ_i is the identity map of $M(i)$ and the maps $N^-(\alpha_i)$ are induced by inclusion maps, we have

$$(M(\alpha_i)\circ\varphi_i)(m_i) = M(\alpha_i)(m_i) = \varphi_j((0, \ldots, 0, m_i, 0, \ldots, 0) + C_{\mathcal{N}}) = (\varphi_j\circ N^-(\alpha_i))(m_i).$$

Thus, $\varphi : \mathcal{N}^- = \Sigma_j^- \Sigma_j^+(\mathcal{M}) \to \mathcal{M}$ is an isomorphism of representations, as claimed. $\qquad\square$

We will now prove the analogous result for the other composition. We also restate this proposition for convenience.

Proposition (Proposition 11.24). *Assume j is a source of a quiver Q' and let $\beta_1, \ldots, \beta_t$ be the arrows in Q' starting at j. Suppose $\mathcal{N}$ is a representation of Q' such that $\bigcap_{i=1}^t \ker(N(\beta_i)) = 0$. Then the representations $\Sigma_j^+ \Sigma_j^-(\mathcal{N})$ and $\mathcal{N}$ are isomorphic.*

Proof. Let $\mathcal{M} = \Sigma_j^-(\mathcal{N})$, then we want to show that $\mathcal{M}^+ = \Sigma_j^+(\mathcal{M})$ is isomorphic to $\mathcal{N}$.

(1) We assume that the intersection of the kernels of the $N(\beta_i)$ is zero, therefore the linear map

$$\varphi_j : N(j) \to C_{\mathcal{N}}, \quad y \mapsto (N(\beta_1)(y), \ldots, N(\beta_t)(y))$$

is injective. It is surjective by definition of $C_{\mathcal{N}}$, hence it is an isomorphism.

(2) We claim that $M^+(j) = C_{\mathcal{N}}$: By definition

$$M^+(j) = \{(n_1, \ldots, n_t) \in N(1) \times \ldots \times N(t) \mid \sum_{i=1}^t M(\bar{\beta_i})(n_i) = 0\}.$$

Now, $M(\bar{\beta_i})(n_i) = N^-(\bar{\beta_i})(n_i) = (0, \ldots, 0, n_i, 0, \ldots, 0) + C_{\mathcal{N}} \in M(j) = N^-(j)$.
So

$$\sum_{i=1}^t M(\bar{\beta_i})(n_i) = (n_1, \ldots, n_t) + C_{\mathcal{N}}.$$

This is zero precisely when $(n_1, \ldots, n_t) \in C_{\mathcal{N}}$. Hence the claim $M^+(j) = C_{\mathcal{N}}$ holds.

(3) Now we define a homomorphism of representations $\varphi : \mathcal{N} \to \mathcal{M}^+$ by setting φ_r as the identity map on $N(r) = M(r)$ if $r \neq j$ and

$$\varphi_j : N(j) \to M^+(j) = C_{\mathcal{N}}, \quad \varphi_j(y) = (N(\beta_1)(y), \ldots, N(\beta_t)(y)).$$

Each of these linear maps is an isomorphism, and we are left to check that φ is a homomorphism of representations.

If $\gamma : r \to s$ is an arrow which is not adjacent to j then φ_r and φ_s are identity maps, and $N(\gamma) = M^+(\gamma)$, and the relevant diagram commutes.

Now let $\beta_i : j \longrightarrow i$ be an arrow, we require that the following diagram commutes:

$$
\begin{array}{ccc}
N(j) & \xrightarrow{\;N(\beta_i)\;} & N(i) \\[2mm]
\varphi_j \downarrow & & \downarrow \varphi_i \\[2mm]
M^+(J) & \xrightarrow[M^+(\beta_i)]{} & M^+(i) = M(i)
\end{array}
$$

Recall that $M^+(\beta_i)$ is the projection onto the i-th component, and φ_i is the identity, and so we get for any $y \in N(j)$ that

$$(M^+(\beta_i) \circ \varphi_j)(y) = M^+(\beta_i)(N(\beta_1)(y), \ldots, N(\beta_t)(y)) = N(\beta_i)(y) = (\varphi_i \circ N(\beta_i))(y),$$

as required. Thus, $\varphi : \mathcal{N} \to \mathcal{M}^+ = \Sigma_j^+ \Sigma_j^-(\mathcal{N})$ is an isomorphism of representations. $\square$

12.2 All Euclidean Quivers Are of Infinite Representation Type

We have already proved in Sect. 11.2 that quivers with underlying graphs of type $\widetilde{A}_n$ and of type $\widetilde{D}_n$ have infinite representation type, over any field K. We will now deal with the three missing Euclidean diagrams $\widetilde{E}_6$, $\widetilde{E}_7$ and $\widetilde{E}_8$. Recall from Corollary 11.26 that the orientation of the arrows does not affect the representation type. So it suffices in each case to consider a quiver with a fixed chosen orientation of the arrows. We take the labelling as in Example 10.4. However, we do not take the orientation as in 10.4. Instead, we take the orientation so that the branch vertex is the only sink. This will make the notation easier (then we can always take the maps to be inclusions).

As a strategy, in each case we will first construct a special representation as in Lemma 11.29 (see Definition 11.30). This will already imply infinite representation type if the underlying field is infinite. This is not yet sufficient for our purposes since we prove Gabriel's theorem for arbitrary fields. Thus we then construct representations of arbitrary dimensions over an arbitrary field and show that they are indecomposable. The details for the general case are analogous to those in the construction of the special representation.

12.2.1 Quivers of Type $\tilde{E}_6$ Have Infinite Representation Type

Let Q be the quiver of Dynkin type E_6 with the following labelling and orientation:

$$3$$
$$\downarrow$$
$$1 \longrightarrow 2 \longrightarrow 4 \longleftarrow 5 \longleftarrow 6$$

Lemma 12.1. *The quiver Q has a special representation $\mathcal{M}$ with $\dim_K M(3) = 2$.*

Proof. We define $\mathcal{M}$ to be a representation for which all maps are inclusion maps, so we do not need to specify names for the arrows. We take $M(4)$ to be a 3-dimensional space, and all other $M(i)$ are subspaces.

$$M(4) = \mathrm{span}\{e, f, g\}$$
$$M(1) = \mathrm{span}\{g\}$$
$$M(2) = \mathrm{span}\{f, g\}$$
$$M(3) = \mathrm{span}\{e + f, f + g\}$$
$$M(5) = \mathrm{span}\{e, f\}$$
$$M(6) = \mathrm{span}\{e\}.$$

According to the definition of a special representation (see Definition 11.30) we must check that any endomorphism of $\mathcal{M}$ is a scalar multiple of the identity. Let $\varphi : \mathcal{M} \to \mathcal{M}$ be a homomorphism of representations. Since all maps are inclusion maps, it follows that for each i, the map φ_i is the restriction of φ_4 to $M(i)$.

Therefore $\varphi_4(g) \in M(1)$ and hence $\varphi_4(g) = c_1 g$ for some $c_1 \in K$. Similarly $\varphi_4(e) \in M(6)$ and $\varphi_4(e) = c_2 e$ for some $c_2 \in K$. Furthermore, $\varphi_4(f) \in M(2) \cap M(5) = \mathrm{span}\{f\}$ and therefore $\varphi_4(f) = c_3 f$ for some $c_3 \in K$.

We have $\varphi_4(e + f) \in M(3)$ so it must be of the form $\varphi_4(e + f) = a(e + f) + b(f + g)$ for $a, b \in K$. On the other hand, by linearity

$$\varphi_4(e + f) = \varphi_4(e) + \varphi_4(f) = c_2 e + c_3 f.$$

Now, e, f, g are linearly independent, and we can equate coefficients. We get $b = 0$ and $a = c_2 = c_3$.

Similarly $\varphi_4(f + g) = a'(e + f) + b'(f + g)$ with $a', b' \in K$, but it is equal to $c_3 f + c_1 g$. We deduce $a' = 0$ and $b' = c_3 = c_1$.

Combining the two calculations above we see that $c_1 = c_2 = c_3$. But this means that $\varphi_4 = c_1 \mathrm{id}_{M(4)}$ and hence φ is a scalar multiple of the identity. $\square$

We have therefore found a special representation of Q, with $M(3)$ two-dimensional. So if we extend the quiver by a new vertex ω and a new arrow $\omega \to 3$ we get the following extended quiver $\widehat{Q}$,

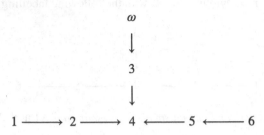

This quiver $\widehat{Q}$ has underlying graph a Euclidean diagram of type $\widetilde{E}_6$. By Lemma 11.29 we can construct pairwise non-isomorphic indecomposable representations $\mathcal{M}_\lambda$ of $\widehat{Q}$ for $\lambda \in K$. In particular, the quiver $\widehat{Q}$ has infinite representation type if the field K is infinite.

We will now define representations of arbitrary large dimension, for the above quiver $\widehat{Q}$ of type $\widetilde{E}_6$. Then we will show that these representations are indecomposable.

Definition 12.2. We fix $m \in \mathbb{N}$, and we define a representation $V = V_m$ of $\widehat{Q}$, where $V(4)$ has dimension $3m$ and all other spaces are subspaces, and all the linear maps are inclusions. That is, let

$$V(4) = \mathrm{span}\{e_1, \ldots, e_m, f_1, \ldots, f_m, g_1, \ldots, g_m\}$$

and define the other spaces as follows,

$V(1) = \mathrm{span}\{g_1, \ldots, g_m\}$

$V(2) = \mathrm{span}\{f_1, \ldots, f_m, g_1, \ldots, g_m\}$

$V(3) = \mathrm{span}\{e_1 + f_1, \ldots, e_m + f_m, f_1 + g_1, \ldots, f_m + g_m\}$

$V(5) = \mathrm{span}\{e_1, \ldots, e_m, f_1, \ldots, f_m\}$

$V(6) = \mathrm{span}\{e_1, \ldots, e_m\}$

$V(\omega) = \mathrm{span}\{e_1 + f_1, (e_2 + f_2) + (f_1 + g_1), \ldots, (e_m + f_m) + (f_{m-1} + g_{m-1})\}.$

Remark 12.3. If we look at the restriction of V_m to the subquiver Q of type E_6 then we see that it is the direct sum of m copies of the special representation defined in the proof of Lemma 12.1. The proof that the representation V_m is indecomposable is thus similar to the proof of Lemma 11.29.

Lemma 12.4. *For every $m \in \mathbb{N}$, the representation V_m of $\widehat{Q}$ is indecomposable.*

Proof. To prove that the representation is indecomposable, we use the criterion in Lemma 9.11. So we take a homomorphism $\varphi : \mathcal{V}_m \to \mathcal{V}_m$ of representations and assume $\varphi^2 = \varphi$. Then it suffices to show that φ is the zero homomorphism or the identity.

Since all maps in the representation $\mathcal{V}_m$ are inclusion maps, any homomorphism of representations $\varphi : \mathcal{V}_m \to \mathcal{V}_m$ is given by a linear map $\varphi_4 : V(4) \to V(4)$ such that $\varphi_4(V(j)) \subseteq V(j)$ for all j of $\widehat{Q}$. We just write φ instead of φ_4, to simplify notation.

We observe that $V(5) \cap V(\omega) = \mathrm{span}\{e_1 + f_1\}$. Since $V(5)$ and $V(\omega)$ are invariant under φ, it follows that $e_1 + f_1$ is an eigenvector of φ. Moreover, since $\varphi^2 = \varphi$ by assumption, the corresponding eigenvalue is 0 or 1. We may assume that $\varphi(e_1 + f_1) = 0$ since otherwise we can replace the representation φ by $\mathrm{id}_{\mathcal{V}_m} - \varphi$. This gives by linearity that $\varphi(e_1) = -\varphi(f_1)$ and this lies in $V(6) \cap V(2)$, which is zero, so that

$$\varphi(e_1) = 0 = \varphi(f_1).$$

Consider now $\varphi(g_1)$, this lies in $V(1)$, and we also have

$$\varphi(g_1) = \varphi(f_1) + \varphi(g_1) = \varphi(f_1 + g_1) \in V(3).$$

But $V(1) \cap V(3) = 0$, therefore $\varphi(g_1) = 0$. In total we know that $\varphi(e_1) = \varphi(f_1) = \varphi(g_1) = 0$.

To prove that φ is the zero map, we show by induction that $\varphi(e_k) = \varphi(f_k) = \varphi(g_k) = 0$ for all $1 \leq k \leq m$.

The case $k = 1$ has been shown above. For the inductive step we have by linearity of φ that

$$\varphi(e_{k+1}) + \varphi(f_{k+1}) = \varphi(e_{k+1} + f_{k+1})$$
$$= \varphi(e_{k+1} + f_{k+1}) + \varphi(f_k) + \varphi(g_k) \quad \text{(by the induction hypothesis)}$$
$$= \varphi((e_{k+1} + f_{k+1}) + (f_k + g_k)).$$

Since φ preserves the spaces spanned by the e_j and the f_j, this element now belongs to $V(5) \cap V(\omega) = \mathrm{span}\{e_1 + f_1\}$. Hence $\varphi(e_{k+1} + f_{k+1}) = \lambda(e_1 + f_1)$ for some scalar $\lambda \in K$. Now using our assumption $\varphi^2 = \varphi$ and that $\varphi(e_1 + f_1) = \varphi(e_1) + \varphi(f_1) = 0$ by induction hypothesis, it follows that $\varphi(e_{k+1} + f_{k+1}) = 0$. From this we deduce by linearity that

$$\varphi(e_{k+1}) = -\varphi(f_{k+1}) \in V(6) \cap V(2) = 0.$$

Furthermore, we can then deduce that

$$\varphi(g_{k+1}) = \varphi(f_{k+1}) + \varphi(g_{k+1}) = \varphi(f_{k+1} + g_{k+1}) \in V(1) \cap V(3) = 0.$$

We have shown that $\varphi(e_{k+1}) = \varphi(f_{k+1}) = \varphi(g_{k+1}) = 0$. This completes the inductive step.

So we have proved that φ is the zero homomorphism and hence that the representation V_m is indecomposable.   □

Since they have different dimensions, the representations V_m for different m are pairwise non-isomorphic, and hence the quiver $\widehat{Q}$ has infinite representation type, over any field K. By the arguments at the beginning of this section, this also shows that every quiver with underlying graph of type $\widetilde{E}_6$ has infinite representation type, over any field K.

12.2.2 Quivers of Type $\widetilde{E}_7$ Have Infinite Representation Type

Let Q be the quiver of Dynkin type E_7 with the following labelling and orientation:

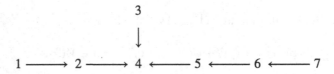

Lemma 12.5. *The quiver Q has a special representation $\mathcal{M}$ with $\dim_K M(1) = 2$.*

Proof. We define a representation $\mathcal{M}$ of this quiver for which all maps are inclusion maps, so we do not specify names for the arrows. We take $M(4)$ to be a 4-dimensional space, and all other spaces are subspaces.

$$M(4) = \text{span}\{e, f, g, h\}$$
$$M(1) = \text{span}\{f - g, e + h\}$$
$$M(2) = \text{span}\{e + f, e + g, e + h\}$$
$$M(3) = \text{span}\{g, h\}$$
$$M(5) = \text{span}\{e, f, g\}$$
$$M(6) = \text{span}\{e, f\}$$
$$M(7) = \text{span}\{e\}.$$

Note that indeed for each arrow in Q the space corresponding to the starting vertex is a subspace of the space corresponding to the end vertex. The only arrow for which this is not immediate is $1 \longrightarrow 2$, and here we have $M(1) \subseteq M(2)$ since $f - g = (e + f) - (e + g)$.

According to Definition 11.30 we have to show that every endomorphism of $\mathcal{M}$ is a scalar multiple of the identity. So let $\varphi : \mathcal{M} \to \mathcal{M}$ be a homomorphism of representations. Then since all maps for $\mathcal{M}$ are inclusions it follows that $\varphi_i : M(i) \to M(i)$ is the restriction of φ_4 to $M(i)$, for each $i = 1, \ldots, 7$.

First, $\varphi_4(e) \in M(7)$ and hence $\varphi_4(e) = c_1 e$ for some $c_1 \in K$. Next $\varphi_4(g) \in M(5) \cap M(3)$, which is spanned by g, so $\varphi_4(g) = c_2 g$ for some $c_2 \in K$.

Furthermore, $\varphi_4(e+f) \in M(6) \cap M(2) = \text{span}\{e+f\}$, so $\varphi_4(e+f) = c_3(e+f)$ for some $c_3 \in K$. Similarly we get $\varphi_4(f - g) \in M(5) \cap M(1) = \text{span}\{f - g\}$ and hence $\varphi_4(f - g) = c_4(f - g)$ with $c_4 \in K$. Finally, $\varphi_4(g - h) \in M(2) \cap M(3) = \text{span}\{g - h\}$ and thus $\varphi_4(g - h) = c_5(g - h)$ for some $c_5 \in K$.

Now consider $\varphi_4(f)$. Using linearity we have two expressions

$$\varphi_4(f) = \varphi_4(e + f) - \varphi_4(e) = c_3(e + f) - c_1 e$$

$$\varphi_4(f) = \varphi_4(f - g) + \varphi_4(g) = c_4(f - g) + c_2 g.$$

Since e, f, g are linearly independent, we can equate coefficients and get $c_3 - c_1 = 0$, $c_3 = c_4$ and $c_2 - c_4 = 0$. Thus, we have $c_1 = c_2 = c_3 = c_4$. This already implies that the basis elements e, g, f are mapped by φ_4 to the same scalar multiple of themselves.

It remains to deal with the fourth basis element h. We observe that

$$\varphi_4(e + h) \in M(1) \cap (M(7) + M(3)) = \text{span}\{e + h\},$$

so $\varphi_4(e + h) = c_6(e + h)$ with $c_6 \in K$. Again using linearity this gives us two expressions for $\varphi_4(h)$, namely

$$\varphi_4(h) = -\varphi_4(g - h) + \varphi_4(g) = (c_2 - c_5)g + c_5 h$$

$$\varphi_4(h) = \varphi_4(e + h) - \varphi_4(e) = (c_6 - c_1)e + c_6 h.$$

By equating coefficients we obtain $c_6 - c_1 = 0$, $c_2 - c_5 = 0$ and $c_5 = c_6$. But this means that h is mapped by φ_4 to the same scalar multiple of itself as e, f, g. Hence the homomorphism $\varphi : \mathcal{M} \to \mathcal{M}$ is a scalar multiple of the identity, as claimed. $\square$

We now have a special representation $\mathcal{M}$ of the quiver Q with $M(1)$ two-dimensional. So if we extend the quiver by a new vertex ω and a new arrow $\omega \longrightarrow 1$ then we obtain the quiver $\widehat{Q}$ as shown below,

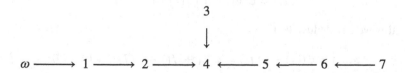

The underlying graph of $\widehat{Q}$ is a Euclidean diagram of type $\widetilde{E}_7$. By Lemma 11.29, the quiver $\widehat{Q}$ has infinite representation type if the field K is infinite.

We will now show that $\widehat{Q}$ has infinite representation type, for arbitrary fields. We construct representations of arbitrary dimension and show that they are indecomposable. Recall from Corollary 11.26 that then every quiver with underlying graph $\widetilde{E}_7$ has infinite representation type.

Definition 12.6. Fix an integer $m \in \mathbb{N}$ and define a representation $V = V_m$ of $\widehat{Q}$ where all maps are inclusions, and all spaces $V(i)$ are subspaces of the $4m$-dimensional K-vector space

$$V(4) = \operatorname{span}\{e_i, f_i, g_i, h_i \mid 1 \leq i \leq m\}.$$

The subspaces assigned to the other vertices of $\widehat{Q}$ are defined as follows:

$$V(\omega) = \operatorname{span}\{f_1 - g_1, (f_i - g_i) + (e_{i-1} + h_{i-1}) \mid 2 \leq i \leq m\}$$

$$V(1) = \operatorname{span}\{f_i - g_i, e_i + h_i \mid 1 \leq i \leq m\}$$

$$V(2) = \operatorname{span}\{e_i + f_i, e_i + g_i, e_i + h_i \mid 1 \leq i \leq m\}$$

$$V(3) = \operatorname{span}\{g_i, h_i \mid 1 \leq i \leq m\}$$

$$V(5) = \operatorname{span}\{e_i, f_i, g_i \mid 1 \leq i \leq m\}$$

$$V(6) = \operatorname{span}\{e_i, f_i \mid 1 \leq i \leq m\}$$

$$V(7) = \operatorname{span}\{e_i \mid 1 \leq i \leq m\}.$$

Lemma 12.7. *For every* $m \in \mathbb{N}$, *the representation* V_m *of the quiver* $\widehat{Q}$ *is indecomposable.*

Proof. We write briefly $V = V_m$. To prove that V is indecomposable, we use Lemma 9.11. So let $\varphi : V \to V$ be a homomorphism with $\varphi^2 = \varphi$. Then we have to show that φ is zero or the identity.

Since all maps in V are inclusions, the morphism φ is given by a single linear map, which we also denote by φ, on $V(4)$ such that all subspaces $V(i)$, where $i \in \{1, \ldots, 7, \omega\}$, are invariant under φ (see Definition 9.4).

First we note that $V(\omega) \cap V(5) = \operatorname{span}\{f_1 - g_1\}$. Therefore $f_1 - g_1$ must be an eigenvector of φ. Again, we may assume that $\varphi(f_1 - g_1) = 0$ (since $\varphi^2 = \varphi$ the eigenvalue is 0 or 1, and if necessary we may replace φ by $\operatorname{id}_V - \varphi$). Then

$$\varphi(f_1) = \varphi(g_1) \in V(6) \cap V(3) = 0.$$

From this we can deduce that

$$\varphi(e_1) = \varphi(e_1) + \varphi(f_1) = \varphi(e_1 + f_1) \in V(7) \cap V(2) = 0$$

and similarly that $\varphi(h_1) = \varphi(e_1 + h_1) \in V(3) \cap V(1) = 0$.

Then we prove by induction on k that if $\varphi(e_k) = \varphi(f_k) = \varphi(g_k) = \varphi(h_k) = 0$ and $k < m$ then also $\varphi(e_{k+1}) = \varphi(f_{k+1}) = \varphi(g_{k+1}) = \varphi(h_{k+1}) = 0$.

By the inductive hypothesis and by linearity we have

$$\varphi(f_{k+1} - g_{k+1}) = \varphi(f_{k+1} - g_{k+1} + e_k + h_k) \in V(5) \cap V(\omega) = \mathrm{span}\{f_1 - g_1\}.$$

Thus, there exists a scalar $\mu \in K$ such that $\varphi(f_{k+1} - g_{k+1}) = \mu(f_1 - g_1)$. Since $\varphi^2 = \varphi$ and $\varphi(f_1) = 0 = \varphi(g_1)$ we conclude that $\varphi(f_{k+1} - g_{k+1}) = 0$. But then we have

$$\varphi(f_{k+1}) = \varphi(g_{k+1}) \in V(6) \cap V(3) = 0.$$

Now we proceed as above, that is, we have

$$\varphi(e_{k+1}) = \varphi(e_{k+1}) + \varphi(f_{k+1}) = \varphi(e_{k+1} + f_{k+1}) \in V(7) \cap V(2) = 0$$

and then also

$$\varphi(h_{k+1}) = \varphi(e_{k+1} + h_{k+1}) \in V(3) \cap V(1) = 0.$$

It follows that $\varphi = 0$ and then Lemma 9.11 implies that the representation $V = V_m$ is indecomposable. $\square$

12.2.3   Quivers of Type $\tilde{E}_8$ Have Infinite Representation Type

Now we consider the quiver Q of type E_8 with labelling and orientation as follows:

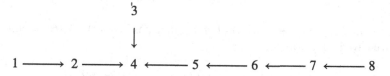

Lemma 12.8. *The quiver Q has a special representation $\mathcal{M}$ with $\dim_K M(8) = 2$.*

Proof. We define a representation $\mathcal{M}$ of this quiver for which all maps are inclusion maps, so we do not specify names of the arrows. We take $M(4)$ to be a 6-dimensional space, and all other spaces are subspaces.

$$M(4) = \mathrm{span}\{e, f, g, h, k, l\}$$
$$M(1) = \mathrm{span}\{e, l\}$$
$$M(2) = \mathrm{span}\{e, f, g, l\}$$

$$M(3) = \operatorname{span}\{h + l, e + g + k, e + f + h\}$$

$$M(5) = \operatorname{span}\{e, f, g, h, k\}$$

$$M(6) = \operatorname{span}\{f, g, h, k\}$$

$$M(7) = \operatorname{span}\{g, h, k\}$$

$$M(8) = \operatorname{span}\{h, k\}.$$

According to Definition 11.30 we have to show that every endomorphism of $\mathcal{M}$ is a scalar multiple of the identity. So let $\varphi : \mathcal{M} \to \mathcal{M}$ be a homomorphism of representations. Then as before each linear map $\varphi_i : M(i) \to M(i)$ is the restriction of φ_4 to $M(i)$. We again use that each subspace $M(i)$ is invariant under φ_4 to get restrictions for φ_4.

First, $\varphi_4(g) = c_1 g$ for some $c_1 \in K$ since $\varphi_4(g) \in M(2) \cap M(7) = \operatorname{span}\{g\}$. Similarly, $\varphi_4(e) \in M(1) \cap M(5) = \operatorname{span}\{e\}$, so $\varphi_4(e) = c_2 e$ for some $c_2 \in K$. Moreover, we have $\varphi_4(k) \in M(8)$, thus $\varphi_4(k) = ck + zh$ for $c, z \in K$. With this, we have by linearity

$$\varphi_4(e + g + k) = c_2 e + c_1 g + (ck + zh).$$

But this must lie in $M(3)$. Since l and f do not occur in the above expression, it follows from the definition of $M(3)$ that $\varphi_4(e + g + k)$ must be a scalar multiple of $e + g + k$ and hence $z = 0$ and $c_1 = c_2 = c$. In particular, $\varphi_4(k) = ck$.

We may write $\varphi_4(l) = ue + vl$ with $u, v \in K$ since it is in $M(1)$, and $\varphi_4(h) = rh + sk$ with $r, s \in K$ since it is in $M(8)$. We have $\varphi_4(h + l) \in M(3)$. In $\varphi_4(h) + \varphi_4(l)$, basis vectors g and f do not occur, and it follows that $\varphi_4(h + l)$ is a scalar multiple of $h + l$. Hence

$$\varphi_4(l) = vl, \quad \varphi_4(h) = rh, \text{ and } v = r.$$

We can write $\varphi_4(f) = af + bg$ with $a, b \in K$ since it is in $M(2) \cap M(6) = \operatorname{span}\{f, g\}$. Now we have by linearity that

$$\varphi_4(e + f + h) = c_2 e + (af + bg) + rh \in M(3).$$

Since the basis vectors l and k do not occur in this expression, it follows that $\varphi_4(e + f + h)$ is a scalar multiple of $e + f + h$. So $b = 0$ and $a = c_2 = r$; in particular, we have $\varphi_4(f) = af$.

In total we have now seen that $c = c_1 = c_2 = a = r = v$, so all six basis vectors of $M(4)$ are mapped to the same scalar multiple of themselves. This proves that φ_4 is a scalar multiple of the identity, and then so is φ. $\square$

We extend now the quiver Q by a new vertex ω and a new arrow $\omega \longrightarrow 8$. Hence we consider the following quiver $\widehat{Q}$ whose underlying graph is a Euclidean diagram of type $\widetilde{E}_8$:

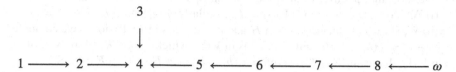

The result in Lemma 12.8 together with Lemma 11.29 already yields that $\widehat{Q}$ has infinite representation type over every infinite field K. But since we prove Gabriel's theorem for arbitrary fields, we need to show that $\widehat{Q}$ has infinite representation type over any field K. To this end we now define representations of arbitrary large dimensions and afterwards show that they are indeed indecomposable. The construction is inspired by the special representation of Q just considered; in fact, the restriction to the subquiver Q is a direct sum of copies of the above special representation.

Definition 12.9. Fix an integer $m \in \mathbb{N}$. We will define a representation $\mathcal{V} = \mathcal{V}_m$ of the above quiver $\widehat{Q}$ where all maps are inclusions, and all spaces $V(i)$ are subspaces of $V(4)$, a $6m$-dimensional vector space over K. We give the bases of the spaces.

$$V(4) = \text{span}\{e_i, f_i, g_i, h_i, k_i, l_i \mid 1 \le i \le m\}$$

$$V(1) = \text{span}\{e_i, l_i \mid 1 \le i \le m\}$$

$$V(2) = \text{span}\{e_i, f_i, g_i, l_i \mid 1 \le i \le m\}$$

$$V(3) = \text{span}\{h_i + l_i, \ e_i + g_i + k_i, e_i + f_i + h_i \mid 1 \le i \le m\}$$

$$V(5) = \text{span}\{e_i, f_i, g_i, h_i, k_i \mid 1 \le i \le m\}$$

$$V(6) = \text{span}\{f_i, g_i, h_i, k_i \mid 1 \le i \le m\}$$

$$V(7) = \text{span}\{g_i, h_i, k_i \mid 1 \le i \le m\}$$

$$V(8) = \text{span}\{h_i, k_i \mid 1 \le i \le m\}$$

$$V(\omega) = \text{span}\{h_1, h_2 + k_1, h_3 + k_2, \ldots, h_m + k_{m-1}\}.$$

Lemma 12.10. *For every* $m \in \mathbb{N}$*, the representation* $\mathcal{V}_m$ *of the quiver* $\widehat{Q}$ *is indecomposable.*

Proof. We briefly write $\mathcal{V} := \mathcal{V}_m$. To show that $\mathcal{V}$ is indecomposable we use the criterion in Lemma 9.11, that is, we show that the only endomorphisms $\varphi : \mathcal{V} \to \mathcal{V}$ of representations satisfying $\varphi^2 = \varphi$ are zero and the identity.

As before, since all maps are given by inclusions, any endomorphism φ on $\mathcal{V}$ is given by a linear map $V(4) \to V(4)$, which we also denote by φ, such that $\varphi(V(i)) \subseteq V(i)$ for all vertices i of $\widehat{Q}$.

The space $V(4)$ is the direct sum of six subspaces, each of dimension m, spanned by the basis vectors with the same letter. We write E for the span of the set $\{e_1, \ldots, e_m\}$, and similarly we define subspaces F, G, H, K and L.

(1) We show that φ leaves each of these six subspaces of $V(4)$ invariant:

(i) We have $\varphi(e_i) \in V(1) \cap V(5) = E$. Similarly, $\varphi(g_i) \in V(7) \cap V(2) = G$.

(ii) We show now that $\varphi(h_i)$ is in H and that $\varphi(l_i)$ is in L: To do so, we compute $\varphi(h_i + l_i) = \varphi(h_i) + \varphi(l_i)$. First, $\varphi(h_i)$ is in $V(8)$, which is $H \oplus K$. Moreover, $\varphi(l_i)$ is in $V(1)$, that is, in $E \oplus L$. Therefore $\varphi(h_i + l_i)$ is in $H \oplus K \oplus E \oplus L$. Secondly, since $h_i + l_i \in V(3)$, its image $\varphi(h_i + l_i)$ is also in $V(3)$. If expressed in terms of the basis of $V(3)$, there cannot be any $e_j + g_j + k_j$ occuring since this involves a non-zero element in G. Similarly no basis vector $e_i + f_i + h_i$ can occur. It follows that $\varphi(h_i + l_i)$ is in $H \oplus L$. This implies that $\varphi(h_i)$ cannot involve any non-zero element in K, so it must lie in H. Similarly $\varphi(l_i)$ must lie in L.

In the following steps, the strategy is similar to that in (ii).

(iii) We show that $\varphi(k_i)$ is in K. To prove this, we compute

$$\varphi(e_i + g_i + k_i) = \varphi(e_i + g_i) + \varphi(k_i).$$

We know $\varphi(e_i + g_i) \in E \oplus G$ and $\varphi(k_i) \in V(8) = H \oplus K$ and therefore $\varphi(e_i + g_i + k_i)$ lies in $E \oplus G \oplus H \oplus K$. On the other hand, $\varphi(e_i + g_i + k_i)$ lies in $V(3)$. It cannot involve a basis element in which some l_j occurs, or some f_j, and it follows that it must be in the span of the elements of the form $e_j + g_j + k_j$. Therefore it follows that $\varphi(k_i) \in K$.

(iv) We claim that $\varphi(f_i)$ is in F. It lies in $V(6) \cap V(2)$, so it is in $F \oplus G$. We compute $\varphi(e_i + f_i + h_i) = \varphi(e_i + h_i) + \varphi(f_i)$. By parts (i) and (ii), we know $\varphi(e_i + h_i)$ is in $E \oplus H$ and therefore $\varphi(e_i + h_i) + \varphi(f_i)$ is in $E \oplus H \oplus F \oplus G$. On the other hand, it lies in $V(3)$ and since it cannot involve a basis vector with a k_j or l_j we deduce that it must be in $E \oplus F \oplus H$. Therefore $\varphi(f_i)$ cannot involve any g_j and hence it belongs of F.

(2) Consider $\varphi(h_1)$. It belongs to H and also to $V(\omega)$, so it is a scalar multiple of h_1, that is, h_1 is an eigenvector of φ. Since $\varphi^2 = \varphi$, the eigenvalue is 0 or 1. As before, we may assume that $\varphi(h_1) = 0$, otherwise we replace φ by $\mathrm{id}_{V(4)} - \varphi$.

(3) We show that $\varphi(l_1) = 0 = \varphi(e_1) = \varphi(f_1)$ and $\varphi(g_1) = \varphi(k_1) = 0$. First we have $\varphi(h_1 + l_1) = \varphi(h_1) + \varphi(l_1) = \varphi(l_1) \in L \cap V(3) = 0$. Next, we have

$$\varphi(e_1 + f_1 + h_1) = \varphi(e_1 + f_1) + \varphi(h_1) = \varphi(e_1 + f_1) \in (E \oplus F) \cap V(3) = 0,$$

hence $\varphi(e_1) = -\varphi(f_1) \in E \cap F = 0$. Similarly

$$\varphi(e_1 + g_1 + k_1) = \varphi(e_1) + \varphi(g_1 + k_1) = \varphi(g_1 + k_1) \in (G \oplus K) \cap V(3) = 0$$

and furthermore $\varphi(g_1) = -\varphi(k_1) \in G \cap K = 0$.

By (2) and (3) we have that all basis vectors with index 1 are mapped to zero by φ.

(4) Now one proves by induction on r that if φ maps each of h_r, l_r, e_r, f_r, g_r and k_r to zero, and if $r < m$ then it also maps all basis vectors h_{r+1}, l_{r+1}, e_{r+1}, f_{r+1}, g_{r+1} and k_{r+1} to zero. The arguments are the same as in step (3), and similar to those in the proofs for $\widetilde{E}_6$ and $\widetilde{E}_7$, so we will omit the details.

Combining (2), (3) and (4) shows that $\varphi = 0$. We have therefore proved that V_m is indecomposable.   □

Since they have different dimensions, the indecomposable representations in Lemma 12.10 are pairwise non-isomorphic. This shows that the quiver $\widehat{Q}$ has infinite representation type over any field K. In Corollary 11.26 we have seen that the representation type is independent of the orientation of the arrows, as long as the underlying graph is a tree, as is the case here. Therefore we have proved that every quiver with underlying graph $\widetilde{E}_8$ has infinite representation type, over any field K.

12.3 Root Systems

Root systems were first discovered in Lie algebras, as collections of eigenvectors of certain diagonalizable linear maps; they form a key tool in the famous Cartan–Killing classification of simple Lie algebras over the complex numbers. Subsequently, root systems have been found in many other contexts, and it is therefore convenient to use an axiomatic definition.

We will only give a brief account of root systems here; for more details we refer to the book by Erdmann and Wildon in this series.[1]

Let E be a finite-dimensional real vector space with an inner product $(-, -)$. Given a non-zero vector $v \in E$, let s_v be the reflection in the hyperplane perpendicular to v. It is given by the formula

$$s_v(x) = x - \frac{2(x, v)}{(v, v)}\, v \quad \text{for every } x \in E.$$

Then $s_v(v) = -v$, and if $(y, v) = 0$ then $s_v(y) = y$. Write $\langle x, v \rangle := \frac{2(x,v)}{(v,v)}$.

Definition 12.11. A subset R of E is a *root system* if it satisfies the following:

(R1) R is finite, it spans E and $0 \notin R$.
(R2) If $\alpha \in R$, the only scalar multiples of α in R are $\pm\alpha$.
(R3) If $\alpha \in R$ then the reflection s_α permutes the elements of R.
(R4) If $\alpha, \beta \in R$ then $\langle \beta, \alpha \rangle \in \mathbb{Z}$.

The elements of the root system R are called *roots*.

---

[1] K. Erdmann, M.J. Wildon, *Introduction to Lie algebras*. Springer Undergraduate Mathematics Series. Springer-Verlag London, Ltd., London, 2006. x+251 pp.

Remark 12.12. Condition (R4) is closely related to the possible angles between two roots. If $\alpha, \beta \in R$ and θ is the angle between α and β then

$$\langle \alpha, \beta \rangle \cdot \langle \beta, \alpha \rangle = 4 \frac{(\alpha, \beta)^2}{|\alpha|^2 |\beta|^2} = 4 \cos^2(\theta) \leq 4$$

and this is an integer by (R4). So there are only finitely many possibilities for the numbers $\langle \beta, \alpha \rangle$.

Definition 12.13. Let R be a root system in E. A **base** of R is a subset B of R such that

(i) B is a vector space basis of E.
(ii) Every $\beta \in R$ can be written as

$$\beta = \sum_{\alpha \in B} k_\alpha \alpha$$

with $k_\alpha \in \mathbb{Z}$ and where all non-zero coefficients k_α have the same sign.

One can show that every root system has a base. With this, $R = R^+ \cup R^-$, where R^+ is the set of all β where the signs are positive, and R^- is the set of all β where signs are negative. Call R^+ the set of 'positive roots', and R^- the set of 'negative roots'.

We fix a base $B = \{\alpha_1, \ldots, \alpha_n\}$ of the root system R. Note that the cardinality n of B is the vector space dimension of E. The **Cartan matrix** of R is the (integral) matrix with (i, j)-entry $\langle \alpha_i, \alpha_j \rangle$.

Root systems are classified by their Cartan matrices. We consider root systems whose Cartan matrices are symmetric, known as 'simply laced' root systems. One can show that for these root systems, if $i \neq j$ then $\langle \alpha_i, \alpha_j \rangle$ is equal to 0 or to -1.

Definition 12.14. Let R be a simply laced root system and let $B = \{\alpha_1, \ldots, \alpha_n\}$ be a base of R. The **Dynkin diagram** of R is the graph Γ_R, with vertices $\{1, 2, \ldots, n\}$, and there is an edge between vertices i and j if $\langle \alpha_i, \alpha_j \rangle \neq 0$.

The classification of root systems via Dynkin diagrams then takes the following form. Recall that the Dynkin diagrams of type A, D, E were given in Fig. 10.1.

Theorem 12.15. *The Dynkin diagrams for simply laced root systems are the unions of Dynkin diagrams of type A_n (for $n \geq 1$), D_n (for $n \geq 4$) and E_6, E_7 and E_8.*

Now we relate this to quivers. Let Q be a connected quiver without oriented cycles, with underlying graph Γ. We have defined the symmetric bilinear form $(-, -)_\Gamma$ on $\mathbb{Z}^n \times \mathbb{Z}^n$, see Definition 10.2. With the same Gram matrix we get a symmetric bilinear form on $\mathbb{R}^n \times \mathbb{R}^n$.

Let Γ be a union of Dynkin diagrams. Then the quadratic form q_Γ corresponding to the above symmetric bilinear form is positive definite; in fact, Proposition 10.10

shows this for a single Dynkin diagram, and the general case can be deduced from the formula (11.1) in the proof of Lemma 11.41.

Hence, $(-, -)_\Gamma$ is an inner product. So we consider the vector space $E = \mathbb{R}^n$ with the inner product $(-, -)_\Gamma$, where n is the number of vertices of Γ.

Proposition 12.16. *Let Q be a quiver whose underlying graph Γ is a union of Dynkin diagrams of type A, D, or E. Let q_Γ be the quadratic form associated to Γ, and let $\Delta_\Gamma = \{x \in \mathbb{Z}^n \mid q_\Gamma(x) = 1\}$ be the set of roots, as in Definition 10.6. Then Δ_Γ is a root system in $E = \mathbb{R}^n$, as in Definition 12.11. It has a base (as in Definition 12.13) consisting of the unit vectors of $\mathbb{Z}^n$. The associated Cartan matrix is the Gram matrix of $(-, -)_\Gamma$.*

Proof. (R1) We have seen that Δ_Γ is finite (see Proposition 10.12 and Remark 11.42). Since the unit vectors are roots (see Exercise 10.3), the set Δ_Γ spans $E = \mathbb{R}^n$. From the definition of q_Γ (see Definition 10.6) we see that $q_\Gamma(0) = 0$, that is, the zero vector is not in Δ_Γ.

(R2) Let $x \in \Delta_\Gamma$ and $\lambda \in \mathbb{R}$ such that $\lambda x \in \Delta_\Gamma$. Then we have $q_\Gamma(\lambda x) = \lambda^2 q_\Gamma(x)$. They are both equal to 1 if and only if $\lambda = \pm 1$.

(R3) We have proved in Lemma 10.9 that a reflection s_i permutes the elements of Δ_Γ but a similar computation shows that for any $y \in \Delta_\Gamma$ we have $q_\Gamma(s_y(x)) = q_\Gamma(x)$: Since $y \in \Delta_\Gamma$ we have $(y, y)_\Gamma = 2q_\Gamma(y) = 2$; then

$$s_y(x) = x - \frac{2(x, y)_\Gamma}{(y, y)_\Gamma} v = x - (x, y)_\Gamma\, y \quad \text{for every } x \in E.$$

It follows that

$$(s_y(x), s_y(x))_\Gamma = (x - (x, y)_\Gamma\, y, x - (x, y)_\Gamma\, y)_\Gamma$$
$$= (x, x)_\Gamma - 2(x, y)_\Gamma^2 + (x, y)_\Gamma^2 (y, y)_\Gamma$$
$$= (x, x)_\Gamma.$$

Thus, $q_\Gamma(s_y(x)) = \frac{1}{2}(s_y(x), s_y(x))_\Gamma = \frac{1}{2}(x, x)_\Gamma = q_\Gamma(x)$ and hence s_y permutes the elements in Δ_Γ.

(R4) We have for $x, y \in \Delta_\Gamma$ that $(y, y) = 2q_\Gamma(y) = 2$ and hence $\langle x, y \rangle = (x, y)_\Gamma \in \mathbb{Z}$.

This proves that Δ_Γ satisfies the axioms of a root system.

We now show that the unit vectors form a base of the root system Δ_Γ (in the sense of Definition 12.13): the unit vectors clearly form a vector space basis of $E = \mathbb{R}^n$; moreover, they satisfy the property (ii) as in Definition 12.13, as we have seen in Lemma 10.13.

As noted, for unit vectors $\varepsilon_i, \varepsilon_j$ we have that $\langle \varepsilon_i, \varepsilon_j \rangle$ is equal to $(\varepsilon_i, \varepsilon_j)_\Gamma$; this says that the Cartan matrix of the root system Δ_Γ is the same as the Gram matrix in Definition 10.2. $\qquad \square$

12.4   Morita Equivalence

A precise definition of Morita equivalence needs the setting of categories and
functors (and it can be found in more advanced texts). Here, we will give an informal
account, for finite-dimensional algebras over a field K, and we give some examples
as illustrations. Roughly speaking, two algebras A and B over the same field K are
Morita equivalent 'if they have the same representation theory'. That is, there should
be a canonical correspondence between finite-dimensional A-modules and finite-
dimensional B-modules, say $M \to F(M)$, and between module homomorphisms:
If M and N are A-modules and $f : M \to N$ is an A-module homomorphism then
$F(f)$ is a B-module homomorphism $F(M) \to F(N)$, such that

(i) it preserves (finite) direct sums, and takes indecomposable A-modules to
     indecomposable B-modules, and simple A-modules to simple B-modules,
(ii) every B-module is isomorphic to $F(M)$ for some A-module M, where M is
      unique up to isomorphism,
(iii) the correspondence between homomorphisms gives a vector space isomor-
       phism

$$\mathrm{Hom}_A(M, N)) \cong \mathrm{Hom}_B(F(M), F(N))$$

for arbitrary A-modules M, N.

Clearly, isomorphic algebras are Morita equivalent. But there are many non-
isomorphic algebras which are Morita equivalent, and we will illustrate this by
considering some examples which appeared in this book.

Example 12.17.

(1) Let $A = M_n(K)$, the algebra of $n \times n$-matrices over the field K. This is
     a semisimple algebra, hence every A-module is a direct sum of simple A-
     modules. Moreover, up to isomorphism there is only one simple A-module,
     namely the natural module K^n. Precisely the same properties hold for K-
     modules, that is, K-vector spaces; every K-module is a direct sum of the only
     simple K-module, which is K itself. So one might ask whether A is Morita
     equivalent to the K-algebra K. This is indeed the case.
         Consider the matrix unit $e = E_{11} \in A$. This is an idempotent, and the algebra
     eAe is one-dimensional, spanned by E_{11}; in particular, eAe is isomorphic to K.
     If M is an A-module, one can take

$$F(M) := eM,$$

which is an eAe-module, by restricting the given action. If $f : M \to N$ is an
A-module homomorphism then $F(f)$ is the restriction of f to eM, and this is
a homomorphism of eAe-modules $eM \to eN$.

If M is an A-module and $M = U \oplus V$ is a direct sum of A-submodules U and V then one checks that $eM = eU \oplus eV$, a direct sum of eAe-submodules. Inductively, one can also see that F preserves arbitrary finite direct sums. If M is indecomposable, then, since A is semisimple, M is simple, and isomorphic to the natural module K^n. We have $e(K^n) = \{(x, 0, \ldots, 0)^t \mid x \in K\}$ so this is 1-dimensional and is indecomposable as a K-module. Hence property (i) is satisfied.

One can see that (ii) holds: since an eAe-module is a vector space, and if it has dimension d then it is isomorphic to eM where M is the direct sum of d copies of the natural module K^n of A. Moreover, since every finite-dimensional A-module is a direct sum of copies of K^n, we see that such M is unique up to isomorphism.

Similarly one can see that (iii) holds.

With a precise definition one can show that Morita equivalence is an equivalence relation, and therefore any two matrix algebras $M_n(K)$ and $M_m(K)$ over a fixed field are Morita equivalent.

This example also shows that dimensions of simple modules are not preserved by a Morita equivalence. In fact, it is an advantage for calculations to have smaller simple modules.

(2) If $A = M_n(D)$ where D is a division algebra over K then similarly A is Morita equivalent to D, with the same correspondence as in (1). But note that if $D \not\cong K$, then K and D are not Morita equivalent. The simple K-module has an endomorphism algebra which is 1-dimensional over K but the endomorphism algebra of the simple D-module, that is of D, is not 1-dimensional over K; so condition (iii) fails.

(3) Consider a finite-dimensional semisimple algebra. If $A = M_n(K) \times M_r(K)$, then one can take $F(M) = eM$ where e is the idempotent

$$e = e_1 + e_2, \quad e_1 = (E_{11}, 0) \quad \text{and} \quad e_2 = (0, E_{11}).$$

Then eAe is Morita equivalent to A. Again, A is semisimple, so indecomposable modules are the same as simple modules. We can see that F gives a bijection between simple modules: The algebra A has simple modules $S_1 := K^n \times 0$ and $S_2 := 0 \times K^r$ (up to isomorphism), and eAe is isomorphic to $K \times K$ and the simple modules are $K \times 0 = F(S_1)$ and $0 \times K = F(S_2)$. This generalizes to the product of finitely many matrix algebras, and also where the factors are of the form as in (2).

Example (3) is a special case of a more general fact. Namely, if A and B are Morita equivalent and C and D are Morita equivalent, then $A \times C$ and $B \times D$ are Morita equivalent. By Lemma 3.30, an indecomposable module of a direct product is an indecomposable module for one of the factors, and the other factor acts as zero. Using this, the two given Morita equivalences can be used to construct one between the direct products.

Given any K-algebra A, we describe a method for constructing algebras Morita equivalent to A (without proofs). Assume that A has n simple modules $S_1, \ldots, S_n$ (up to isomorphism). Suppose e is an idempotent of A such that the modules eS_i are all non-zero. Then one can show that eAe is Morita equivalent to A. (It is easy to check that eS_i is a simple module for eAe.)

A special case of this method is used to construct an algebra which is Morita equivalent to A and is called *a basic algebra* of A. Namely, start with the algebra $\bar{A} := A/J(A)$. This is semisimple, see Theorem 4.23, so it is isomorphic to a product of finitely many matrix algebras. Take an idempotent ε of $\bar{A}$ as we have described in Example 12.17(3). Then $\varepsilon\bar{A}\varepsilon$ is a direct product of finitely many division algebras D_i, and the modules for $\varepsilon\bar{A}\varepsilon$ are then as vector spaces also equal to the D_i. One can show that there is an idempotent e of A such that $e + J(A) = \varepsilon$, and that the idempotent e produces such a basic algebra of A. Note that then the simple eAe-modules are the same as the simple $\varepsilon\bar{A}\varepsilon$-modules.

In the case when $\bar{A}$ is a direct product $M_{n_1}(K) \times \ldots \times M_{n_r}(K)$, then we have that all simple eAe-modules are 1-dimensional. This is the nicest setting, and it has inspired the following definition:

Definition 12.18. A finite-dimensional K-algebra B is said to be basic if all simple B-modules are 1-dimensional. Equivalently, the factor algebra $B/J(B)$ (which is semisimple) is isomorphic to a direct product of copies of K.

These conditions are equivalent, recall that by Theorem 4.23 the simple B-modules are the same as the simple $B/J(B)$-modules. Since $B/J(B)$ is semisimple, it is a direct product of matrix algebras over K, and the simple modules are 1-dimensional if and only if all blocks have size 1 and are just K.

Example 12.19.

(1) Let A be the product of matrix algebras over division algebras. Then the basic algebra of A is basic in the above sense if and only if all the division algebras are equal to K.
(2) Let $A = KQ$ where Q is any quiver without oriented cycles. Then A is basic, see Theorem 3.26.

To summarize, a basic algebra of an algebra is Morita equivalent to A, and in the nicest setting it has all simple modules 1-dimensional. There is a theorem of Gabriel which gives a completely general description of such algebras. This theorem might also give another reason why quivers play an important role.

Recall that if Q is a quiver then $KQ^{\geq t}$ is the span of all paths of length $\geq t$. We call an ideal I in a path algebra KQ admissible if $KQ^{\geq m} \subseteq I \subseteq KQ^{\geq 2}$ for some $m \geq 2$.

Theorem 12.20 (Gabriel). *Assume A is a finite-dimensional K-algebra which is basic. Then $A \cong KQ/I$, where Q is a quiver, and I is an admissible ideal of KQ. Moreover, the quiver Q is unique.*

The condition that I contains all paths of length $\geq m$ for some m makes sure that KQ/I is finite-dimensional, and one can show that Q is unique by using that I is contained in the span of paths of length ≥ 2. The ideal I in this theorem is not unique.

Example 12.21.

(1) Let $A = K[X]/(X^r)$ for $r \geq 1$. We have seen that $K[X]$ is isomorphic to the path algebra KQ where Q is the quiver with one vertex and a loop α, hence A is isomorphic to KQ/I where $I = (\alpha^r)$ is the ideal generated by α^r. Moreover, A is basic, see Proposition 3.23.
(2) Let $A = KG$ where G is the symmetric group S_3, with elements $r = (1\ 2\ 3)$ and $s = (1\ 2)$.

   (i) Assume that K has characteristic 3. Then by Example 8.20 we know that the simple A-modules are 1-dimensional. They are the trivial module and the sign module. Therefore A is basic.

   (ii) Now assume that $K = \mathbb{C}$, then we have seen in Example 6.7 that the simple A-modules have dimensions $1, 1, 2$ and hence A is not basic. In Exercise 6.4 we obtained the Artin–Wedderburn decomposition of $\mathbb{C}S_3$, with orthogonal idempotents e_+, e_-, f and f_1. We can take as isomorphism classes of simple A-modules Ae_+ and Ae_- (the trivial and the sign representation) and in addition Af. So we can write down the basic algebra eAe of A by taking $e = e_+ + e_- + f$.

Example 12.22. As before, let $A = KG$ be the group algebra for the symmetric group $G = S_3$ and assume that K has characteristic 3. We describe the quiver Q and a factor algebra KQ/I isomorphic to A. Note that the group algebra $K\langle s \rangle$ of the subgroup of order 2 is a subalgebra, and it is semisimple (by Maschke's theorem). It has two orthogonal idempotents $\varepsilon_1, \varepsilon_2$ with $1_A = \varepsilon_1 + \varepsilon_2$. Namely, take $\varepsilon_1 = -1 - s$ and $\varepsilon_2 = -1 + s$ (use that $-2 = 1$ in K). These are also idempotents of A, still orthogonal. Then one checks that A as a left A-module is the direct sum $A = A\varepsilon_1 \oplus A\varepsilon_2$. For the following, we leave the checking of the details as an exercise.

For $i = 1, 2$ the vector space $A\varepsilon_i$ has a basis of the form $\{\varepsilon_i, r\varepsilon_i, r^2\varepsilon_i\}$. We can write $A\varepsilon_1 = \varepsilon_1 A\varepsilon_1 \oplus \varepsilon_2 A\varepsilon_1$ as vector spaces, and $A\varepsilon_2 = \varepsilon_2 A\varepsilon_2 \oplus \varepsilon_1 A\varepsilon_2$. One checks that for $i = 1, 2$ the space $\varepsilon_i A\varepsilon_i$ has basis $\{\varepsilon_i, (1 + r + r^2)\varepsilon_i\}$. Moreover, $\varepsilon_2 A\varepsilon_1$ is 1-dimensional and spanned by $\varepsilon_2 r\varepsilon_1$, and $\varepsilon_1 A\varepsilon_2$ also is 1-dimensional and spanned by $\varepsilon_1 r\varepsilon_2$. Furthermore, one checks that

(i) $(\varepsilon_1 r\varepsilon_2)(\varepsilon_2 r\varepsilon_1) = (1 + r + r^2)\varepsilon_1$ and $(\varepsilon_2 r\varepsilon_1)(\varepsilon_1 r\varepsilon_2) = (1 + r + r^2)\varepsilon_2$
(ii) $(\varepsilon_i r\varepsilon_j)(\varepsilon_j r\varepsilon_i)(\varepsilon_i r\varepsilon_j) = 0$ for $\{i, j\} = \{1, 2\}$.

We take the quiver Q defined as

Note that the path algebra KQ is infinite-dimensional since Q has oriented cycles. Then we have a surjective algebra homomorphism $\psi : KQ \to A$ where we let

$$\psi(e_i) = \varepsilon_i, \quad \psi(\alpha) = \varepsilon_2 r \varepsilon_1, \quad \psi(\beta) = \varepsilon_1 r \varepsilon_2$$

and extend this to linear combinations of paths in KQ. Let I be the kernel of ψ. One shows, using (i) and (ii) above, that $I = (\alpha\beta\alpha, \ \beta\alpha\beta)$, the ideal generated by $\alpha\beta\alpha$ and $\beta\alpha\beta$. Hence I is an admissible ideal of KQ, and KQ/I is isomorphic to A.

Appendix A
Induced Modules for Group Algebras

In this appendix we provide the details of the construction of induced modules for group algebras of finite groups, as used in Sect. 8.2 for determining the representation type of group algebras. Readers familiar with tensor products over algebras can entirely skip this appendix. When defining induced modules we will take a hands-on approach and only develop the theory as far as we need it in this book to make it self-contained. For more details on induced modules for group algebras we refer the reader to textbooks on group representation theory.

We start by introducing tensor products of vector spaces. For simplicity we restrict to finite-dimensional vector spaces (however, the construction carries over almost verbatim to arbitrary vector spaces).

Definition A.1. Let K be a field and let V and W be two finite-dimensional K-vector spaces. We choose bases $\{v_1, \ldots, v_n\}$ of V and $\{w_1, \ldots, w_m\}$ of W, and introduce new symbols $v_i \otimes w_j$ for $1 \leq i \leq n$ and $1 \leq j \leq m$. The *tensor product* of V and W over K is the K-vector space having the symbols $v_i \otimes w_j$ as a basis, that is,

$$V \otimes_K W := \mathrm{span}\{v_i \otimes w_j \mid 1 \leq i \leq n, 1 \leq j \leq m\}.$$

Remark A.2.

(1) By definition, for the dimensions we have

$$\dim_K(V \otimes_K W) = (\dim_K V)(\dim_K W).$$

(2) An arbitrary element of $V \otimes_K W$ has the form $\sum_{i=1}^{n} \sum_{j=1}^{m} \lambda_{ij}(v_i \otimes w_j)$ with scalars $\lambda_{ij} \in K$. To ease notation, we often abbreviate the double sum to

© Springer International Publishing AG, part of Springer Nature 2018
K. Erdmann, T. Holm, *Algebras and Representation Theory*, Springer
Undergraduate Mathematics Series, https://doi.org/10.1007/978-3-319-91998-0

$\sum_{i,j} \lambda_{ij}(v_i \otimes w_j)$. Addition and scalar multiplication in $V \otimes_K W$ are given by

$$\left(\sum_{i,j} \lambda_{ij}(v_i \otimes w_j)\right) + \left(\sum_{i,j} \mu_{ij}(v_i \otimes w_j)\right) = \sum_{i,j}(\lambda_{ij} + \mu_{ij})(v_i \otimes w_j)$$

and

$$\lambda\left(\sum_{i,j} \lambda_{ij}(v_i \otimes w_j)\right) = \sum_{i,j} \lambda \lambda_{ij}(v_i \otimes w_j),$$

respectively.

The next result collects some fundamental 'calculation rules' for tensor products of vector spaces, in particular, it shows that the tensor product is bilinear. Moreover it makes clear that the vector space $V \otimes_K W$ does not depend on the choice of the bases used in the definition.

Proposition A.3. *We keep the notation from Definition A.1. For arbitrary elements* $v \in V$ *and* $w \in W$*, expressed in the given bases as* $v = \sum_i \lambda_i v_i$ *and* $w = \sum_j \mu_j w_j$*, we set*

$$v \otimes w := \sum_{i,j} \lambda_i \mu_j (v_i \otimes w_j) \in V \otimes_K W. \qquad (A.1)$$

Then the following holds.

(a) For all $v \in V$*,* $w \in W$ *and* $\lambda \in K$ *we have* $v \otimes (\lambda w) = (\lambda v) \otimes w = \lambda(v \otimes w)$*.*
(b) Let $x_1, \ldots, x_r \in V$ *and* $y_1, \ldots, y_s \in W$ *be arbitrary elements. Then we have*

$$\left(\sum_{i=1}^r x_i\right) \otimes \left(\sum_{j=1}^s y_j\right) = \sum_{i=1}^r \sum_{j=1}^s x_i \otimes y_j.$$

(c) Let $\{b_1, \ldots, b_n\}$ *and* $\{c_1, \ldots, c_m\}$ *be arbitrary bases of* V *and* W*, respectively. Then* $\{b_k \otimes c_\ell \mid 1 \le k \le n, 1 \le \ell \le m\}$ *is a basis of* $V \otimes_K W$*. In particular, the vector space* $V \otimes_K W$ *does not depend on the choice of the bases of* V *and* W*.*

Proof. (a) Since the scalars from the field K commute we get from (A.1) that

$$v \otimes (\lambda w) = \left(\sum_i \lambda_i v_i\right) \otimes \left(\sum_j \lambda \mu_j w_j\right) = \sum_{i,j} \lambda_i \lambda \mu_j (v_i \otimes w_j)$$

$$= \sum_{i,j} \lambda \lambda_i \mu_j (v_i \otimes w_j) = \left(\sum_i \lambda \lambda_i v_i\right) \otimes \left(\sum_j \mu_j w_j\right)$$

$$= (\lambda v) \otimes w.$$

Completely analogously one shows that $(\lambda v) \otimes w = \lambda(v \otimes w)$.

(b) We write the elements in the given bases as $x_i = \sum_k \lambda_{ik} v_k \in V$ and $y_j = \sum_\ell \mu_{j\ell} w_\ell \in W$. Then, again by (A.1), we have

$$\left(\sum_i x_i\right) \otimes \left(\sum_j y_j\right) = \left(\sum_k \left(\sum_i \lambda_{ik}\right) v_k\right) \otimes \left(\sum_\ell \left(\sum_j \mu_{j\ell}\right) w_\ell\right)$$

$$= \sum_{k,\ell} \left(\sum_i \lambda_{ik}\right)\left(\sum_j \mu_{j\ell}\right)(v_k \otimes w_\ell)$$

$$= \sum_{i,j} \sum_{k,\ell} \lambda_{ik} \mu_{j\ell}(v_k \otimes w_\ell)$$

$$= \sum_{i,j} \left(\left(\sum_k \lambda_{ik} v_k\right) \otimes \left(\sum_\ell \mu_{j\ell} w_\ell\right)\right) = \sum_{i,j} x_i \otimes y_j.$$

(c) We express the elements of the given bases in the new bases as $v_i = \sum_k \lambda_{ik} b_k$ and $w_j = \sum_\ell \mu_{j\ell} c_\ell$. Applying parts (b) and (a) we obtain

$$v_i \otimes w_j = \sum_{k,\ell} (\lambda_{ik} b_k \otimes \mu_{j\ell} c_\ell) = \sum_{k,\ell} \lambda_{ik} \mu_{j\ell}(b_k \otimes c_\ell).$$

This means that $\{b_k \otimes c_\ell \mid 1 \le k \le n, 1 \le \ell \le m\}$ is a spanning set for the vector space $V \otimes_K W$ (since they generate all basis elements $v_i \otimes w_j$). On the other hand, this set contains precisely $nm = \dim_K(V \otimes_K W)$ elements, so it must be a basis. $\square$

Now we can start with the actual topic of this section, namely defining induced modules for group algebras.

Let K be a field, G be a finite group and $H \subseteq G$ a subgroup. Suppose we are given a finite-dimensional KH-module W, with basis $\{w_1, \ldots, w_m\}$. Our aim is to construct from W a (finite-dimensional) KG-module in a way which extends the given KH-action on W.

First we can build the tensor product of the K-vector spaces KG and W, thus

$$KG \otimes_K W = \text{span}\{g \otimes w_i \mid g \in G, i = 1, \ldots, m\}.$$

We can turn $KG \otimes_K W$ into a KG-module by setting

$$x \cdot (g \otimes w) = xg \otimes w, \tag{A.2}$$

for every $x, g \in G$ and $w \in W$, and then extending linearly. (We leave it as an exercise to verify that the axioms for a module are indeed satisfied.) However, note that this is not yet what we want, since the given KH-action on W is completely ignored. Therefore, in $KG \otimes_K W$ we consider the K-subspace

$$\mathcal{H} := \text{span}\{gh \otimes w - g \otimes hw \mid g \in G, h \in H, w \in W\}.$$

It is immediate from (A.2) that $\mathcal{H} \subseteq KG \otimes_K W$ is a KG-submodule.

Definition A.4. We keep the notations from above. The factor module

$$KG \otimes_H W := (KG \otimes_K W)/\mathcal{H}$$

is called the KG-module induced from the KH-module W. For short we write

$$g \otimes_H w := g \otimes w + \mathcal{H} \in KG \otimes_H W;$$

by definition we then have

$$gh \otimes_H w = g \otimes_H hw \text{ for all } g \in G, h \in H \text{ and } w \in W. \tag{A.3}$$

We now want to collect some elementary properties of induced modules. We do not aim at giving a full picture but restrict to the few facts which we use in the main text of the book.

Proposition A.5. *We keep the above notations. Let $T \subseteq G$ be a system of representatives of the left cosets of H in G. Then the set*

$$\{t \otimes_H w_i \mid t \in T, i = 1, \ldots, m\}$$

is a basis of the K-vector space $KG \otimes_H W$. In particular, for the dimension of the induced module we have

$$\dim_K(KG \otimes_H W) = |G : H| \cdot \dim_K W,$$

where $|G : H|$ is the index of the subgroup H in G (that is, the number of cosets).

Proof. By definition, G is a disjoint union $G = \bigcup_{t \in T} tH$ of left cosets. So every $g \in G$ can be written as $g = t\tilde{h}$ for some $t \in T$ and $\tilde{h} \in H$. Hence every element $g \otimes_H w_i$ is of the form $g \otimes_H w_i = t\tilde{h} \otimes_H w_i = t \otimes_H \tilde{h}w_i$, by (A.3). Since the elements $g \otimes_H w_i$ for $g \in G, i = 1, \ldots, m$, clearly form a spanning set of the vector space $KG \otimes_H W$, we conclude that $\{t \otimes_H w_i \mid t \in T, i = 1, \ldots, m\}$ also forms a spanning set and that

$$\dim_K(KG \otimes_H W) \leq |T| \cdot \dim_K W = |G : H| \cdot \dim_K W.$$

Conversely, we claim that $\dim_K(KG \otimes_H W) \geq |G : H| \cdot \dim_K W$. Note that this is sufficient for proving the entire proposition since then $\dim_K(KG \otimes_H W) = |G : H| \cdot \dim_K W$ and hence the spanning set $\{t \otimes_H w_i \mid t \in T, i = 1, \ldots, m\}$ from above must be linearly independent.

To prove the claim, we exhibit a generating set for the subspace $\mathcal{H}$. Take an arbitrary generating element $\alpha := gh \otimes w - g \otimes hw$ from $\mathcal{H}$. By linearity of the tensor product in both arguments (see Proposition A.3) we can assume that $w = w_i$

is one of the basis elements of W. Again write $g = t\tilde{h}$ with $t \in T$ and $\tilde{h} \in H$. Then

$$\alpha = gh \otimes w - g \otimes hw = t\tilde{h}h \otimes w - t\tilde{h} \otimes hw$$
$$= t\tilde{h}h \otimes w - t \otimes \tilde{h}hw + t \otimes \tilde{h}hw - t\tilde{h} \otimes hw.$$

But this implies that the set

$$\{ty \otimes w_i - t \otimes yw_i \mid t \in T, y \in H, i = 1, \ldots, m\}$$

is a spanning set of the subspace $\mathcal{H}$ of $KG \otimes_K W$. Moreover, for $y = 1$ the identity element of the subgroup H the expression becomes 0 and hence can be left out. This means that the vector space $\mathcal{H}$ can be spanned by at most $|T| \cdot (|H| - 1) \cdot \dim_K W$ elements. For the dimension of the induced module we thus obtain

$$\dim_K (KG \otimes_H W) = \dim_K (KG \otimes_K W)/\mathcal{H}$$
$$= \dim_K (KG \otimes_K W) - \dim_K \mathcal{H}$$
$$\geq |G| \cdot \dim_K W - |T| \cdot (|H| - 1) \cdot \dim_K W$$
$$= |G : H| \cdot \dim_K W,$$

where the last equation follows from Lagrange's theorem $|G| = |G : H| \cdot |H|$ from elementary group theory. As remarked above, this completes the proof of the proposition.   □

We collect some further properties needed in the main text of the book in the following proposition.

Proposition A.6. *We keep the notations from above.*

(a) *The map $i : W \to KG \otimes_H W$, $w \mapsto 1 \otimes_H w$, is an injective KH-module homomorphism.*
(b) *Identifying W with the image under the map i from part (a), the induced module $KG \otimes_H W$ has the desired property that the given KH-module action on W is extended.*
(c) *In the case $G = H$ there is an isomorphism $W \cong KG \otimes_G W$ of KG-modules.*

Proof. (a) Let $w = \sum_{j=1}^{m} \lambda_j w_j \in \ker(i)$, written as a linear combination of a basis $\{w_1, \ldots, w_m\}$ of W. Then we have

$$0 = i(w) = 1 \otimes_H w = \sum_{j=1}^{m} \lambda_j (1 \otimes_H w_j).$$

We can choose our system T of coset representatives to contain the identity element, then Proposition A.5 in particular says that the elements $1 \otimes_H w_j$, for $j = 1, \ldots, m$, are part of a basis of $KG \otimes_H W$. So in the equation above we can deduce that all $\lambda_j = 0$. This implies that $w = 0$ and hence i is injective.

It is easy to check that i is a K-linear map. Moreover, using (A.3), the following holds for every $h \in H$ and $w \in W$:

$$i(hw) = 1 \otimes_H hw = h \otimes_H w = h(1 \otimes_H w) = hi(w). \tag{A.4}$$

Thus i is a KH-module homomorphism.

(b) This follows directly from (A.4).

(c) By part (a), the map $i : W \to KG \otimes_G W$ is an injective KG-module homomorphism. On the other hand, $\dim_K(KG \otimes_G W) = |G : G| \cdot \dim_K W = \dim_K W$ by Proposition A.5; so i is also surjective and hence an isomorphism. $\square$

Appendix B
Solutions to Selected Exercises

Usually there are many different ways to solve a problem. The following are possible approaches, but are not unique.

Chapter 1

Exercise 1.12. Let $u = a + bi + cj + dk \in \mathbb{H}$ where $a, b, c, d \in \mathbb{R}$. The product formula given in Example 1.8 yields that

$$u^2 = (a^2 - b^2 - c^2 - d^2) + 2abi + 2acj + 2adk.$$

So u is a root of $X^2 + 1$ if and only if $a^2 - b^2 - c^2 - d^2 = -1$ and $2ab = 2ac = 2ad = 0$. There are two cases. If $a \neq 0$ then $b = c = d = 0$ and hence $a^2 = -1$, contradicting $a \in \mathbb{R}$. If $a = 0$ then the only condition remaining is $b^2 + c^2 + d^2 = 1$. There are clearly infinitely many solutions; in fact, the solutions correspond bijectively to the points on the unit sphere in $\mathbb{R}^3$.

Exercise 1.13. We check the three conditions in Definition 1.14. By definition all subsets are K-subspaces of $M_3(K)$. The first subset is not a subalgebra since it is not closed under multiplication, for example

$$\begin{pmatrix} 0 & 1 & 0 \\ 0 & 0 & 1 \\ 0 & 0 & 0 \end{pmatrix}^2 = \begin{pmatrix} 0 & 0 & 1 \\ 0 & 0 & 0 \\ 0 & 0 & 0 \end{pmatrix}$$

which does not lie in the first subalgebra. The fifth subset is also not a subalgebra since it does not contain the identity matrix, which is the identity element of $M_3(K)$. The other four subsets are subalgebras: they contain the identity element of $M_3(K)$, and one checks that they are closed under taking products.

© Springer International Publishing AG, part of Springer Nature 2018
K. Erdmann, T. Holm, *Algebras and Representation Theory*, Springer
Undergraduate Mathematics Series, https://doi.org/10.1007/978-3-319-91998-0

Exercise 1.14. Let $\psi = \phi^{-1}$, so that ψ is a map from B to A and $\psi \circ \phi = \mathrm{id}_A$ and $\phi \circ \psi = \mathrm{id}_B$.

Let $b, b' \in B$, then $b = \phi(a)$ for a unique $a \in A$, and $b' = \phi(a')$ for a unique $a' \in A$, and then $\psi(b) = a$ and $\psi(b') = a'$. We check the three conditions in Definition 1.22.

(i) Let $\lambda, \mu \in K$. We must show that $\psi(\lambda b + \mu b') = \lambda \psi(b) + \mu \psi(b')$. In fact,

$$\psi(\lambda b + \mu b') = \psi(\lambda \phi(a) + \mu \phi(a')) = \psi(\phi(\lambda a + \mu a'))$$
$$= (\psi \circ \phi)(\lambda a + \mu a') = \mathrm{id}_A(\lambda a + \mu a')$$
$$= \lambda a + \mu a' = \lambda \psi(b) + \mu \psi(b'),$$

where we have used in the second step that ϕ is K-linear.

(ii) To check that ψ commutes with taking products,

$$\psi(bb') = \psi(\phi(a)\phi(a')) = \psi(\phi(aa')) = (\psi \circ \phi)(aa')$$
$$= \mathrm{id}_A(aa') = aa' = \psi(b)\psi(b').$$

In the second step we have used that ϕ commutes with taking products.

(iii) We have $\psi(1_B) = \psi(\phi(1_A)) = \psi \circ \phi(1_A) = \mathrm{id}_A(1_A) = 1_A$.

Exercise 1.15. (i) If $a^2 = 0$ then $\phi(a)^2 = \phi(a^2) = \phi(0) = 0$, using that ϕ is an algebra homomorphism. Conversely, if $\phi(a^2) = \phi(a)^2 = 0$, then a^2 is in the kernel of ϕ which is zero since ϕ is injective.

(ii) Let a be a (left) zero divisor, that is, there exists an $a' \in A \setminus \{0\}$ with $aa' = 0$. It follows that $0 = \phi(0) = \phi(aa') = \phi(a)\phi(a')$. Moreover, $\phi(a') \neq 0$ since ϕ is injective. Hence $\phi(a)$ is a zero divisor. Conversely, let $\phi(a)$ be a (left) zero divisor, then there exists a $b \in B \setminus \{0\}$ such that $\phi(a)b = 0$. Since ϕ is surjective there exists an $a' \in A$ with $b = \phi(a')$; note that $a' \neq 0$ because $b \neq 0$. Then $0 = \phi(a)b = \phi(a)\phi(a') = \phi(aa')$. This implies $aa' = 0$ since ϕ is injective, and hence a is a (left) zero divisor. The same proof works for right zero divisors.

(iii) Let A be commutative, and let $b, b' \in B$. We must show that $bb' - b'b = 0$. Since ϕ is surjective, there are $a, a' \in A$ such that $b = \phi(a)$ and $b' = \phi(a')$. Therefore

$$bb' - b'b = \phi(a)\phi(a') - \phi(a')\phi(a) = \phi(aa' - a'a) = \phi(0) = 0$$

using that ϕ is an algebra homomorphism, and that A is commutative.

For the converse, interchange the roles of A and B, and use the inverse ϕ^{-1} of ϕ (which is also an algebra isomorphism, see Exercise 1.14).

(iv) Assume A is a field, then B is commutative by part (iii). We must show that every non-zero element in B is invertible. Take $0 \neq b \in B$, then $b = \phi(a)$ for

$a \in A$ and a is non-zero. Since A is a field, there is an $a' \in A$ such that $aa' = 1_A$, and then

$$b\phi(a') = \phi(a)\phi(a') = \phi(aa') = \phi(1_A) = 1_B.$$

That is, b has inverse $\phi(a') \in B$. We have proved if A is a field then so is B. For the converse, interchange the roles of A and B, and use the inverse ϕ^{-1} of ϕ.

Exercise 1.16. (i) One checks that each A_i is a K-subspace of $M_3(K)$, contains the identity matrix, and that each A_i is closed under multiplication; hence it is a K-subalgebra of $M_3(K)$. Moreover, some direct computations show that A_1, A_2, A_3, A_5 are commutative. But A_4 is not commutative, for example

$$\begin{pmatrix} 1 & 0 & 0 \\ 0 & 0 & 0 \\ 0 & 0 & 1 \end{pmatrix} \begin{pmatrix} 1 & 1 & 0 \\ 0 & 0 & 0 \\ 0 & 0 & 1 \end{pmatrix} \neq \begin{pmatrix} 1 & 1 & 0 \\ 0 & 0 & 0 \\ 0 & 0 & 1 \end{pmatrix} \begin{pmatrix} 1 & 0 & 0 \\ 0 & 0 & 0 \\ 0 & 0 & 1 \end{pmatrix}.$$

(ii) We compute $\bar{A}_1 = 0$,

$$\bar{A}_2 = \left\{ \begin{pmatrix} 0 & y & z \\ 0 & 0 & 0 \\ 0 & 0 & 0 \end{pmatrix} \mid y, z \in K \right\}, \quad \bar{A}_3 = \left\{ \begin{pmatrix} 0 & 0 & z \\ 0 & 0 & 0 \\ 0 & 0 & 0 \end{pmatrix} \mid z \in K \right\}$$

$$\bar{A}_4 = \left\{ \begin{pmatrix} 0 & y & 0 \\ 0 & 0 & 0 \\ 0 & 0 & 0 \end{pmatrix} \mid y \in K \right\}, \quad \bar{A}_5 = \left\{ \begin{pmatrix} 0 & 0 & y \\ 0 & 0 & z \\ 0 & 0 & 0 \end{pmatrix} \mid y, z \in K \right\}.$$

(iii) By part (iii) of Exercise 1.15, A_4 is not isomorphic to any of A_1, A_2, A_3, A_5. Note that the sets $\bar{A}_i$ computed in (ii) are even K-subspaces; hence we can compare dimensions. By part (ii) of Exercise 1.15 we conclude that A_1 is not isomorphic to A_2, A_3, A_5, that A_2 is not isomorphic to A_3 and that A_3 is not isomorphic to A_5.

The remaining algebras A_2 and A_5 are isomorphic; in fact the map

$$\phi : A_2 \to A_5, \quad \begin{pmatrix} x & y & z \\ 0 & x & 0 \\ 0 & 0 & x \end{pmatrix} \mapsto \begin{pmatrix} x & 0 & y \\ 0 & x & z \\ 0 & 0 & x \end{pmatrix}$$

is bijective by definition, and one checks that it is a K-algebra homomorphism.

Exercise 1.18. $T_n(K)$ has a K-basis given by matrix units $\{E_{ij} \mid 1 \leq i \leq j \leq n\}$. On the other hand, we denote by α_i the arrow from vertex $i + 1$ to i (where $1 \leq i \leq n - 1$). Then the path algebra KQ has a K-basis

$$\{e_i \mid 1 \leq i \leq n\} \cup \{\alpha_i \alpha_{i+1} \ldots \alpha_{j-1} \mid 1 \leq i < j \leq n\}.$$

We define a bijective K-linear map $\phi : T_n(K) \to KQ$ by mapping a basis to a basis as follows: $\phi(E_{ii}) = e_i$ for $1 \leq i \leq n$ and $\phi(E_{ij}) = \alpha_i \alpha_{i+1} \ldots \alpha_{j-1}$ for all $1 \leq i < j \leq n$. It remains to check that ϕ is a K-algebra homomorphism. To show that ϕ preserves products, it suffices to check it on basis elements, see Remark 1.23. Note that $\phi(E_{ij})$ is a path starting at vertex j, and $\phi(E_{k\ell})$ is a path ending at vertex k. Thus, if $j \neq k$ then

$$\phi(E_{ij})\phi(E_{k\ell}) = 0 = \phi(0) = \phi(E_{ij}E_{k\ell}).$$

So suppose $j = k$, and we have to show that $\phi(E_{ij})\phi(E_{j\ell}) = \phi(E_{ij}E_{j\ell})$. One checks that this formula holds if $i = j$ or $j = \ell$. Otherwise, we have

$$\phi(E_{ij})\phi(E_{j\ell}) = \alpha_i \alpha_{i+1} \ldots \alpha_{j-1} \alpha_j \alpha_{j+1} \ldots \alpha_{\ell-1} = \phi(E_{i\ell}) = \phi(E_{ij}E_{j\ell}).$$

Finally, to check condition (iii) of Definition 1.22, we have

$$\phi(1_{T_n(K)}) = \phi(\sum_{i=1}^{n} E_{ii}) = \sum_{i=1}^{n} \phi(E_{ii}) = \sum_{i=1}^{n} e_i = 1_{KQ}.$$

Exercise 1.24. Recall that in the proof of Proposition 1.29 the essential part was to find a basis $\{1, \tilde{b}\}$ for which $\tilde{b}$ squares to either 0 or ± 1. For $D_2(\mathbb{R})$ we can choose $\tilde{b}$ as a diagonal matrix with entries 1 and -1; clearly, $\tilde{b}^2$ is the identity matrix and hence $D_2(\mathbb{R}) \cong \mathbb{R}[X]/(X^2 - 1)$. For A choose $\tilde{b} = \begin{pmatrix} 0 & 1 \\ 0 & 0 \end{pmatrix}$; then $\tilde{b}^2 = 0$ and hence $A \cong \mathbb{R}[X]/(X^2)$. Finally, for B we choose $\tilde{b} = \begin{pmatrix} 0 & 1 \\ -1 & 0 \end{pmatrix}$; then $\tilde{b}^2$ equals the negative of the identity matrix and hence $B \cong \mathbb{R}[X]/(X^2 + 1)$. In particular, there are no isomorphisms between any of the three algebras.

Exercise 1.26. One possibility is to imitate the proof of Proposition 1.29; it still gives the three possibilities $A_0 = \mathbb{C}[X]/(X^2)$, $A_1 = \mathbb{C}[X]/(X^2 - 1)$ and $A_{-1} = \mathbb{C}[X]/(X^2 + 1)$. Furthermore, the argument that $A_0 \not\cong A_1$ also works for $\mathbb{C}$. However, the $\mathbb{C}$-algebras A_1 and A_{-1} are isomorphic. In fact, the map

$$\mathbb{C}[X]/(X^2 - 1) \to \mathbb{C}[X]/(X^2 + 1), \ g(X) + (X^2 - 1) \mapsto g(iX) + (X^2 + 1)$$

is well-defined (that is, independent of the coset representatives) and a bijective $\mathbb{C}$-algebra homomorphism.

Alternatively, we apply Exercise 1.25. In each of (a) and (b) we get a unique algebra (they are A_{-1} and A_0 with the previous notation). Furthermore, we know that there are no irreducible polynomials of degree 2 in $\mathbb{C}[X]$ and therefore there is no $\mathbb{C}$-algebra of dimension 2 in (c).

Exercise 1.27. We fix the basis $\{1_A, \alpha\}$ of $A = \mathbb{Q}[X]/(X^2 - p)$ where 1_A is the coset of 1 and α is the coset of X. That is, $\alpha^2 = p1_A$. Similarly we fix the basis of $B = \mathbb{Q}[X]/(X^2 - q)$ consisting of 1_B and β with $\beta^2 = q1_B$. Suppose $\psi : A \to B$ is a $\mathbb{Q}$-algebra isomorphism, then

$$\psi(1_A) = 1_B, \quad \psi(\alpha) = c1_B + d\beta,$$

where $c, d \in \mathbb{Q}$ and $d \neq 0$ (otherwise ψ is not injective). We then have

$$p1_B = p\psi(1_A) = \psi(p1_A) = \psi(\alpha^2) = \psi(\alpha)^2$$
$$= (c1_B + d\beta)^2 = c^2 1_B + d^2\beta^2 + 2cd\beta$$
$$= (c^2 + d^2 q)1_B + 2cd\beta.$$

Comparing coefficients yields $2cd = 0$ and $p = (c^2 + d^2 q)$. Since $d \neq 0$ we must have $c = 0$. Then it follows that $p = d^2 q$. Now write $d = \frac{m}{n}$ with coprime $m, n \in \mathbb{Z}$. Then we have $n^2 p = m^2 q$, which forces $p = q$ (otherwise p divides m^2 and then since p is prime, p also divides n^2, contradicting m and n being coprime). But we assumed that p and q are different, so this shows that A and B are not isomorphic.

Exercise 1.28. (a) The elements of B are $\begin{pmatrix} 0 & 0 \\ 0 & 0 \end{pmatrix}, \begin{pmatrix} 1 & 0 \\ 0 & 1 \end{pmatrix}, \begin{pmatrix} 0 & 1 \\ 1 & 1 \end{pmatrix}, \begin{pmatrix} 1 & 1 \\ 1 & 0 \end{pmatrix}$. It is then easily checked that B is a $\mathbb{Z}_2$-subspace of $M_2(\mathbb{Z}_2)$ and that it is closed under matrix multiplication. Moreover, every non-zero matrix in B has determinant 1 and hence is invertible. Furthermore, B is 2-dimensional and therefore is commutative (see Proposition 1.28). That is, B is a field.

(b) Let A be a 2-dimensional $\mathbb{Z}_2$-algebra. One can imitate the proof of Proposition 1.29 (we will not write this down). Alternatively, we apply Exercise 1.25. For each of (a) and (b) we get a unique algebra, namely $\mathbb{Z}_2[X]/(X(X + 1)) \cong \mathbb{Z}_2 \times \mathbb{Z}_2$ in (a) and $\mathbb{Z}_2[X]/(X^2)$ in (b). Then to get an algebra in (c) we need to look for irreducible polynomials over $\mathbb{Z}_2$ of degree 2. By inspection, we see that there is just one such polynomial, namely $X^2 + X + 1$. In total we have proved that every 2-dimensional $\mathbb{Z}_2$-algebra is isomorphic to precisely one of

$$\mathbb{Z}_2[X]/(X^2 + X) , \ \mathbb{Z}_2[X]/(X^2) \ \text{ or } \ \mathbb{Z}_2[X]/(X^2 + X + 1).$$

Exercise 1.29. (a) We use the axioms of a K-algebra from Definition 1.1.
(i) For every $x, y \in A$ and $\lambda, \mu \in K$ we have

$$l_a(\lambda x + \mu y) = a(\lambda x + \mu y) = a(\lambda x) + a(\mu y)$$
$$= \lambda(ax) + \mu(ay) = \lambda l_a(x) + \mu l_a(y).$$

(ii) For every $x \in A$ we have

$$l_{\lambda a + \mu b}(x) = (\lambda a + \mu b)x = \lambda(ax) + \mu(bx)$$
$$= \lambda l_a(x) + \mu l_b(x) = (\lambda l_a + \mu l_b)(x).$$

Since $x \in A$ was arbitrary it follows that $l_{\lambda a + \mu b} = \lambda l_a + \mu l_b$.

(iii) For every $x \in A$ we have $l_{ab}(x) = (ab)x = a(bx) = l_a(l_b(x))$, hence $l_{ab} = l_a \circ l_b$.

(b) If l_a is the zero map for some $a \in A$, then $ax = l_a(x) = 0$ for every $x \in A$. In particular, $a = a 1_A = 0$, hence ψ is injective.

(c) Choosing a fixed basis of A yields an algebra isomorphism $\mathrm{End}_K(A) \cong M_n(K)$, see Example 1.24. So from part (b) we get an injective algebra homomorphism $\tilde{\psi} : A \to M_n(K)$, that is, A is isomorphic to the image of $\tilde{\psi}$, a subalgebra of $M_n(K)$ (see Theorem 1.26).

(d) We fix the basis $\{e_1, e_2, \alpha\}$ of KQ. From the multiplication table of KQ in Example 1.13 one works out the matrices of the maps l_{e_1}, l_{e_2} and l_α as

$$l_{e_1} \leftrightarrow \begin{pmatrix} 1 & 0 & 0 \\ 0 & 0 & 0 \\ 0 & 0 & 1 \end{pmatrix} \quad l_{e_2} \leftrightarrow \begin{pmatrix} 0 & 0 & 0 \\ 0 & 1 & 0 \\ 0 & 0 & 0 \end{pmatrix} \quad l_\alpha \leftrightarrow \begin{pmatrix} 0 & 0 & 0 \\ 0 & 0 & 0 \\ 0 & 1 & 0 \end{pmatrix}.$$

By parts (b) and (c), the path algebra KQ is isomorphic to the subalgebra of $M_3(K)$ consisting of all K-linear combinations of the matrices corresponding to l_{e_1}, l_{e_2} and l_α, that is, to the subalgebra

$$\left\{ \begin{pmatrix} a & 0 & 0 \\ 0 & b & 0 \\ 0 & c & a \end{pmatrix} \mid a, b, c \in K \right\}.$$

Chapter 2

Exercise 2.12. According to Theorem 2.10 the 1-dimensional A-modules are in bijection with scalars $\alpha \in K$ satisfying $\alpha^4 = 2$.

(i) For $K = \mathbb{Q}$ there is no such scalar, that is, there is no 1-dimensional $\mathbb{Q}[X]/(X^4 - 2)$-module.

(ii) For $K = \mathbb{R}$ there are precisely two 1-dimensional $\mathbb{R}[X]/(X^4 - 2)$-modules, given by $X \cdot v = \pm\sqrt[4]{2}v$ (where v is a basis vector of the 1-dimensional module).

(iii) For $K = \mathbb{C}$ there are four 1-dimensional $\mathbb{C}[X]/(X^4 - 2)$-modules, given by $X \cdot v = \pm\sqrt[4]{2}v$ and $X \cdot v = \pm i\sqrt[4]{2}v$.

(iv) There is no $\alpha \in \mathbb{Z}_3$ with $\alpha^4 = 2$; hence there is no 1-dimensional $\mathbb{Z}_3[X]/(X^4 - 2)$-module.

(v) In $\mathbb{Z}_7 = \{\bar{0}, \bar{1}, \bar{2}, \bar{3}, \bar{4}, \bar{5}, \bar{6}\}$ there are precisely two elements with fourth power equal to $\bar{2}$, namely $\bar{2}$ and $\bar{5}$. So there are precisely two 1-dimensional $\mathbb{Z}_7[X]/(X^4 - 2)$-modules, given by $X \cdot v = \pm\bar{2}v$.

Exercise 2.14. (a) Suppose that $U \subseteq K^n$ is a $T_n(K)$-submodule. If $U = 0$ then $U = V_0$ and we are done. If $U \neq 0$ we set

$$i := \max\{k \mid \exists (u_1, \ldots, u_n)^t \in U \text{ with } u_k \neq 0\}.$$

We now show that $U = V_i$ (proving the claim in (a)). By definition of i we have $U \subseteq V_i$. Conversely, we take an element $u = (u_1, \ldots, u_i, 0, \ldots, 0)^t \in U$ with $u_i \neq 0$ (which exists by definition of i). Then for every $1 \leq r \leq i$ we have that $e_r = u_i^{-1} E_{ri} u \in U$ since U is a $T_n(K)$-submodule. So $V_i = \mathrm{span}\{e_1, \ldots, e_i\} \subseteq U$.

(b) Let $\phi : V_{i,j} \to V_{r,s}$ be a $T_n(K)$-module homomorphism. Note that we have $V_{i,j} = \mathrm{span}\{e_{j+1} + V_j, \ldots, e_i + V_j\}$ and similarly $V_{r,s} = \mathrm{span}\{e_{s+1} + V_s, \ldots, e_r + V_s\}$. Then we consider the image of the first basis vector, $\phi(e_{j+1} + V_j) = \sum_{\ell=s+1}^r \lambda_\ell (e_\ell + V_s)$, and we get

$$\phi(e_{j+1} + V_j) = \phi(E_{j+1,j+1}(e_{j+1} + V_j)) = E_{j+1,j+1}\phi(e_{j+1} + V_j)$$

$$= \sum_{\ell=s+1}^r \lambda_\ell E_{j+1,j+1}(e_\ell + V_s).$$

Note that if $j < s$ then $E_{j+1,j+1}e_\ell = 0$ for all $\ell = s+1, \ldots, r$. Thus, if $j < s$ then $e_{j+1} + V_j \in \ker(\phi)$ and ϕ is not an isomorphism.

Completely analogously, there can't be an isomorphism ϕ if $j > s$ (since then ϕ^{-1} is not injective, by the argument just given). So we have shown that if ϕ is an isomorphism, then $j = s$. Moreover, if ϕ is an isomorphism, the dimensions of $V_{i,j}$ and $V_{r,s}$ must agree, that is, we have $i - j = r - s = r - j$ and hence also $i = r$. This shows that the $T_n(K)$-modules $V_{i,j}$ with $0 \leq j < i \leq n$ are pairwise non-isomorphic. An easy counting argument shows that there are $\frac{n(n+1)}{2}$ such modules.

(c) The annihilator $\mathrm{Ann}_{T_n(K)}(e_i)$ consists precisely of those upper triangular matrices with i-th column equal to zero. So the factor module $T_n(K)/\mathrm{Ann}_{T_n(K)}(e_i)$ is spanned by the cosets $E_{1i} + \mathrm{Ann}_{T_n(K)}(e_i), \ldots, E_{ii} + \mathrm{Ann}_{T_n(K)}(e_i)$. Then one checks that the map

$$\psi : V_i \to T_n(K)/\mathrm{Ann}_{T_n(K)}(e_i), \quad e_j \mapsto E_{ji} + \mathrm{Ann}_{T_n(K)}(e_i)$$

(for $1 \leq j \leq i$) is a $T_n(K)$-module isomorphism.

Exercise 2.16. (a) E is non-empty since $(0, 0) \in E$. Let $(m_1, m_2), (n_1, n_2) \in E$ and $\lambda, \mu \in K$. Then

$$\alpha_1(\lambda m_1 + \mu n_1) + \alpha_2(\lambda m_2 + \mu n_2) = \lambda \alpha_1(m_1) + \mu \alpha_1(n_1) + \lambda \alpha_2(m_2) + \mu \alpha_2(n_2)$$
$$= \lambda(\alpha_1(m_1) + \alpha_2(m_2)) + \mu(\alpha_1(n_1) + \alpha_2(n_2))$$
$$= \lambda \cdot 0 + \mu \cdot 0 = 0,$$

that is, $\lambda(m_1, m_2) + \mu(n_1, n_2) = (\lambda m_1 + \mu n_1, \lambda m_2 + \mu n_2) \in E$.

(b) By definition, the map $\varphi : M_1 \times M_2 \rightarrow M$, $(m_1, m_2) \mapsto \alpha_1(m_1) + \alpha_2(m_2)$ has kernel E, and it is surjective since $M = \text{im}(\alpha_1) + \text{im}(\alpha_2)$ by assumption. The rank-nullity theorem from linear algebra then gives

$$\dim_K M_1 + \dim_K M_2 = \dim_K M_1 \times M_2$$
$$= \dim_K \text{im}(\varphi) + \dim_K \ker(\varphi)$$
$$= \dim_K M + \dim_K E,$$

proving the claim.

(c) By part (a), E is a K-subspace. Moreover, for $a \in A$ and $(m_1, m_2) \in E$ we have

$$\alpha_1(am_1) + \alpha_2(am_2) = a\alpha_1(m_1) + a\alpha_2(m_2)$$
$$= a(\alpha_1(m_1) + \alpha_2(m_2)) = a \cdot 0 = 0,$$

that is, $a(m_1, m_2) = (am_1, am_2) \in E$ and E is an A-submodule.

Exercise 2.17. (a) Let $(\beta_1(w), \beta_2(w)) \in C$ and $(\beta_1(v), \beta_2(v)) \in C$, with $w, v \in W$. For any $\lambda, \mu \in K$ we have

$$\lambda(\beta_1(w), \beta_2(w)) + \mu(\beta_1(v), \beta_2(v)) = (\lambda\beta_1(w) + \mu\beta_1(v), \lambda\beta_2(w) + \mu\beta_2(v))$$
$$= (\beta_1(\lambda w + \mu v), \beta_2(\lambda w + \mu v)),$$

which is in C since W is a K-vector space, that is, $\lambda w + \mu v \in W$. Then F is the factor space, which is known from linear algebra to be a K-vector space.

(b) The map $\psi : W \rightarrow C$, $w \mapsto (\beta_1(w), \beta_2(w))$ is surjective by definition of C, and it is injective since $\ker(\beta_1) \cap \ker(\beta_2) = 0$ by assumption. For the dimensions we obtain

$$\dim_K F = \dim_K M_1 \times M_2 - \dim_K C = \dim_K M_1 + \dim_K M_2 - \dim_K W.$$

(c) By part (a), C is a K-subspace. For every $a \in A$ and $(\beta_1(w), \beta_2(w)) \in C$ we have

$$a((\beta_1(w), \beta_2(w)) = (a\beta_1(w), a\beta_2(w)) = (\beta_1(aw), \beta_2(aw)) \in C,$$

hence C is an A-submodule. Then F is the factor module as in Proposition 2.18.

Exercise 2.18. (a) By definition of the maps β_i we have

$$\ker(\beta_1) \cap \ker(\beta_2) = \{(m_1, m_2) \in E \mid m_1 = 0, m_2 = 0\} = \{(0, 0)\}.$$

(b) By construction

$$C = \{(\beta_1(w), \beta_2(w)) \mid w = (m_1, m_2) \in E\} = \{(m_1, m_2) \mid (m_1, m_2) \in E\} = E.$$

(c) Let $\varphi : M_1 \times M_2 \to M$ be the map $(m_1, m_2) \mapsto \alpha_1(m_1) + \alpha_2(m_2)$, as in Exercise 2.16. Since $M = \operatorname{im}(\alpha_1) + \operatorname{im}(\alpha_2)$, this map is surjective. The kernel is E, by definition of E. We have proved in part (b) that $E = C$ and therefore by the isomorphism theorem of vector spaces,

$$F = (M_1 \times M_2)/C = (M_1 \times M_2)/\ker(\varphi) \cong \operatorname{im}(\varphi) = M.$$

Exercise 2.22. We prove part (b) (all other parts are straightforward). Let $W \subseteq Ae_1$ be a 1-dimensional submodule, take $0 \neq w \in W$. We express it in terms of the basis of Ae_1, as $w = ce_1 + d\alpha$ with $c, d \in K$. Then $e_1 w$ and $e_2 w$ are both scalar multiples of w, and $e_1 w = ce_1$ but $e_2 w = d\alpha$. It follows that one of c, d must be zero and the other is non-zero. If $c \neq 0$ then $\alpha w = c\alpha \neq 0$ and is not a scalar multiple of w. Hence $c = 0$ and w is a non-zero scalar multiple of α.

Now suppose we have non-zero submodules U and V of Ae_1 such that $Ae_1 = U \oplus V$. Then U and V must be 1-dimensional. By part (b) it follows that $U = \operatorname{span}\{\alpha\} = V$ and we do not have a direct sum, a contradiction.

Exercise 2.23. We apply Example 2.22 with $R = A$. We have seen that a map from A to a module taking $r \in A$ to rm for m a fixed element in a module is an A-module homomorphism. Hence we have the module homomorphism

$$\psi : K[X] \mapsto (g)/(f), \quad \psi(r) = r(g + (f)) = rg + (f),$$

which is surjective. We have $\psi(r) = 0$ if and only if rg belongs to (f), that is, rg is a multiple of $f = gh$. This holds if and only if r is a multiple of h (recall that the ring $K[X]$ is a UFD). By the isomorphism theorem, $K[X]/(h) = K[X]/\ker(\psi) \cong \operatorname{im}(\psi) = (g)/(f)$.

Chapter 3

Exercise 3.3. (a) It suffices to prove one direction; the other one then follows by using the inverse of the isomorphism. So suppose that V is a simple A-module, and let $U \subseteq W$ be an A-submodule. By Proposition 2.26 the preimage $\phi^{-1}(U)$ is an A-submodule of V. We assume V is simple, and we conclude that $\phi^{-1}(U) = 0$ or

$\phi^{-1}(U) = V$. Since ϕ is surjective this implies that $U = 0$ or $U = W$. Hence W is a simple A-module.

(b) For $i \in \{0, 1, \ldots, n - 1\}$ let $\psi : V_{i+1} \rightarrow \phi(V_{i+1})/\phi(V_i)$ be the composition of the restriction $\phi_{|V_{i+1}} : V_{i+1} \rightarrow \phi(V_{i+1})$ with the canonical map onto $\phi(V_{i+1})/\phi(V_i)$; this is a surjective A-module homomorphism and $\ker(\psi) = V_i$ (since $\phi_{|V_{i+1}}$ is an isomorphism). By the isomorphism theorem we have $V_{i+1}/V_i \cong \phi(V_{i+1})/\phi(V_i)$. Since V_{i+1}/V_i is simple, $\phi(V_{i+1})/\phi(V_i)$ is also a simple A-module by part (a). Hence we have a composition series of W, as claimed.

Exercise 3.6. (a)/(b) The statement in (a) is a special case of the following solution of (b). Let $V_n := \mathrm{span}\{v_1, \ldots, v_n\}$ be an n-dimensional vector space on which we let x and y act as follows: $x \cdot v_1 = 0$, $x \cdot v_i = v_{i-1}$ for $2 \le i \le n$, and $y \cdot v_i = v_{i+1}$ for $1 \le i \le n - 1$, $y \cdot v_n = 0$. (This action is extended to a basis of KQ by letting products of powers of x and y act by successive action of x and y.) Let $0 \ne U \subseteq V_n$ be a KQ-submodule, and $\alpha = \sum_{i=1}^{n} \lambda_i v_i \in U \setminus \{0\}$. Set $m = \max\{i \mid \lambda_i \ne 0\}$. Then $v_1 = \lambda_m^{-1} x^{m-1} \cdot \alpha \in U$. But then for every $j \in \{1, \ldots, n\}$ we also get that $v_j = y^{j-1} v_1 \in U$, that is, $U = V_n$ and V_n is simple.

(c) We consider an infinite-dimensional vector space $V_\infty := \mathrm{span}\{v_i \mid i \in \mathbb{N}\}$ and make this a KQ-module by setting $x \cdot v_1 = 0$, $x \cdot v_i = v_{i-1}$ for all $i \ge 2$ and $y \cdot v_i = v_{i+1}$ for all $i \in \mathbb{N}$. Then the argument in (b) carries over verbatim (in part (b) we have not used that $y \cdot v_n = 0$). Thus, V_∞ is a simple KQ-module.

Exercise 3.10. By Example 3.25, every simple A-module S has dimension at most 2. If S is 1-dimensional, then the endomorphism algebra is also 1-dimensional, hence $\mathrm{End}_\mathbb{R}(S) \cong \mathbb{R}$, by Proposition 1.28. Now let S be of dimension 2. Then for every $v \in S \setminus \{0\}$ we have $S = \mathrm{span}\{v, Xv\}$ (where we write X also for its coset in A); in fact, otherwise $\mathrm{span}\{v\}$ is a non-trivial A-submodule of S. So the action of X on S is given by a matrix $\begin{pmatrix} 0 & a \\ 1 & b \end{pmatrix}$ for some $a, b \in \mathbb{R}$. Writing endomorphisms as matrices we have

$$\mathrm{End}_\mathbb{R}(S) \cong \left\{ \varphi \in M_2(\mathbb{R}) \mid \varphi \begin{pmatrix} 0 & a \\ 1 & b \end{pmatrix} = \begin{pmatrix} 0 & a \\ 1 & b \end{pmatrix} \varphi \right\}.$$

By direct calculation, φ is of the form $\varphi = \begin{pmatrix} \alpha & a\gamma \\ \gamma & \alpha + b\gamma \end{pmatrix}$, where $\alpha, \gamma \in \mathbb{R}$; in particular, $\mathrm{End}_\mathbb{R}(S)$ has dimension 2. On the other hand, Schur's lemma (cf. Theorem 3.33) implies that $\mathrm{End}_\mathbb{R}(S)$ is a division algebra. In the classification of 2-dimensional $\mathbb{R}$-algebras in Proposition 1.29 there is only one division algebra, namely $\mathbb{R}[X]/(X^2 + 1) \cong \mathbb{C}$.

Exercise 3.16. We use the matrix units E_{ij} where $1 \le i, j \le n$. Every element of $M_n(D)$ can be written as $\sum_{i,j=1}^{n} z_{ij} E_{ij}$ with $z_{ij} \in D$. Suppose that

$z = \sum_{i,j=1}^{n} z_{ij} E_{ij} \in Z(A)$. Then for every $k, \ell \in \{1, \ldots, n\}$ we have on the one hand

$$E_{k\ell} z = \sum_{i,j=1}^{n} z_{ij} E_{k\ell} E_{ij} = \sum_{j=1}^{n} z_{\ell j} E_{kj}$$

and on the other hand

$$z E_{k\ell} = \sum_{i,j=1}^{n} z_{ij} E_{ij} E_{k\ell} = \sum_{i=1}^{n} z_{ik} E_{i\ell}.$$

Note that the first matrix has the ℓ-th row of z in row k, the second one has the k-th column of z in column ℓ. However, the two matrices must be equal since $z \in Z(A)$. So we get $z_{\ell k} = 0$ for all $\ell \neq k$ and $z_{\ell\ell} = z_{kk}$ for all k, ℓ. This just means that z is a multiple of the identity matrix by a 'scalar' from D. However, D need not be commutative so only the $Z(D)$-multiples of the identity matrix form the centre.

Exercise   3.18.   We   apply   Proposition   3.17   to   the   chain $M_0 \subset M_1 \subset \ldots \subset M_{r-1} \subset M_r = M$. Since all inclusions are strict we obtain for the lengths that

$$0 \leq \ell(M_0) < \ell(M_1) < \ldots < \ell(M_{r-1}) < \ell(M_r) = \ell(M),$$

and hence $r \leq \ell(M)$.

Chapter 4

Exercise 4.3. By the division algorithm of polynomials, since g and h are coprime, there are polynomials q_1, q_2 such that $1_{K[X]} = g q_1 + h q_2$ and therefore we have

$$1_{K[X]} + (f) = (g q_1 + (f)) + (h q_2 + (f))$$

and it follows that $K[X]/(f)$ is the sum of the two submodules $(g)/(f)$ and $(h)/(f)$. We show that the intersection is zero: Suppose r is a polynomial such that $r + (f)$ is in the intersection of $(g)/(f)$ and $(h)/(f)$. Then

$$r + (f) = g p_1 + (f) = h p_2 + (f)$$

for polynomials p_1, p_2. Since $r + (f) = g p_1 + (f)$ there is some polynomial k such that $r - g p_1 = f k = g h k$ and therefore $r = g(p_1 + hk)$, that is, g divides r. Similarly h divides r. Since $K[X]$ is a UFD, and since g, h are coprime it follows that gh divides r, and $f = gh$ and therefore $r + (f)$ is zero in $K[X]/(f)$.

Exercise 4.5. (a) This holds since $v_1 + v_2 + v_3$ is fixed by every $g \in G$.

(b) Consider the 2-dimensional subspace $C := \text{span}\{v_1 - v_2, v_2 - v_3\}$. One checks that for any field K this is a KS_3-submodule of V. We claim that C is even a simple module. In fact, let $\tilde{C} \subseteq C$ be a non-zero submodule, and $c := \alpha(v_1 - v_2) + \beta(v_2 - v_3) \in \tilde{C} \setminus \{0\}$, where $\alpha, \beta \in K$. Then $\tilde{C} \ni c - (1\,3)c = (\alpha + \beta)(v_1 - v_3)$. So if $\alpha \neq -\beta$ then $v_1 - v_3 \in \tilde{C}$, but then also $(1\,2)(v_1 - v_3) = v_2 - v_3 \in \tilde{C}$ and hence $\tilde{C} = C$. So we can suppose that $\alpha = -\beta$; then $\tilde{C} \ni c - (1\,2)c = 3\alpha(v_1 - v_2)$. Since K has characteristic $\neq 3$ (and $\alpha \neq 0$, otherwise $c = 0$) this implies that $v_1 - v_2 \in \tilde{C}$ and then $\tilde{C} = C$ as above.

Finally, we show that $V = U \oplus C$. For dimension reasons it suffices to check that $U \cap C = 0$; let $\alpha(v_1 + v_2 + v_3) = \beta(v_1 - v_2) + \gamma(v_2 - v_3) \in U \cap C$, with $\alpha, \beta, \gamma \in K$. Comparing coefficients yields $\alpha = \beta = -\gamma = -\beta + \gamma$; this implies that $3\alpha = 0$, hence $\alpha = 0$ (since K has characteristic different from 3) and then also $\beta = \gamma = 0$. So we have shown that $V = U \oplus C$ is a direct sum of simple KS_3-submodules, that is, V is a semisimple KS_3-module.

(c) Let K have characteristic 3, so $-2 = 1$ in K. Then $v_1 + v_2 + v_3 = (v_1 - v_2) - (v_2 - v_3) \in C$ (recall that C is a submodule for every field). Hence $U \subseteq C$. Suppose (for a contradiction) that V is semisimple. Then the submodule C is also semisimple, by Corollary 4.7. But then, by Theorem 4.3, U must have a complement, that is, a KS_3-module $\tilde{U}$ such that $C = U \oplus \tilde{U}$; note that $\tilde{U}$ must be 1-dimensional, say $\tilde{U} = \text{span}\{\tilde{u}\}$ where $\tilde{u} = \alpha(v_1 - v_2) + \beta(v_2 - v_3)$ with $\alpha, \beta \in K$. Since $\tilde{U}$ is a KS_3-submodule we have

$$\tilde{u} + (1\,2)\tilde{u} = \beta(v_1 + v_2 + v_3) \in U \cap \tilde{U}.$$

But $U \cap \tilde{U} = 0$, so $\beta = 0$. Similarly, $\tilde{u} + (2\,3)\tilde{u} = -\alpha(v_1 + v_2 + v_3)$ yields $\alpha = 0$. So $\tilde{u} = 0$, a contradiction. Thus the KS_3-module V is not semisimple if K has characteristic 3.

Exercise 4.6. Let $x \in I$, and suppose (for a contradiction) that $x \notin J(A)$. Since $J(A)$ is the intersection over all maximal left ideals of A, there exists a maximal left ideal M with $x \notin M$. Since M is maximal, we have $M + Ax = A$. In particular, we can write $1_A = m + ax$ with $m \in M$ and $a \in A$, so $1_A - ax = m \in M$. On the other hand, $ax \in I$ since I is a left ideal and hence $(ax)^r = 0$ by assumption on I. It follows that

$$(1_A + ax + (ax)^2 + \ldots + (ax)^{r-1})(1_A - ax) = 1_A,$$

that is, $1_A - ax$ is invertible in A. But from above $1_A - ax = m \in M$, so also $1_A \in M$ and thus $M = A$, a contradiction since M is a maximal left ideal of A.

Exercise 4.7. A_1 is isomorphic to $K \times K \times K$, hence is semisimple (see Example 4.19). For A_2 and A_3, consider the left ideal $I = \begin{pmatrix} 0 & 0 & * \\ 0 & 0 & 0 \\ 0 & 0 & 0 \end{pmatrix}$; this is nilpotent, namely $I^2 = 0$, and therefore $I \subseteq J(A_2)$ and $I \subseteq J(A_3)$ by Exercise 4.6. In particular, $J(A_2) \neq 0$ and $J(A_3) \neq 0$ and then by Theorem 4.23 (g), the algebras A_2 and A_3 are not semisimple. We have a map from A_4 to the algebra $B = \begin{pmatrix} * & 0 & 0 \\ 0 & * & * \\ 0 & * & * \end{pmatrix}$

defined by $\begin{pmatrix} a & 0 & b \\ 0 & c & 0 \\ d & 0 & e \end{pmatrix} \mapsto \begin{pmatrix} c & 0 & 0 \\ 0 & a & b \\ 0 & d & e \end{pmatrix}$. One checks that this is an algebra isomorphism.

The algebra B is isomorphic to $K \times M_2(K)$ which is semisimple (see Example 4.19), hence so is A_4.

Exercise 4.9. (a) According to Proposition 4.14 we have to check that f is square-free in $\mathbb{R}[X]$ if and only if f is square-free in $\mathbb{C}[X]$. Recall that every irreducible polynomial in $\mathbb{R}[X]$ has degree at most 2; let

$$f = (X - a_1) \ldots (X - a_r)(X - z_1)(X - \overline{z}_1) \ldots (X - z_s)(X - \overline{z}_s)$$

with $a_i \in \mathbb{R}$ and $z_i \in \mathbb{C} \setminus \mathbb{R}$. This is the factorisation in $\mathbb{C}[X]$, the one in $\mathbb{R}[X]$ is obtained from it by taking together $(X - z_i)(X - \overline{z}_i) \in \mathbb{R}[X]$ for each i. Then f is square-free in $\mathbb{R}[X]$ if and only if the a_i and z_j are pairwise different if and only if f is square-free in $\mathbb{C}[X]$.

(b) This is also true. It is clear that if f is square-free in $\mathbb{R}[X]$ then f is also square-free in $\mathbb{Q}[X]$. For the converse, let $f = f_1 \ldots f_r$ be square-free with pairwise coprime irreducible polynomials $f_i \in \mathbb{Q}[X]$. Let $z \in \mathbb{C}$ be a root of some f_i (here we use the fundamental theorem of algebra). Then the minimal polynomial of z (over $\mathbb{Q}$) divides f_i. Since f_i is irreducible in $\mathbb{Q}[X]$ it follows that f_i is the minimal polynomial of z (up to a non-zero scalar multiple). This means in particular that different f_i cannot have common factors in $\mathbb{R}[X]$. Moreover, since we are in characteristic 0, an irreducible polynomial in $\mathbb{Q}[X]$ cannot have multiple roots, thus no f_i itself can produce multiple factors in $\mathbb{R}[X]$. So f is square-free in $\mathbb{R}[X]$.

Chapter 5

Exercise 5.1. According to the Artin–Wedderburn theorem there are precisely four such algebras: $\mathbb{C}^9$, $\mathbb{C}^5 \times M_2(\mathbb{C})$, $M_3(\mathbb{C})$ and $\mathbb{C} \times M_2(\mathbb{C}) \times M_2(\mathbb{C})$.

Exercise 5.2. Here, again by the Artin–Wedderburn theorem there are five possible algebras, up to isomorphism, namely: $\mathbb{R}^4$, $M_2(\mathbb{R})$, $\mathbb{R} \times \mathbb{R} \times \mathbb{C}$, $\mathbb{C} \times \mathbb{C}$ and $\mathbb{H}$.

Exercise 5.4. (a) By the Artin–Wedderburn theorem, $A \cong M_{n_1}(D_1) \times \ldots \times M_{n_r}(D_r)$ with positive integers n_i and division algebras D_i. For the centres we get $Z(A) \cong \prod_{i=1}^{r} Z(M_{n_i}(D_i))$. We have seen in Exercise 3.16 that the centre of a matrix algebra $M_{n_i}(D_i)$ consists precisely of the $Z(D_i)$-multiples of the identity matrix, that is $Z(M_{n_i}(D_i)) \cong Z(D_i)$ as K-algebras. Note that $Z(D_i)$ is by definition a commutative algebra, and it is still a division algebra (the inverse of an element in $Z(D_i)$ also commutes with all elements in D_i); thus $Z(D_i)$ is a field. Hence we get that $Z(A) \cong \prod_{i=1}^{r} Z(D_i)$ is isomorphic to a direct product of fields.

(b) By part (a), there are fields $K_1, \dots, K_r$ (containing K), such that $Z(A) \cong K_1 \times \dots \times K_r$. Since fields do not contain any non-zero zero divisors, an element

$$x := (x_1, \dots, x_r) \in K_1 \times \dots \times K_r$$

can only be nilpotent (that is, satisfy $x^\ell = 0$ for some $\ell \in \mathbb{N}$) if all $x_i = 0$, that is, if $x = 0$. Clearly, this property is invariant under isomorphisms, so holds for $Z(A)$ as well.

Exercise 5.11. (a) By definition of ε_i we have that $(X - \lambda_i + I)\varepsilon_i$ is a scalar multiple of $\prod_{i=1}^{r}(X - \lambda_i) + I$, which is zero in A.

(b) The product $\varepsilon_i \varepsilon_j$ for $i \neq j$ has a factor $\prod_{t=1}^{r}(X - \lambda_t) + I$ and hence is zero in A. Moreover, using part (a), we have

$$\varepsilon_i^2 = (1/c_i) \prod_{j \neq i}(X - \lambda_j + I)\varepsilon_i = (1/c_i) \prod_{j \neq i}(\lambda_i - \lambda_j)\varepsilon_i = \varepsilon_i.$$

(c) It follows from (b) that the elements $\varepsilon_1, \dots, \varepsilon_r$ are linearly independent over K (if $\lambda_1 \varepsilon_1 + \dots + \lambda_r \varepsilon_r = 0$ then multiplication by ε_i yields $\lambda_i \varepsilon_i = 0$ and hence $\lambda_i = 0$). Observe that $\dim_K A = r$ and hence $\varepsilon_1, \dots, \varepsilon_r$ are a K-basis for A. So we can write $1_A = b_1 \varepsilon_1 + \dots + b_r \varepsilon_r$ for $b_i \in K$. Now use $\varepsilon_i = \varepsilon_i \cdot 1_A$ and deduce that $b_i = 1$ for all i.

Chapter 6

Exercise 6.5. Let $U \subseteq V$ be a non-zero $\mathbb{C}S_n$-submodule. Take a non-zero element $u \in U$; by a suitable permutation of the coordinates we can assume that u is of the form $(0, \dots, 0, u_i, \dots, u_n)$ where $u_j \neq 0$ for all $j = i, i+1, \dots, n$. Since the coordinates must sum to zero, u must have at least two non-zero entries; so there exists an index $j > i$ such that $u_j \neq 0$. Then we consider the element $\tilde{u} := u - \frac{u_i}{u_j}(i\ j)u$, which is in U (since U is a submodule). By construction, the i-th coordinate of $\tilde{u}$ is zero, that is, $\tilde{u} \in U$ has fewer non-zero entries than u. We repeat the argument until we reach an element in U with precisely two non-zero entries. Since the coordinates sum to 0, this element is of the form $(0, \dots, 0, u_{n-1}, -u_{n-1})$. By scaling and then permuting coordinates we conclude that the $n - 1$ linearly independent vectors of the form $(0, \dots, 0, 1, -1, 0, \dots, 0)$ are in U. Since V has dimension $n - 1$ this implies that $U = V$ and hence V is simple.

Exercise 6.8. (i) Since -1 commutes with every element in G, the cyclic group $\langle -1 \rangle$ generated by -1 is a normal subgroup in G. The factor group $G/\langle -1 \rangle$ is of order 4, hence abelian. This implies that $G' \subseteq \langle -1 \rangle$. On the other hand, as G is not abelian, $G' \neq 1$, and we get $G' = \langle -1 \rangle$.

(ii) The number of one-dimensional simple $\mathbb{C}G$-modules is equal to $|G/G'| = 4$. Then the only possible solution of

$$8 = |G| = \sum_{i=1}^{k} n_i^2 = 1 + 1 + 1 + 1 + \sum_{i=5}^{k} n_i^2$$

is that $k = 5$ and $n_5 = 2$. So $\mathbb{C}G$ has five simple modules, of dimensions $1, 1, 1, 1, 2$.

(iii) By part (ii), the Artin–Wedderburn decomposition of $\mathbb{C}G$ has the form

$$\mathbb{C}G \cong \mathbb{C} \times \mathbb{C} \times \mathbb{C} \times \mathbb{C} \times M_2(\mathbb{C}).$$

This is actually the same as for the group algebra of the dihedral group D_4 of order 8. Thus, the group algebras $\mathbb{C}G$ and $\mathbb{C}D_4$ are isomorphic algebras (but, of course, the groups G and D_4 are not isomorphic).

Exercise 6.9. (i) No. Every group G has the 1-dimensional trivial $\mathbb{C}G$-module; hence there is always at least one factor $\mathbb{C}$ in the Artin–Wedderburn decomposition of $\mathbb{C}G$.

(ii) No. Such a group would have order $|G| = 1^2 + 2^2 = 5$. But every group of prime order is cyclic (by Lagrange's theorem from elementary group theory), and hence abelian. But then every simple $\mathbb{C}G$-module is 1-dimensional (see Theorem 6.4), so a factor $M_2(\mathbb{C})$ cannot occur.

(iii) Yes. $G = S_3$ is such a group, see Example 6.7.

(iv) No. Such a group would have order $|G| = 1^2 + 1^2 + 3^2 = 11$. On the other hand, the number of 1-dimensional $\mathbb{C}G$-modules divides the group order (see Corollary 6.8). But this would give that 2 divides 11, a contradiction.

Chapter 7

Exercise 7.3. Let $M = U_1 \oplus \ldots \oplus U_r$, and let $\gamma : M \to M$ be an isomorphism. Then $\gamma(M) = M$ since γ is surjective. We verify the conditions in Definition 2.15. (i) By Theorem 2.24, each $\gamma(U_i)$ is a submodule of $\gamma(M)$, and then $\gamma(U_1) + \ldots + \gamma(U_r)$ is also a submodule of $\gamma(M)$, using Exercise 2.3 (and induction on r). To prove equality, take an element in $\gamma(M)$, that is, some $\gamma(m)$ with $m \in M$. Then $m = u_1 + \ldots + u_r$ with $u_i \in U_i$ and we have

$$\gamma(m) = \gamma(u_1) + \ldots + \gamma(u_r) \in \gamma(U_1) + \ldots + \gamma(U_r).$$

(ii) Now let $x \in \gamma(U_i) \cap \sum_{j \neq i} \gamma(U_j)$, then $x = \gamma(u_i) = \sum_{j \neq i} \gamma(u_j)$ for elements $u_i \in U_i$ and $u_j \in U_j$. We have

$$0 = -\gamma(u_i) + \sum_{j \neq i} \gamma(u_j) = \gamma(-u_i + \sum_{j \neq i} u_j).$$

Since γ is injective, it follows that $-u_i + \sum_{j \neq i} u_j = 0$ and then u_i is in the intersection of U_i and $\sum_{j \neq i} U_j$. This is zero, by the definition of a direct sum. Then also $x = \gamma(u_i) = 0$. This shows that $\gamma(M) = \gamma(U_1) \oplus \ldots \oplus \gamma(U_r)$.

Exercise 7.4. (a) Analogous to the argument in the proof of Exercise 2.14 (b) one shows that for any fixed t with $j + 1 \leq t \leq i$ we have $\phi(e_t + V_j) = \lambda_t(e_t + V_j)$, where $\lambda_t \in K$. If we use the action of $E_{j+1,t}$ similarly, we deduce that $\lambda_t = \lambda_{j+1}$ and hence ϕ is a scalar multiple of the identity. This gives $\text{End}_{T_n(K)}(V_{i,j}) \cong K$, the 1-dimensional K-algebra.

(b) By part (a), the endomorphism algebra of $V_{i,j}$ is a local algebra. Then Fitting's Lemma (see Corollary 7.16) shows that each $V_{i,j}$ is an indecomposable $T_n(K)$-module.

Exercise 7.6. (a) The left ideals of $A = K[X]/(f)$ are given by $(h)/(f)$, where h divides f. In particular, the maximal left ideals are given by the irreducible divisors h of f. By definition, A is a local algebra if and only if A has a unique maximal left ideal. By the previous argument this holds if and only if f has only one irreducible divisor (up to scalars), that is, if $f = g^m$ is a power of an irreducible polynomial.

(b) Applying part (a) gives the following answers: (i) No; (ii) Yes (since $X^p - 1 = (X - 1)^p$ in characteristic p); (iii) Yes (since $X^3 - 6X^2 + 12X - 8 = (X - 2)^3$).

Exercise 7.9. As a local algebra, A has a unique maximal left ideal (see Exercise 7.1) which then is the Jacobson radical $J(A)$ (see Definition 4.21). Now let S be a simple A-module. So $S = As$ for any non-zero $s \in S$ (see Lemma 3.3) and by Lemma 3.18 we know that $S \cong A/\text{Ann}_A(s)$. Then $\text{Ann}_A(s)$ is a maximal left ideal of A (by the submodule correspondence), thus $\text{Ann}_A(s) = J(A)$ as observed above, since A is local. Hence any simple A-module is isomorphic to $A/J(A)$, which proves uniqueness.

Exercise 7.12. (a) The Artin–Wedderburn theorem gives that up to isomorphism the semisimple algebra A has the form

$$A \cong M_{n_1}(D_1) \times \ldots \times M_{n_r}(D_r)$$

with division algebras D_i over K. For each $1 \leq i \leq r$ and $1 \leq j \leq n_i$ we consider the left ideal

$$I_{i,j} := \prod_{\ell=1}^{i-1} M_{n_\ell}(D_\ell) \times U_j \times \prod_{\ell=i+1}^{r} M_{n_\ell}(D_\ell),$$

where $U_j \subseteq M_{n_i}(D_i)$ is the left ideal of all matrices with zero entries in the j-th column. Clearly, each $I_{i,j}$ is a left ideal of A, that is, an A-submodule, with factor module $A/I_{i,j} \cong D_i^{n_i}$. The latter is a simple A-module (see Lemma 5.8, and its proof), and thus each $I_{i,j}$ is a maximal left ideal of A. However, if A is also a local algebra, then it must have a unique maximal left ideal. So in the Artin–Wedderburn decomposition above we must have $r = 1$ and $n_1 = 1$, that is, $A = D_1$ is a division algebra over K.

Conversely, any division algebra over K is a local algebra, see Example 7.15.

(b) For the group G with one element we have $KG \cong K$, which is local and semisimple. So let G have at least two elements and let $1 \neq g \in G$. If m is the order of g then $g^m = 1$ in G, which implies that in KG we have

$$(g - 1)(g^{m-1} + \ldots + g^2 + g + 1) = 0,$$

that is, $g - 1$ is a non-zero zero divisor. In particular, $g - 1$ is not invertible and hence KG is not a division algebra. But then part (a) implies that KG is not local.

Chapter 8

Exercise 8.1. (a) (i) Recall that the action of the coset of X on V_α is given by the linear map α, and T is a different name for this map. Since the coset of X commutes with every element of A, applying α commutes with the action of an arbitrary element in A. Hence T is an A-module homomorphism. Since $T = \alpha$, it has minimal polynomial g^t, that is, $g^t(T) = 0$ and if h is a polynomial with $h(T) = 0$ then g^t divides h.

(ii) By assumption, V_α is cyclic, so we can fix a generator w of V_α as an A-module. The cosets of monomials X^i span A, and they take w to $\alpha^i(w)$ and therefore V_α is spanned by elements of the form $\alpha^i(w)$. Since V_α is finite-dimensional, there exists an $m \in \mathbb{N}$ such that V_α has a K-basis of the form $\{\alpha^i(w) \mid 0 \leq i \leq m\}$.

Let $\phi : V_\alpha \to V_\alpha$ be an A-module homomorphism, then

$$\phi(w) = \sum_{i=0}^{m} a_i \alpha^i(w) = \sum_{i=0}^{m} a_i T^i(w) = (\sum_{i=0}^{m} a_i T^i)(w)$$

for unique elements $a_i \in K$. But the A-module homomorphisms ϕ and $h(T) := \sum_{i=0}^{m} a_i T^i$ are both uniquely determined by the image of w since w generates V_α and it follows that $\phi = h(T)$. (To make this explicit: an arbitrary element in V_α is of the form zw for $z \in A$. Then $\phi(zw) = z\phi(w) = z(h(T)(w)) = h(T)(zw)$ and therefore the two maps ϕ and $h(T)$ are equal.)

(iii) By (ii) we have $\phi = h(T)$, where h is a polynomial. Assume $\phi^2 = \phi$, then $h(T)(h(T) - \mathrm{id}_V)$ is the zero map on V_α. By part (i), the map T has minimal polynomial g^t, and therefore g^t divides $h(h - 1)$. We have that

g is irreducible and clearly h and $h - 1$ are coprime, then since $K[X]$ is a unique factorisation domain, it follows that either g^t divides h or g^t divides $h - 1$. In the first case, $h(T)$ is the zero map on V_α and $\phi = 0$ and in the second case $h(T) - \mathrm{id}_V$ is the zero map on V_α and $\phi = \mathrm{id}_V$.

We have shown that the zero map and the identity map are the only idempotents in the endomorphism algebra of V_α. Then Lemma 7.3 implies that V_α is indecomposable as an A-module.

(b) By assumption,

$$
J_n(\lambda) = \begin{pmatrix} \lambda & & & & \\ 1 & \lambda & & & \\ & 1 & \ddots & & \\ & & \ddots & \ddots & \\ & & & 1 & \lambda \end{pmatrix}
$$

is the matrix of α with respect to some K-basis $\{w_1, \ldots, w_n\}$ of V. This means that $\alpha(w_i) = \lambda w_i + w_{i+1}$ for $1 \le i \le n - 1$ and $\alpha(w_n) = \lambda w_n$. Since α describes the action of the coset of X, this implies that $A w_1$ contains $w_1, \ldots, w_n$ and hence V_α is a cyclic A-module, generated by w_1. The minimal polynomial of $J_n(\lambda)$, and hence of α, is g^n, where $g = X - \lambda$. This is irreducible, so we can apply part (a) and V_α is indecomposable by (iii).

Exercise 8.6. (a) Recall (from basic algebra) that the conjugacy class of an element g in some group G is the set $\{xgx^{-1} \mid x \in G\}$, and that the size of the conjugacy class divides the order of G. If $|G| = p^n$ then each conjugacy class has size some power of p. Since G is the disjoint union of conjugacy classes, the number of conjugacy classes of size 1 must be a multiple of p. It is non-zero since the identity element is in such a class. Hence there must be at least one non-identity element g with conjugacy class of size 1. Then g is in the centre $Z(G)$, hence part (a) holds.

(b) Assume $Z(G)$ is a proper subgroup of G; it is normal, so we have the factor group. Assume (for a contradiction) that it is cyclic, say generated by the coset $xZ(G)$ for $x \notin Z(G)$. So every element of G belongs to a coset $x^r Z(G)$ for some r. Take elements $y_1, y_2 \in G$, then $y_1 = x^r z_1$ and $y_2 = x^s z_2$ for some $r, s \in \mathbb{N}_0$ and $z_1, z_2 \in Z(G)$. We see directly that y_1 and y_2 commute. So G is abelian, and then $Z(G) = G$, a contradiction.

(c) Since G is not cyclic, we must have $n \ge 2$. Assume first that $n = 2$: Then G must be abelian. Indeed, otherwise the centre of G would be a subgroup of order p, by (a) and Lagrange's theorem. Then the factor group $G/Z(G)$ has order p and must be cyclic, and this contradicts (b). So if G of order p^2 is not cyclic then it is abelian and can only be $C_p \times C_p$ by the structure of finite abelian groups. Thus the claim holds, with $N = \{1\}$ the trivial normal subgroup.

As an inductive hypothesis, we assume the claim is true for any group of order p^m where $m \le n$. Now take a group G of order p^{n+1} and assume G is not cyclic. If

G is abelian then it is a direct product of cyclic groups with at least two factors, and then we can see directly that the claim holds. So assume G is not abelian. Then by (a) we have $G/Z(G)$ has order p^k for $k \leq n$, and the factor group is not cyclic, by (b). By the inductive hypothesis it has a factor group $\bar{N} = (G/Z(G))/(N/Z(G))$ isomorphic to $C_p \times C_p$. By the isomorphism theorem for groups, this group is isomorphic to G/N, where N is normal in G, that is, we have proved (c).

Exercise 8.9. Let $\{v_1, \ldots, v_r\}$ be a basis of V and $\{w_1, \ldots, w_s\}$ be a basis of W as K-vector spaces. Then the elements $(v_1, 0), \ldots, (v_r, 0), (0, w_1), \ldots, (0, w_s)$ form a K-vector space basis of $V \oplus W$. (Recall that we agreed in Sect. 8.2 to use the symbol $\oplus$ also for external direct sums, that is, elements of $V \oplus W$ are written as tuples, see Definition 2.17.)

Let T be a system of representatives of the left cosets of H in G. Then we get K-vector space bases for the modules involved from Proposition A.5 in the appendix. Indeed, a basis for $KG \otimes_H (V \oplus W)$ is given by the elements

$$t \otimes_H (v_1, 0), \ldots, t \otimes_H (v_r, 0), t \otimes_H (0, w_1), \ldots, t \otimes_H (0, w_s).$$

On the other hand, a basis for $(KG \otimes_H V) \oplus (KG \otimes_H W)$ is given by

$$(t \otimes_H v_1, 0), \ldots, (t \otimes_H v_r, 0), (0, t \otimes_H w_1), \ldots, (0, t \otimes_H w_s).$$

From linear algebra it is well known that mapping a basis onto a basis and extending linearly yields a bijective K-linear map. If we choose to map

$$t \otimes_H (v_i, 0) \mapsto (t \otimes_H v_i, 0) \text{ for } 1 \leq i \leq r$$

and

$$t \otimes_H (0, w_j) \mapsto (0, t \otimes_H w_j) \text{ for } 1 \leq j \leq s$$

then we obtain a bijective K-linear map of the form

$$KG \otimes_H (V \oplus W) \to (KG \otimes_H V) \oplus (KG \otimes_H W), \quad g \otimes_H (v, w) \mapsto (g \otimes_H v, g \otimes_H w)$$

for all $g \in G$, $v \in V$ and $w \in W$. Since the KG-action on induced modules is by multiplication on the first factor and on direct sums diagonally, it is readily seen that the above map is also a KG-module homomorphism. So we have the desired isomorphism of KG-modules.

Chapter 9

Exercise 9.8. Let $v \in M(j)$ be a non-zero element. Then we define a representation $\mathcal{T}_v$ of Q by setting $T_v(j) = \text{span}\{v\}$ and $T_v(i) = 0$ for all $i \neq j$. Moreover, for every arrow γ of Q we set $T_v(\gamma) = 0$. Then $\mathcal{T}_v$ clearly is a subrepresentation of $\mathcal{M}$ and it is isomorphic to the simple representation $\mathcal{S}_j$.

Now let $v_1, \ldots, v_d$ be a K-basis of $M(j)$, that is, $M(j) = \text{span}\{v_1\} \oplus \ldots \oplus \text{span}\{v_d\}$ is a direct sum of K-vector spaces. Then $\mathcal{M} = \mathcal{T}_{v_1} \oplus \ldots \oplus \mathcal{T}_{v_d} \cong \mathcal{S}_j \oplus \ldots \oplus \mathcal{S}_j$ is a direct sum of subrepresentations.

The last statement on the converse is clear.

Exercise 9.9. (a) (i) To define $\mathcal{Y}$, take the spaces as given in the question. If γ is any arrow of Q then, by assumption, γ starts at a vertex different from j and we take $Y(\gamma) = M(\gamma)$. If $\gamma = \alpha_i$ for some i then the image of $Y(\gamma)$ is in $Y(j)$, by the definition of $Y(j)$. Otherwise, if γ ends at a vertex $k \neq j$ then $Y(k) = M(k)$ and $Y(\gamma)$ maps into $Y(k)$. This shows that $\mathcal{Y}$ is a subrepresentation of $\mathcal{M}$.

To construct $\mathcal{X}$, for $X(j)$ we choose a subspace of $M(j)$ such that $X(j) \oplus Y(j) = M(j)$. Note that this may be zero. For $k \neq j$ we take $X(k) = 0$ and for each arrow γ of Q we take $X(\gamma)$ to be the zero map. Then $\mathcal{X}$ is a subrepresentation of $\mathcal{M}$. With these, for each vertex k of Q we have $M(k) = X(k) \oplus Y(k)$ as vector spaces. This proves the claim.

(ii) The representation $\mathcal{X}$ satisfies the conditions of Exercise 9.8. Hence it is isomorphic to a direct sum of copies of $\mathcal{S}_j$, the number of copies is equal to $\dim_K X(j) = \dim_K M(j) - \dim_K Y(j)$, by basic linear algebra.

(b) Suppose $\mathcal{M} = \mathcal{X} \oplus \mathcal{S}_j$ for a subrepresentation $\mathcal{X}$ of $\mathcal{M}$. Then for each i, we have $S_j(\alpha_i) = 0$ and the map $M(\alpha_i)$ has image contained in $X(j)$. So $\sum_{i=1}^{t} \text{im}(M(\alpha_i)) \subseteq X(j)$ which is a proper subspace of $M(j) = X(j) \oplus S_j(j) = X(j) \oplus K$.

Exercise 9.10. (a) (i) Take $\mathcal{Y}$ with $Y(k)$ as defined in the question. Let γ be an arrow. It does not end at j, by assumption. If $\gamma : k \to l$ and $k \neq j$ then take $Y(\gamma) = N(\gamma)$, it maps into $Y(l) = N(l)$. If $k = j$ then we take $Y(\gamma)$ to be the restriction of $N(\gamma)$ to $Y(j)$, it maps to $Y(l) = N(l)$. So $\mathcal{Y}$ is a subrepresentation of $\mathcal{N}$. Define $\mathcal{X}$ with $X(j)$ as in the question, and $X(i) = 0$ for $i \neq j$, and take $X(\gamma)$ to be the zero map, for all arrows γ. This is a subrepresentation of $\mathcal{N}$, and moreover, $\mathcal{N} = \mathcal{X} \oplus \mathcal{Y}$.

(ii) This follows by applying Exercise 9.8.

(b) Let $\mathcal{N} = \mathcal{U} \oplus \mathcal{V}$, where $\mathcal{V} \cong \mathcal{S}_j$, then $V(j) = K$ is contained in the kernel of each $N(\beta_i)$, so the intersection of these kernels is non-zero.

Exercise 9.11. (i) Take the last quiver in Example 9.21, with $i = 1_1$. Then we set

$$P_i(i) = \text{span}\{e_i\}, \quad P_i(1_2) = \text{span}\{\gamma\}, \quad P_i(2) = \text{span}\{\alpha, \beta\gamma\}.$$

For each arrow, the corresponding map in the representation is given by left multiplication with this arrow, for instance $P_i(\beta)$ is given by $\gamma \mapsto \beta\gamma$.

In general, for each vertex j of Q, define $P_i(j)$ to be the span of all paths from i to j (which may be zero). Assume $\beta : j \to t$ is an arrow in Q. Then define $P_i(\beta) : P_i(j) \to P_i(t)$ by mapping each element p in $P_i(j)$ to βp, which lies in $P_i(t)$.

(ii) Assume Q has no oriented cycles. We give three alternative arguments for proving indecomposability.

Let $\mathcal{P}_i = \mathcal{U} \oplus \mathcal{V}$ for subrepresentations $\mathcal{U}$ and $\mathcal{V}$. The space $P_i(i) = \text{span}\{e_i\}$ is only 1-dimensional, by assumption, and must be equal to $U(i) \oplus V(i)$, and it follows that one of $U(i)$ and $V(i)$ must be zero, and the other is spanned by e_i. Say $U(i) = \text{span}\{e_i\}$ and $V(i) = 0$. Take some vertex t. Then $P_i(t)$ is the span of all paths from i to t. Suppose $P_i(t)$ is non-zero, then any path p in $P_i(t)$ is equal to pe_i. By the definition of the maps in $\mathcal{P}_i$, such a path belongs to $U(t)$. This means that $U(t) = P_i(t)$, and then $V(t) = 0$ since by assumption $P_i(t) = U(t) \oplus V(t)$. This is true for all vertices t and hence $\mathcal{V}$ is the zero representation.

Alternatively we can also show that KQe_i is an indecomposable KQ-module: Suppose $KQe_i = U \oplus V$ for submodules U and V. Then $e_i = u_i + v_i$ for unique $u_i \in U$ and $v_i \in V$. Since $e_i^2 = e_i$ we have $u_i + v_i = e_i = e_i^2 = e_i u_i + e_i v_i$. Note that $e_i u_i \in U$ and $e_i v_i \in V$ since U and V are submodules. By uniqueness, $e_i u_i = u_i$ and $e_i v_i = v_i$; these are elements of $e_i KQe_i$ which is 1-dimensional (since Q has no oriented cycles). If u_i and v_i were non-zero then they would be linearly independent (they lie in different summands), a contradiction. So say $v_i = 0$. Then $e_i = u_i$ belongs to U and since e_i generates the module, $U = KQe_i$ and $V = 0$.

As a second alternative, we can also prove indecomposability by applying our previous work. By Exercise 5.9 we know that $\text{End}_{KQ}(KQe_i)$ is isomorphic to $(e_i KQe_i)^{op}$. Since there are no oriented cycles, this is just the span of e_i. Then we see directly that any endomorphism φ of $P_i = KQe_i$ with $\varphi^2 = \varphi$ is zero or the identity. Hence the module is indecomposable, by Lemma 7.3.

Chapter 10

Exercise 10.10. (a) Using the explicit formula for C_Γ in Example 10.15, we find that the orbit of ε_n is $\{\varepsilon_n, \varepsilon_{n-1}, \ldots, \varepsilon_1, -\alpha_{1,n}\}$.
(b) Using the formula for $C_\Gamma(\alpha_{r,s})$ in Example 10.15 we see that the orbit of $\alpha_{r,n}$ is given by

$$\{\alpha_{r,n}, \alpha_{r-1,n-1}, \ldots, \alpha_{1,n-r+1}, \ -\alpha_{n-r+1,n}, -\alpha_{n-r,n-1}, \ldots, -\alpha_{1,r}\}.$$

This has $n + 1$ elements and $\alpha_{r,n}$ is the unique root of the form $\alpha_{t,n}$. Moreover, there are n such orbits, each containing $n + 1$ elements. Since in total there are $n(n + 1)$ roots for Dynkin type A_n (see Example 10.8), these are all orbits and the claim follows.

(c) We have $C_\Gamma^{n+1}(\varepsilon_i) = \varepsilon_i$ for all $i = 1, \ldots, n$ (as one sees from the calculation for (a)). The ε_i form a basis for $\mathbb{R}^n$, and C_Γ is linear. It follows that C_Γ^{n+1} fixes each element in $\mathbb{R}^n$.

Exercise 10.11. In Exercise 10.10 we have seen that C_Γ^{n+1} is the identity. Moreover, for $1 \le k < n + 1$ the map C_Γ^k is not the identity map, as we see from the computation in the previous exercise, namely all C_Γ-orbits have size $n + 1$.

Exercise 10.14. (a) The positive roots for D_4 are

$$\varepsilon_i, (1, 0, 1, 0), (0, 1, 1, 0), (0, 0, 1, 1), (1, 1, 1, 0), (0, 1, 1, 1), (1, 0, 1, 1),$$

$$(1, 1, 1, 1), (1, 1, 2, 1).$$

(b) The positive roots for type D_5 are as follows: first, append a zero to each of the roots for D_4. Then any other positive root for D_5 has 5-th coordinate equal to 1. One gets

$$\varepsilon_5, (0, 0, 0, 1, 1), (0, 0, 1, 1, 1), (0, 1, 1, 1, 1), (1, 0, 1, 1, 1), (1, 1, 1, 1, 1),$$

$$(1, 1, 2, 1, 1), (1, 1, 2, 2, 1).$$

We get 20 roots in total.

Chapter 11

Exercise 11.2. By Exercises 9.9 and 9.10 we have that $\mathcal{M} = \mathcal{X} \oplus \mathcal{Y}$ is a direct sum of subrepresentations. Since $\mathcal{M}$ is indecomposable, one of $\mathcal{X}$ and $\mathcal{Y}$ must be the zero representation. Suppose that $\mathcal{Y}$ is the zero representation, then $\mathcal{M} = \mathcal{X}$, but as shown in Exercises 9.9 and 9.10, the representation $\mathcal{X}$ is isomorphic to a direct sum of copies of the simple representation $\mathcal{S}_j$. This is a contradiction to the assumption that $\mathcal{M}$ is indecomposable and not isomorphic to $\mathcal{S}_j$. Therefore, $\mathcal{X}$ is the zero representation and $\mathcal{M} = \mathcal{Y}$. For part (a), by definition in Exercise 9.9 we then have $M(j) = Y(j) = \sum_{i=1}^{t} \text{im}(M(\alpha_i))$, as claimed. For part (b), by definition in Exercise 9.10 we then have $0 = X(j) = \bigcap_{i=1}^{t} \ker(M(\alpha_i))$, as claimed.

Exercise 11.8. (a) Vertex 1 is a source, so by Exercise 9.10 we can write $\mathcal{M} = \mathcal{X} \oplus \mathcal{Y}$ where $X(1) = \ker(M(\alpha_1))$ and $\mathcal{X}$ is a direct sum of copies of the simple representation $\mathcal{S}_1$. We assume $\mathcal{M}$ is indecomposable and not simple. It follows that $\mathcal{X} = 0$, that is, $M(\alpha_1)$ is injective. Similarly $M(\alpha_2)$ is injective. The vertex 2 is a sink, so by Exercise 9.9 we can write $\mathcal{M} = \mathcal{X} \oplus \mathcal{Y}$ where $Y(2) = \text{im}(M(\alpha_1)) + \text{im}(M(\alpha_2))$, and where $\mathcal{X}$ is isomorphic to a direct sum of copies of $\mathcal{S}_2$. Since $\mathcal{M}$ is indecomposable and not simple, it follows that $\mathcal{X} = 0$, and therefore $M(2) = Y(2) = \text{im}(M(\alpha_1)) + \text{im}(M(\alpha_2))$.

(b) An indecomposable representation $\mathcal{M}$ with $M(1) = 0$ is the extension by zero of an indecomposable representation of the subquiver of Dynkin type A_2 obtained by removing vertex 1. The indecomposable representations of a quiver of type A_2 are the simple representations and one other with dimension vector $(1, 1)$. Therefore the indecomposable representations $\mathcal{M}$ of Q with $M(1) = 0$ are

$$S_2, \ S_3 \ \text{and} \ 0 \longrightarrow K \xleftarrow{\text{id}} K.$$

Similarly the indecomposable representations $\mathcal{M}$ of Q with $M(3) = 0$ are

$$S_1, \ S_2 \ \text{and} \ K \xrightarrow{\text{id}} K \longleftarrow 0.$$

(c) This follows directly from part (a).

Exercise 11.10. Exercise 11.8 describes the indecomposable representations with $M(1) = 0$ or $M(3) = 0$; moreover, we have seen that if $M(1)$ and $M(3)$ are both non-zero then $M(2)$ must also be non-zero. In Exercise 11.9 we see that up to isomorphism there is only one indecomposable representation of Q with $M(i) \neq 0$ for $i = 1, 2, 3$. In total we have six indecomposable representations, and their dimension vectors are in fact the positive roots of the Dynkin diagram of type A_3 (which we have computed in Example 10.8).

Exercise 11.13. (i) $\Rightarrow$ (ii): This holds by Proposition 11.22.

(ii) $\Rightarrow$ (i): We show that if (i) does not hold, then also (ii) does not hold. Assume $\sum_{i=1}^{t} \text{im}(M(\alpha_i))$ is a proper subspace of $M(j)$. By Exercise 9.9 we have $\mathcal{M} = \mathcal{X} \oplus \mathcal{Y}$, where $\mathcal{X}$ is isomorphic to a direct sum of copies of S_j, and $Y(j) = \sum_{i=1}^{t} \text{im}(M(\alpha_i))$. By our assumption, $\mathcal{X}$ is non-zero, and since $\mathcal{M}$ is indecomposable, it follows that $\mathcal{M} \cong S_j$. But then $\Sigma_j^+(\mathcal{M}) = 0$ (by Proposition 11.12) and $\Sigma_j^- \Sigma_j^+(\mathcal{M}) = 0 \not\cong \mathcal{M}$, so (ii) does not hold.

(i) $\Rightarrow$ (iii): This holds by Proposition 11.37.

(iii) $\Rightarrow$ (iv): Assume that (iv) does not hold, that is, $\mathcal{M} \cong S_j$. Then we have $s_j(\underline{\dim}\mathcal{M}) = s_j(\underline{\dim}S_j) = -\varepsilon_j$; on the other hand, $\Sigma_j^+(\mathcal{M}) = 0$ and therefore $\underline{\dim}\Sigma_j^+(\mathcal{M}) = 0 \neq -\varepsilon_j$. So (iii) does not hold.

(iv) $\Rightarrow$ (v): This is Theorem 11.25 part (a) (use that $\mathcal{M}$ is indecomposable by assumption).

(v) $\Rightarrow$ (iv): Assume $\Sigma_j^+(\mathcal{M})$ is indecomposable. Suppose we had $\mathcal{M} \cong S_j$, then $\Sigma_j^+(\mathcal{M}) = 0$, a contradiction. So $\mathcal{M}$ is not isomorphic to S_j.

(iv) $\Rightarrow$ (i): This holds by Exercise 9.9.

Exercise 11.14. We only formulate the analogue; the proof is very similar to the proof of Exercise 11.13 and is left to the reader.

Let Q' be a quiver and j a source in Q'. We denote by $\beta_1, \ldots, \beta_t$ the arrows of Q' starting at j. Let $\mathcal{N}$ be an indecomposable representation of Q'. Show that the following statements are equivalent:

(i) $\bigcap_{i=1}^{t} \ker(N(\beta_i)) = 0$.

(ii) $\Sigma_j^+ \Sigma_j^-(\mathcal{N}) \cong \mathcal{N}$.

(iii) $\underline{\dim}\,\Sigma_j^-(\mathcal{N}) = s_j(\underline{\dim}\mathcal{N})$.

(iv) $\mathcal{N}$ is not isomorphic to the simple representation $\mathcal{S}_j$.

(v) $\Sigma_j^-(\mathcal{N})$ is indecomposable.

Exercise 11.18. If one of conditions (i) or (ii) of Exercise 11.16 does not hold then the representation is decomposable, by Exercise 11.16. So assume now that both (i) and (ii) hold. We only have to show that the hypotheses of Exercise 11.17 are satisfied. By (i) we know

$$\dim_K \ker(M(\alpha_1)) + \dim_K \ker(M(\alpha_2)) \le \dim_K M(1) = 4$$

and by rank-nullity

$$4 - \dim_K \ker(M(\alpha_i)) = \dim_K M(1) - \dim_K \ker(M(\alpha_i)) = \dim_K \mathrm{im}(M(\alpha_i)) \le 2$$

for $i = 1, 2$. Hence $2 \le \dim_K \ker(M(\alpha_i))$ for $i = 1, 2$. These force

$$2 = \dim_K \ker(M(\alpha_i)) = \dim_K \mathrm{im}(M(\alpha_i))$$

and also $M(1) = \ker(M(\alpha_1)) \oplus \ker(M(\alpha_2))$.

Exercise 11.19. (a) We observe that $\sigma_2 \sigma_1 Q = Q$, and the statements follow from Theorem 11.25, and Proposition 11.37; note that the properties of $\Sigma_j^{\pm}$ hold not just for Dynkin quivers, but also for the Kronecker quiver Q as well.

(b) Since $\mathcal{M}$ is not simple, it is not isomorphic to $\mathcal{S}_1$ and hence $\Sigma_1^-(\mathcal{M})$ is not zero. So $\Sigma_2^- \Sigma_1^-(\mathcal{M})$ is zero precisely for $\Sigma_1^-(\mathcal{M}) = \mathcal{S}_2$. Let $(a, b) = \underline{\dim}\mathcal{M}$. The dimension vector of $\Sigma_1^-(\mathcal{M})$ is $s_1(\underline{\dim}\mathcal{M}) = (-a + 2b, b)$ and this is equal to $(0, 1)$ precisely for $(a, b) = (2, 1)$. Explicitly, we can take $\mathcal{M}$ to have $M(1) = K^2$ and $M(2) = K$, and for $x = (x_1, x_2) \in K^2$ the maps are $M(\alpha_1)(x) = x_1$ and $M(\alpha_2)(x) = x_2$, that is, the projections onto the first and second coordinate.

Exercise 11.20. (a) Recall from Exercise 10.9 that C_Γ has matrix $\begin{pmatrix} -1 & 2 \\ -2 & 3 \end{pmatrix}$, with respect to the standard basis of $\mathbb{R}^2$. Then $a_1 = -a + 2b$ and $b_1 = -2a + 3b$ and therefore we have $a_1 - b_1 = a - b > 0$. Furthermore, $a - a_1 = 2(a - b) > 0$.

(b) This follows by iterating the argument in (a).

(c) This follows by applying Exercise 11.19.

(d) Clearly, either $\mathcal{N} = \mathcal{S}_1$, or otherwise $\Sigma_2^-(\mathcal{N}) = \mathcal{S}_2$. By Theorem 11.25 we can apply $(\Sigma_1^+ \Sigma_2^+)^r$ to $\mathcal{N}$ and get $\mathcal{M}$, that is, we see that such a representation exists. Then

$$\underline{\dim}\mathcal{M} = \underline{\dim}(\Sigma_1^+ \Sigma_2^+)^r(\mathcal{N}) = (s_1 s_2)^r \underline{\dim}\mathcal{N}$$

by Proposition 11.37. Note that $s_1 s_2 = (s_2 s_1)^{-1} = C_\Gamma^{-1}$ and this is given by the matrix $\begin{pmatrix} 3 & -2 \\ 2 & -1 \end{pmatrix}$. Then one checks that the dimension vector of $\mathcal{M}$ is of the form $(a, a - 1)$; indeed, if $\mathcal{N} = \mathcal{S}_1$ then $\underline{\dim}\mathcal{M} = \binom{2r+1}{2r}$ and if $\Sigma_2^-(\mathcal{N}) = \mathcal{S}_2$ then $\underline{\dim}\mathcal{M} = \binom{2r+2}{2r+1}$.

Finally, we get uniqueness by using the argument as in Proposition 11.46.

Exercise 11.22. (a) This is straightforward.

(b) Let $\varphi : V_\mathcal{M} \to V_\mathcal{M}$ be an A-module homomorphism such that $\varphi^2 = \varphi$; according to Lemma 7.3 we must show that φ is the zero map or the identity. Write φ as a matrix, in block form, as

$$T = \begin{pmatrix} T_1 & T' \\ T'' & T_2 \end{pmatrix}$$

so that T_i is the matrix of a linear transformation of $M(i)$, and we may view T' as a linear map $M(2) \to M(1)$. This is an A-module endomorphism of $V_\mathcal{M}$ if and only if T commutes with x and y, equivalently, if and only if for $i = 1, 2$ we have

(i) $T'M(\alpha_i) = 0$ and $M(\alpha_i)T' = 0$
(ii) $T_2 M(\alpha_i) = M(\alpha_i)T_1$.

Assuming these, we claim that $T' = 0$. Say $M(\alpha_1)$ is surjective, then

$$T'(M(2)) = T'(M(\alpha_1)(M(1))) = T'M(\alpha_1)(M(1)) = 0$$

by (i), that is, $T' = 0$. Recall that $\varphi^2 = \varphi$, since $T' = 0$ we deduce from this that $T_1^2 = T_1$ and $T_2^2 = T_2$. Let $\tau := (T_1, T_2)$, by (ii) this defines a homomorphism of representations $\mathcal{M} \to \mathcal{M}$ and we have $\tau^2 = \tau$. We assume $\mathcal{M}$ is indecomposable, so by Lemma 9.11 it follows that τ is zero or the identity. If $\tau = 0$ then we see that $T^2 = 0$. But we should have $T^2 = T$ since $\varphi^2 = \varphi$ and therefore $T = 0$ and $\varphi = 0$. If $\tau = \text{id}_\mathcal{M}$ then $T^2 = T$ implies that $2T'' = T''$ and therefore $T'' = 0$ and then $\varphi = \text{id}_{V_\mathcal{M}}$. We have proved that $V_\mathcal{M}$ is indecomposable.

(c) Assume $\varphi : V_\mathcal{M} \to V_\mathcal{N}$ is an isomorphism and write it as a matrix in block form, as

$$T = \begin{pmatrix} T_1 & T' \\ T'' & T_2 \end{pmatrix}.$$

We write $x_\mathcal{M}$ and $y_\mathcal{M}$ for the matrices describing the action of X and Y on $V_\mathcal{M}$ and similarly $x_\mathcal{N}$ and $y_\mathcal{N}$ for the action on $V_\mathcal{N}$. Since φ is an A-module homomorphism we have $x_\mathcal{N} T = T x_\mathcal{M}$, and $y_\mathcal{N} T = T y_\mathcal{M}$. That is, the following two conditions are satisfied for $i = 1, 2$:

(i) $N(\alpha_i)T' = 0$ and $T'M(\alpha_i) = 0$
(ii) $N(\alpha_i)T_1 = T_2 M(\alpha_i)$.

Say $M(\alpha_1)$ is surjective, then it follows by the argument as in part (b) that $T' = 0$. Using this, we see that since T is invertible, we must have that both T_1 and T_2 are invertible. Therefore $\tau = (T_1, T_2)$ gives an isomorphism of representations $\tau : \mathcal{M} \to \mathcal{N}$. This proves part (c).

Exercise 11.24. (a) An A-module M can be viewed as a module of KQ, by inflation. So we get from this a representation $\mathcal{M}$ of Q as usual. That is, $M(i) = e_i M$ and the maps are given by multiplication with the arrows. Since $\beta\gamma$ is zero in A we have $M(\beta) \circ M(\gamma)$ is zero, similarly for $\alpha\gamma$ and $\delta\gamma$.

(b) (i) Vertex 5 is a source, so we can apply Exercise 9.10, which shows that $\mathcal{M} = \mathcal{X} \oplus \mathcal{Y}$ where $X(5) = \ker(M(\gamma))$, and $\mathcal{X}$ is isomorphic to a direct sum of copies of the simple representation $\mathcal{S}_5$.

(ii) Let $\mathcal{U}$ be the subrepresentation with $U(5) = M(5)$ and $U(3) = \operatorname{im}(M(\gamma))$, and where $U(\gamma) = M(\gamma)$. This is a subrepresentation of $\mathcal{M}$ since $M(\alpha), M(\beta)$ and $M(\delta)$ map the image of $M(\gamma)$ to zero, by part (a). The dimension vector of this subrepresentation is $(0, 0, d, 0, d)$ since $M(\gamma)$ is injective. From the construction it is clear that $\mathcal{U}$ decomposes as a direct sum of d copies of a representation with dimension vector $(0, 0, 1, 0, 1)$, and that this representation is indecomposable (it is the extension by zero of an indecomposable representation for a quiver of type A_2).

Now choose a subspace C of $M(3)$ such that $C \oplus U(3) = M(3)$, and then we have a subrepresentation $\mathcal{V}$ of $\mathcal{M}$ with $V(i) = M(i)$ for $i = 1, 2, 4$ and $V(3) = C$, $V(5) = 0$ and where $V(\omega) = M(\omega)$ for $\omega = \alpha, \beta, \delta$. Then $\mathcal{M} = \mathcal{U} \oplus \mathcal{V}$.

(c) Let M be an indecomposable A-module, and let $\mathcal{M}$ be the corresponding indecomposable representation of Q, satisfying the relations as in (a).

Suppose first that $M(5) = 0$. Then $\mathcal{M}$ is the extension by zero of an indecomposable representation of a quiver of Dynkin type D_4, so there are finitely many of these, by Gabriel's theorem.

Suppose now that $M(5) \neq 0$. If $M(\gamma)$ is not injective then $\mathcal{M} \cong \mathcal{S}_5$, by part (b)(i) (and since $\mathcal{M}$ is indecomposable by assumption). So we can assume now that $M(\gamma)$ is injective. Then by part (b)(ii) (and because $\mathcal{M}$ is indecomposable), $\mathcal{M}$ has dimension vector $(0, 0, 1, 0, 1)$, and hence is unique up to isomorphism.

In total we have finitely many indecomposable representations of Q satisfying the relations in A, and hence we have finitely many indecomposable A-modules.

(d) Gabriel's theorem is a result about path algebras of quivers; it does not make any statement about representations of a quiver where the arrows satisfy relations.

Index

© Springer International Publishing AG, part of Springer Nature 2018
K. Erdmann, T. Holm, *Algebras and Representation Theory*, Springer
Undergraduate Mathematics Series, https://doi.org/10.1007/978-3-319-91998-0